Hailstorms of the Central High Plains

Volume 2
Case Studies of the National Hail Research Experiment

Edited by
Charles A. Knight and Patrick Squires

Colorado Associated University Press
Boulder, Colorado
1982

The National Center for Atmospheric
Research is sponsored by the National Science
Foundation. Any opinions, findings, and conclusions or recommendations expressed in this
publication are those of the author(s) and do
not necessarily reflect the views of the
National Science Foundation.

Published by Colorado Associated University Press in
association with the National Center for Atmospheric
Research.

Colorado Associated University Press
Boulder, Colorado 80309

The National Center for Atmospheric Research (NCAR) is
operated by the nonprofit University Corporation for
Atmospheric Research (UCAR) under the sponsorship of
the National Science Foundation.

International Standard Book Number: 0-87081-097-9
Library of Congress Card Catalog Number: 81-66733

Printed in the United States of America.

Preface

In planning this report on research carried out in NHRE, the editors faced two difficult choices: whether to have a second volume for the case studies at all, and whether to try to furnish a final chapter of summary, overview, and recommendations. The decision to publish Volume 2 was not clear because much of the material would also be submitted to journals for publication. Three factors led to the decision to prepare Volume 2. First, the case studies represented a major part of the research carried on in NHRE, and so formed an essential component of this report. Second, we perceive a real value in having these case studies all together in one place. The contrasts and similarities between them illustrate how the kind of analyses described in Volume 1 must be supplemented in order to approach the goal of an adequate understanding of thunderstorm precipitation. Third, we have an opportunity to be somewhat more complete in this volume than the journals would allow, in terms of the amount of data presented.

The second decision was not to write a concluding chapter, either to Volume 1 or to both volumes as a set. This also was not a clear-cut decision. It was made with a strong sense of the complexities of thunderstorm precipitation. At this time there are not many simple, positive conclusions to this work, and it does not seem appropriate to devote a chapter to the proposition that more research is needed, however amply that conclusion may be justified.

The 1976 field season of NHRE was the first to bring to bear on thunderstorm problems a Doppler radar network coordinated with several aircraft. The new and more complete data sets that resulted provide documentation of complexities that, by and large, had been foreseen and expected, such as complicated growth trajectories or the likelihood of individual precipitation elements not being confined to single convective cells or thermals, as was suggested some time ago by Ludlam. They also documented the incorrectness of a number of assumptions that had been widely used for convenience though most serious students of the field could hardly have been convinced of their actual validity, such as the assumption that maximum hail size can be predicted from a predicted maximum updraft velocity.

It is a commonly held view that the technologies of multiple-Doppler radar studies and new aircraft capabilities have moved this field back to a stage in which the appropriate emphasis is on new observations. The three case studies in this volume may be seen as illustrating this proposition, in that none of them fits comfortably into any present storm classification scheme. Yet the storms studied are quite different from each other, in both size and organization. Parts of the discussions of all three cases concern the problem of generalization of these specific results, but an attractive scheme has not yet emerged.

In order to avoid repetition of footnotes, certain conventions used in this volume are summarized here. Times are all Mountain Daylight Time (MDT = Z + 6 hours). Heights are relative to mean sea level unless otherwise stated (the surface is at about 1.6 km). Radar reflectivity factor values are always the equivalent ones, calculated using the dielectric constant of liquid water rather than ice, in spite of the fact that above the freezing level ice was always the dominant form of precipitation. To emphasize this we use Z_e and dBZ_e as the symbols for the reflectivity factor, though dBZ is perhaps more conventional.

Specific acknowledgements that would be appropriate in journal papers have been omitted. At least a hundred people contributed in important ways to the data gathering and analysis: the NHRE staff, the Atmospheric Technology Division of NCAR, the Wave Propagation Laboratory of NOAA, and the groups at the University of Wyoming and the South Dakota School of Mines and Technology were major institutional contributors. A complete listing is given in the preface to Volume 1.

CONTENTS

The 22 June 1976 Case Study

The 22 July 1976 Case Study

Chapter 16. Radar Echo Structure and Evolution

Chapter 17. Low-Level Airflow and Mesoscale Influences

Chapter 18. Storm Airflow, Updraft Structure, and Mass Flux from Triple-Doppler Measurements

Chapter 19. Storm Structure Deduced from a Penetrating Aircraft

Chapter 20: Hail Growth

The 22 June 1976 Case Study

CHAPTER 13
The 22 June 1976 Case Study: Large-Scale Influences, Radar Echo Structure, and Mesoscale Circulations

James C. Fankhauser

13.1 INTRODUCTION

This chapter is the first of three giving a description and discussion of a large multicellular storm that formed on 22 June 1976 within 30 km of the NHRE field headquarters, located near the town of Grover in northeastern Colorado. It remained nearly stationary during its 90-min lifetime and produced copious rainfall and widespread hail. The positions of fixed facilities and the surface topography are given in Fig. 13.1.

This chapter and Chapter 14, which presents the detailed structure and evolution of internal airflow patterns, are intended to lay the groundwork for interpretations of precipitation formation mechanisms covered in Chapter 15. The emphasis here will be on significant larger-scale influences on storm initiation, the manner in which mesoscale features modulate storm development and, finally, the way these combine to produce the evolving radar echo structures.

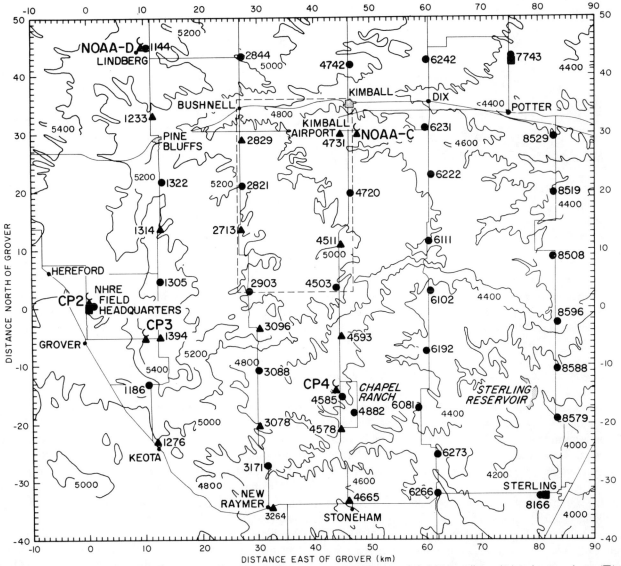

Figure 13.1 *Topographic map of the area of NHRE field operations in 1976. Contour interval is 200 ft (≈ 60 m). Symbols identify locations of Doppler radars (▲), surface mesonetwork stations (●); PAM installations (▲), and rawinsonde sites (■). Dashed rectangle shows position of a dense network for precipitation measurement at the surface.*

13.2 SYNOPTIC FEATURES

Severe storm forecasters have recognized for some time that coupling between zones of upper tropospheric divergence and low-level convergence is favorable for aiding the release of latent thermodynamic instability (see, e.g., Beebe and Bates, 1955), the mechanism being large-scale ascent of convectively unstable air. Similarly favorable conditions for realizing potential instability exist when warm moist air is mechanically lifted in flowing over an upward-sloping terrain. Both of these mechanisms are enhanced in regions of localized boundary layer convergence.

Figure 13.2 shows the 500-mb constant pressure analysis for 0000 GMT, 23 June 1976. In the foregoing context, a

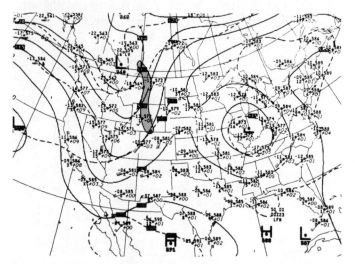

Figure 13.2 Contours of geopotential height at 60-m intervals (solid) and isotherms at 5 °C increments (dashed) at a pressure of 50 kPa for 0000 GMT, 23 June 1976. Stippling shows region of cyclonic vorticity advection $\geq 2.5 \times 10^{-8}$ s^{-1} at 25 kPa.

prominent feature favoring convective development is the distinctly diffluent airflow overlying eastern Colorado and western Nebraska. A band of positive vorticity advection extending from eastern Montana into northern Colorado was found at 250 mb at a position somewhat ahead of the 500-mb trough axis. A maximum core centered over south-central Wyoming was undoubtedly responsible for the surface low pressure center depicted near the same location on the satellite photograph taken at 1800 GMT on 22 June (Fig. 13.3). As indicated by superimposed surface data, cyclonic circulation around this deepening low produced southeasterly surface airflow in the area of northeastern Colorado where, as shown in Fig. 13.1, the terrain rises some 500 m in 80 km.

From the standpoint of later convective developments, the most striking feature in the satellite photography is the narrow cloud line extending from extreme western Nebraska through Colorado into southwestern Kansas. Surface dew points, in degrees Fahrenheit, are in the middle 50's (10 °C to 20 °C) east of this cloud band, while stations near the foot of

the Rocky Mountains have dew points near 40 °F (4 °C to 5 °C). Thus, the cloud line seems to identify a rather sharp moisture gradient frequently observed in the High Plains region. Mesoscale data to be presented later also show this to be a zone of surface wind convergence. Massive cloud cover near and somewhat ahead of the surface cold front in Wyoming indicates widespread thunderstorm activity already present there by midday. A cloud-free zone occupied by decidedly drier air extends northward from the Colorado plains into southeastern Wyoming, separating the frontal activity from the distinct cloud band to the east.

The vertical structure of the environment is given by two soundings in Fig. 13.4. These were released around 1500 MDT from special rawinsonde sites located near Potter, Nebraska, and Sterling, Colorado (see Fig. 13.1 for site locations). Thermodynamic parameters below 615 mb in the Potter sounding have been modified with data gathered around 1620 by Queen Air N304D, which made a spiral descent from cloud base altitude to near the surface (≈ 850 mb), within the inflow to the storm of principal interest. Thus, the Potter sounding may be taken to represent conditions in the near environment, while the Sterling sounding is more indicative of general ambient conditions.

In the inflow sector the aircraft observed a well-mixed subcloud regime, with a dry adiabatic lapse rate and a nearly uniform mixing ratio ($q \approx 8$ g kg^{-1}) extending from the surface to 616 mb (4.2 km MSL). Average potential temperature (θ) and mixing ratio in the lowest 50-mb layer were 318.6 K and 8.4 g kg^{-1}, respectively. The lifting condensation level (LCL) for air parcels with these properties is 616 mb, and the pseudo-adiabat passing through the LCL has an equivalent potential temperature (θ_e) of 345.5 K. Moist adiabatic ascent above the LCL gives a stability index of -4.7 °C at 500 mb.

Low-level conditions on the Sterling sounding show a temperature lapse rate slightly less than dry adiabatic and a gradual decrease in mixing ratio with height. This results in a corresponding decrease of θ_e with height (inset), which is characteristic of a convectively unstable atmosphere. Coldest and driest air aloft is found on the Potter sounding between 440 and 430 mb, where minimum values are near 327 K.

The Potter site was best situated to observe winds representative of the storm's environment, as plotted in Fig. 13.4. Mean winds below cloud base altitude were southeasterly with a velocity of 15 m s^{-1}. Between cloud base and 400 mb (7.5 km) winds veered to southwesterly and increased to 20 m s^{-1}. Above 400 mb the flow was from the west-southwest and nearly constant with height averaging 20 m s^{-1}. Wind shear computed through the layer from cloud base to the 200-mb level is 2.7×10^{-3} s^{-1}, similar to values reported by Marwitz (1972b) for a series of multicellular storms. A shear that is, perhaps, more relevant from the standpoint of storm organization and dynamics (see, e.g., Newton, 1966; Browning, 1977) is one that considers the difference between winds in the subcloud and cloud-bearing layers. If the mean airflow in these layers is used, a value of

Figure 13.3 GOES-E visual satellite image at 1800 GMT, 22 June 1976, with frontal positions and selected surface data superimposed. Surface station data include wind (1 full barb = 10 kt), temperature (upper left), and dew point (lower left), with the latter both in °F. Small dashed box shows location of area in Fig. 13.1. Cloud line extending south and southeastward from that region presumably marks boundary between moist air to the east and drier air to the west and southwest, along which storms shown in Figs. 13.5 and 13.6 eventually form.

$3.9 \times 10^{-3}\,\text{s}^{-1}$ results, with the shear vector oriented from 280°.

Most of the ambient conditions discussed above may be compared with averages derived from representative soundings for a sample of 57 hail days, which constitute the experimental units in the NHRE randomized seeding experiment (Fankhauser and Mohr, 1977). Such a comparison shows that subcloud thermodynamic parameters on this date were potentially warmer by about 5 K and slightly moister than their average low-level conditions. Static stability as measured by either the 500-mb stability index or the net positive energy associated with unmixed parcel ascent also shows that the situation was quite unstable compared with the average. The magnitude of the wind shear between mean airflow in the subcloud and cloud layers quoted above falls between the values given for average shear associated with multicellular storms and the more strongly sheared environments of persistent unicell storms (supercells). One final feature of note is the strong southeasterly subcloud airflow.

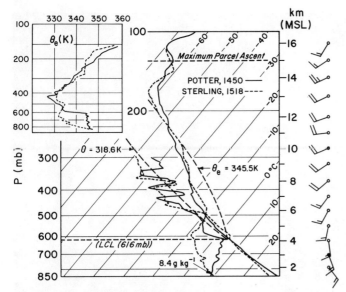

Figure 13.4 Thermodynamic diagram showing the vertical distribution of temperature and dew point from soundings at Potter, Nebraska (solid), and Sterling, Colorado (dashed), on the afternoon of 22 June 1976. Dry and moist adiabats and mixing ratio corresponding to the designated lifting condensation level (LCL) and level of maximum parcel ascent are based on aircraft measurements between the surface and cloud base altitude, as plotted on the Potter sounding at pressures greater than 616 mb. Winds given on the right at 1-km intervals (1 full barb = 10 m s⁻¹) are also taken from the Potter sounding. Inset gives equivalent potential temperature (θ_e) vs. pressure for both soundings.

Early storm development is illustrated in Fig. 13.5 by PPI radar echoes observed in the altitude range of 4 to 6 km. First echoes formed around 1330 within the cloud line noted in Fig. 13.3, which, according to subsequent satellite photographs and surface time-lapse photography, had moved westward to a position indicated by echoes at 1400. Factors undoubtedly influencing the preferred region of formation are (1) the north-south area of elevated terrain located 10-15 km east of the Grover site (Fig. 13.1), and (2) localized surface convergence at the boundary between the moist southeasterly flow (seen at all times over the southeastern portions of the grids in Fig. 13.5) and a drier air mass with south-southwesterly winds lying just to the west of the surface network (evident only at 1530 in data shown at Grover [GRO] in Fig. 13.5).

Discrete echo systems, identified by alphabetical sequence according to their order of appearance, formed at approximately 10-min intervals near or over these boundary layer features and moved rapidly northward at speeds of 15 to 20 m s⁻¹ while evolving in a fairly typical multicellular manner (see, e.g., Renick, 1971; Marwitz; 1972b; and Chalon et al., 1976). Each system formed successively further southward, and as many as seven are seen to coexist at 1500. Individual storms were typically composed of three to six evolving high-reflectivity cores or "cells," and total storm lifetime was about 45 min to an hour. The geographically fixed boundary-layer forcing and the predominantly northward storm motion produced only minor eastward displace-

ment for the line as a whole during the 2-h period shown.

First evidence of cool outflowing air from downdrafts produced by the earliest storms appeared in mesonetwork data at about 1520. Three distinct outflow surges from storms G₂, J, and K₁ are indicated at 1530 by heavy barbed lines, which intersect to form cusps near the south edges of storms G₂ and J. At and prior to this time all storm systems had free access to moist southeasterly airflow, which had surface properties generally similar to those at low levels on the Potter sounding (Fig. 13.4). After about 1530, however, storm development patterns underwent a major change, as outflow air spread rapidly eastward over the northern part of the grid. Concurrently, storms comprising the northern portion of the line diminished in intensity, and the large echo complex identified as K in Fig. 13.5 became the dominant feature. It is this large and persistent system that receives our major attention in analyses to follow.

The visual appearance of the storm at 1600 is given by the photographic composite in Fig. 13.6a. This figure was constructed from a series of photographs obtained in rapid sequence from an altitude of about 4 km by the pilot of the NCAR/NOAA sailplane in descending orbit over the location northwest of the storm indicated in Fig. 13.5e. Figure 13.6b was photographed from an altitude of ≈ 9 km and near the same location as that of the sailplane, but 26 min later, by an observer on the NCAR Sabreliner (N307D) during simulated cloud-top seeding trials (Biter and Solak, 1979).

During the hour prior to 1600 the sailplane had participated with Queen Air N10UW in cloud penetrations to obtain precipitation physics and air motion measurements within the cumulus congestus clouds seen flanking the southwest portion of the main cloud mass. This part of the study was in line with the first objective of the field measurements program mentioned in Section 13.1 and has been summarized by Breed (1978).

Echo system K, and probably L as well, is embedded within the largely glaciated portion of the cloud mass located in the right-central portion of Fig. 13.6a, and the precipitation faintly visible below cloud base is undoubtedly associated with these. The crisp, freshly-growing congestus apparent on the western and southwestern flanks in Fig. 13.6a and primarily on the southern and southwestern flanks in Fig. 13.6b is of particular interest since, as we shall see in the following section, this is the genesis region for discrete new echo growth, which plays a significant role in determining the storm's overall behavior. Cumulus turrets can be seen in Fig. 13.6a, extending well to the south with diminishing vertical development. This line presumably delineates the boundary between dry and more moist air masses mentioned earlier.

13.3 RADAR ECHO STRUCTURE

Constant altitude radar echo patterns (i.e., CAPPI's) for

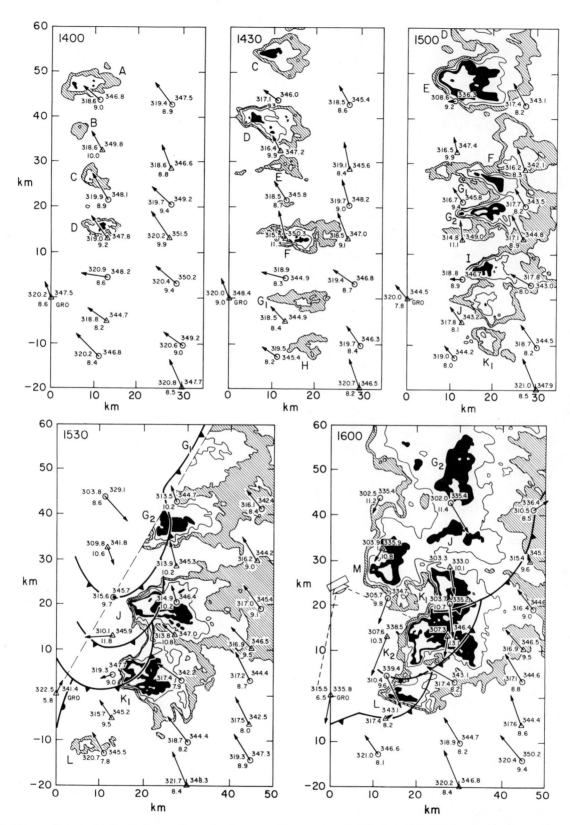

Figure 13.5 PPI radar echo contours in the altitude range of 4 to 6 km shown at 10-dB intervals increasing from 15 dBZ$_e$. Discrete multicellular storms are labeled alphabetically according to their order of development. Storm system K, first seen at the southern end of the line at 1500, is the subject of later analyses. Wind vectors (15 m s^{-1} is equivalent to 10 km) and thermodynamic data are plotted at surface mesonetwork sites with potential temperature at upper left, mixing ratio at lower left, and equivalent potential temperature at upper right. Heavy barbed lines denote the boundary between ambient south-southeasterly winds and the outflow from downdrafts of earlier storms. Border labels indicate the distance (in kilometers) from the Grover, Colorado, field headquarters, where the radar is located. Symbol at 1600 near y = 25 and x = 0 km shows approximate position and field of view for the photographs in Fig. 13.6.

Figure 13.6 Photographs taken toward the southeast from aircraft at the general location designated in Fig. 13.5: (a) composite from the NCAR/NOAA sailplane at about 1600 and near 4 km altitude; (b) from the NCAR Sabreliner (N307D) at about 1630 and about 9-km altitude. Both photographs depict southwest flank of the main storm, which is embedded within glaciated cloud mass on the left.

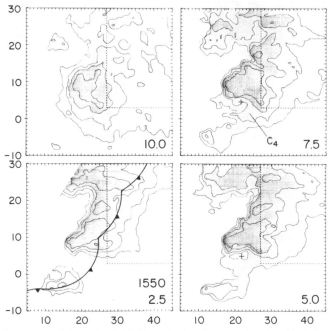

Figure 13.7a Constant altitude PPI (CAPPI) radar echo contours at indicated levels (km) for 1550 (MDT). Graduated shading represents 10-dB steps in radar reflectivity factor increasing from 10 dBZ_e. Early positions of radar reflectivity maxima tracked as cells are identified by the letter C and subscript. Cross (+) designates location of weak echo region or echo-free vault at 5.0- and 7.5-km altitudes. Borders are labeled in kilometers from Grover radar, and interior dashed boundary locates dense precipitation network shown in Fig. 13.1. Heavy barbed line delineates the inflow-outflow interface at the surface. Heavy dashed line, which appears at times west and north of the outflow boundary in other parts of Fig. 13.7 identifies axis of low-level radar return.

four levels are given at 10-min intervals in Fig. 13.7 for a period covering most of the lifetime of storm system K. Shading gradation from lightest to darkest represents 10-dB step increases in equivalent radar reflectivity factor, Z_e, beginning at 20 dBZ_e. The CAPPI configurations were produced from computer software developed by Mohr and Vaughan (1979) for transforming radar data observed in conventional azimuth-elevation angle mode to more easily interpreted Cartesian coordinates. The close proximity of the storm to the radar and large increments in elevation angle (required to obtain frequent complete scans) prohibited reliable interpolation of sparse data at altitudes greater than about 12 km. For this reason levels chosen for display emphasize radar echo history only in the lower and middle portions of the storm (i.e., between 2.5 and 10.0 km). Later analyses will demonstrate that changes occurring at these altitudes are adequate for explaining most of the important features in the storm's evolution, but it should be noted that radar echo

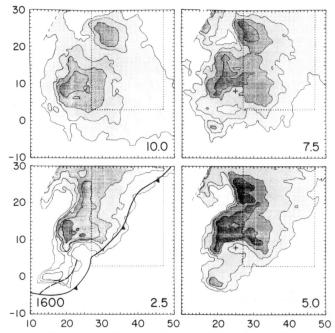

Figure 13.7b Same as Fig. 13.7a, except for 1600. Note shift in x coordinates.

≥ 20 dBZ_e was observed at altitudes ≥ 15 km throughout most of the period shown. Also, infrared radiation (IR) satellite data (Reynolds, 1980) indicate that cloud-top temperatures were as cold as −66°C, a temperature found only above 15 km on the profiles in Fig. 13.4. Thus, the indications are that the storm consistently penetrated some 2 to 3 km above the tropopause height, which is shown to be around 13 km in Fig. 13.4.

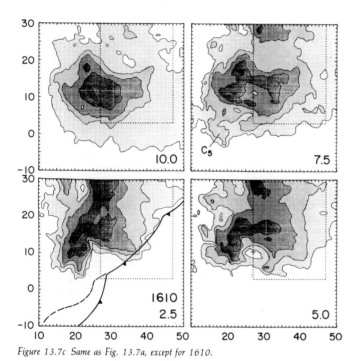

Figure 13.7c Same as Fig. 13.7a, except for 1610.

Figure 13.7d Same as Fig. 13.7a, except for 1620.

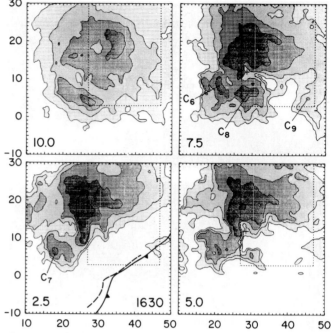

Figure 13.7e Same as Fig. 13.7a, except for 1630.

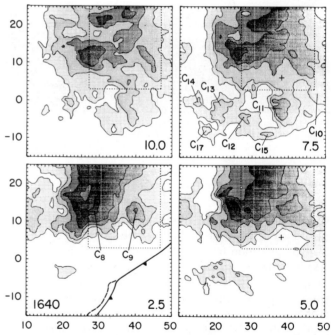

Figure 13.7f Same as Fig. 13.7a, except for 1640. Note shift in y coordinates.

13.3.1 Quasi-Steady Characteristics

One impression to be drawn from the CAPPI sequence in Fig. 13.7 is a certain degree of stationarity in storm location, shape, and intensity. This is particularly evident between 1600 and 1630, when the three-dimensional structure of

7

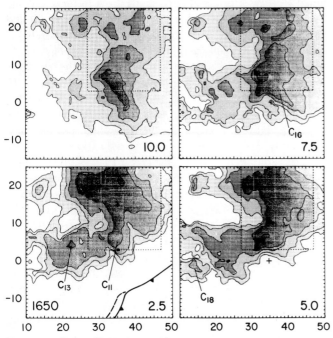

Figure 13.7g Same as Fig. 13.7a, except for 1650.

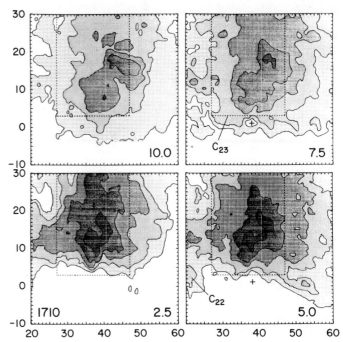

Figure 13.7i Same as Fig. 13.7a, except for 1710. Note shift in x and y coordinates.

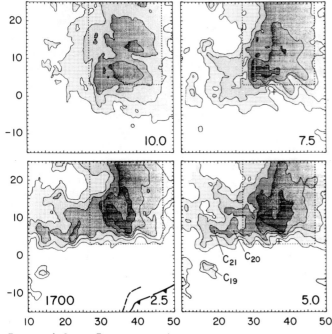

Figure 13.7h Same as Fig. 13.7a, except for 1700.

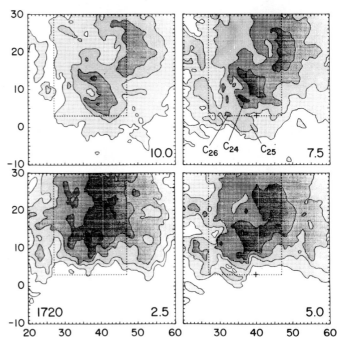

Figure 13.7j Same as Fig. 13.7a, except for 1720.

radar reflectivity factor displays many of the characteristics of supercell storms, as conceived by Browning (1964) and summarized by Marwitz (1972a) and Chisholm and Renick (1972). These characteristics include:

(1) A nearly circular symmetry in the shape of the outer radar reflectivity contours at 10 km, expanding from early dimensions of 15-20 km to 30 or 40 km by 1630,

and forming an extensive overhang to the south of the echo at lower levels.

(2) A persistent weak echo vault with horizontal dimensions of 5 to 8 km (designated by a + at 5.0 and 7.5 km) penetrating upward to altitudes of 6 to 10 km beneath the upper overhang. Such a vault is indicative of broad, strong updrafts in which there is only a short

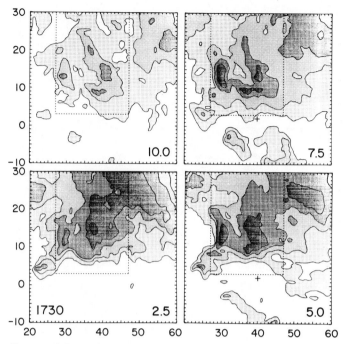

Figure 13.7k Same as Fig. 13.7a, except for 1730.

residence time for particle growth. Although it fluctuates somewhat in size, relative position, and depth of vertical penetration, echo profiles in Fig. 13.8 demonstrate that a vault is always present during the 30-min period.

(3) An alar-shaped high-reflectivity core extending from mid-cloud levels to the ground bordering the vault on its north side, with a narrow southward protrusion on the west. Like the vault, the zone of highest reflectivity factor undergoes some minor variation in intensity but retains the same basic shape and position relative to the vault.

The combination of these radar echo features in a relatively unchanging state has been interpreted by Browning (1964) and others as indicating quasi-steady and well-organized internal airflow patterns. Further evidence for a quasi-persistent circulation during this period is shown in Fig. 13.8 by vertical sections taken along constant x and y planes, which intersect near the point of the vault's highest penetration at any given time. The data are presented at 5-min intervals (approximately twice the basic scan interval) and are again generated with the software of Mohr and Vaughan (1979).

It can be seen that at all times a significant amount of precipitation echo is suspended above the 6- to 8-km altitude in the overhang, which extends some 10 km south and as much as 20 km east of the primary heavy precipitation reaching the ground. From arguments involving precipitation growth and terminal fall velocity it is logical to conclude that a large part of suspended precipitation is generated and main-

tained aloft by strong and persistent inflow in these sectors. Analysis to be presented later will show that the south and east sectors are favored locations for underlying updraft.

The changes in vault configuration illustrated in Figs. 13.7b through e and in Fig. 13.8 are not unlike the situation described by Nelson and Braham (1975), where an unsteady but persistent bounded weak echo region was observed for more than 40 min. As in their case, fluctuations in the vault's vertical extent and shifts in its horizontal position are presumably a result of corresponding oscillations in updraft strength, which in turn are probably due to local precipitation loading. Browning (1977) suggests that one requirement for maintaining steady broad updrafts, and a correspondingly persistent vaulted structure, is that the magnitude of the sub-cloud inflow to the storm must be properly matched to overall updraft strength. Later analysis will show that, while the horizontal speed of the inflow is quite strong in the present case (15 to 20 m s^{-1}), it is less than the strongest updrafts (30 to 40 m s^{-1}) estimated from multiple-Doppler radar measurements (discussed in Chapter 14).

Steep gradients in radar reflectivity factor, typically found between a vaulted region and the core of maximum Z_e, are seen only occasionally in Fig. 13.7 (e.g., on the 7.5-km CAPPI at 1620). Thus, as will be discussed further in Chapter 15, precipitation size-sorting, resulting from variable terminal velocity of different sized particles, does not seem to be a dominant feature of the vaulted structure observed here.

Radar echo characteristics associated with a quasi-steady airflow are less obvious after about 1630, when CAPPI configurations in Fig. 13.7 begin to display considerably more irregularity and complexity. Maximum reflectivities remain quite high, however, until a gradual decline in overall intensity is detected between 1700 and 1730 (compare, e.g., the size of the 50-dBZ$_e$ contour at 7.5 km in Figs. 13.7h and k). Vertical sections like those in Fig. 13.8 show that echo overhang persisted after 1630 along the southern and eastern edges of the storm, but the vaulted structure, while discernible at times (see, e.g., the 7.5 km contours in Fig. 13.7), was neither as large nor as persistent. Therefore, the positions of crosses after 1630 do not necessarily identify vaults but rather the location of weak echo regions in the context of Chisholm's (1973) definition, where organized updraft is still apt to exist.

13.3.2 Cellular Structure

Attention, which to now has centered on the broader-scale radar reflectivity features with relatively long-term stability, will turn here to a cellular structure that exhibits considerably more variation at comparatively smaller time and space scales. Because of the inevitable subjectivity involved, it is essential to define what is meant by the term "cell." We are concerned with discrete and significantly large areas of precipitation having definable histories as radar entities for

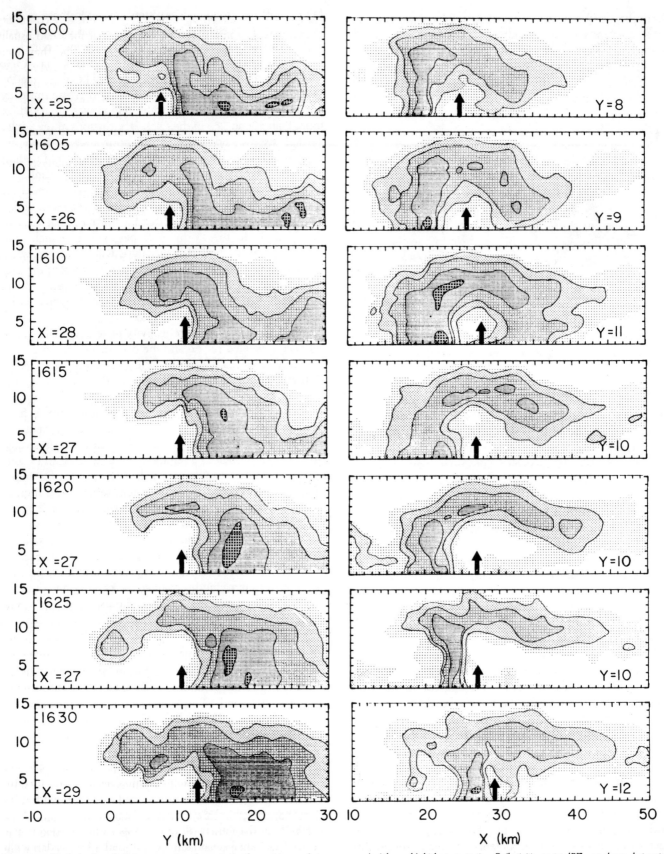

Figure 13.8 Vertical radar reflectivity cross sections demonstrating the persistence and sta-tionarity of a weak echo vault during the 30-min period shown. At each time the x-z (right) and y-z (left) planes are chosen so as to intersect at or near the point (vertically pointing ar-rows) of the vault's highest penetration. Reflectivities ≥ 20 dBZ$_e$ are shown, but contours are drawn at 10-dB intervals only for reflectivity ≥ 30 dBZ$_e$ to emphasize suspended echo "overhang" in sectors south and east of the vaulted region.

reasonably long periods of time. A distinction is made between a cell as a radar-detectable phenomenon and a cell as defined in *The Thunderstorm* (Byers and Braham, 1949), that is, a dynamic entity characterized by a concentrated region of strong vertical air motion, or updraft. It is implicit, however, that cells as radar units occur as a direct consequence of some dynamic counterpart. To the extent possible this will be verified in Chapter 14.

As mentioned above, no analysis of cellular evolution can be completely objective, but criteria involving minimum and maximum radar reflectivity, echo size, and duration can be set to achieve a reasonably consistent treatment. For the purposes of the analyses to follow, radar reflectivity maxima were tracked in three-dimensional space from the earliest appearance of a discrete 20 dBZ_e echo contour to the time of their descent through cloud base altitude (≈ 4 km). Only those entities attaining a reflectivity of ≥ 50 dBZ_e were considered, and temporal continuity for at least 10 min (4 or 5 scans) was required. While these criteria may not be particularly meaningful in the context of precipitation physics, the approach assured a consistent analysis of the more dominant echo features and eliminated from consideration a multitude of finer-scale elements that had little or no apparent influence on the storm's overall evolution and persistence.

In the practical analysis, cells were usually first identified on CAPPI's, typically in the altitude range of 6 to 10 km at some time near the middle of their life history, and were subsequently traced forward and backward in time to insure continuity. In fact, a number of cells were analyzed in this manner before the above criteria were established, and were either excluded or retained in the final sample according to whether the adopted standards were met. It should be emphasized that at all times a great deal of echo existed that was never an obvious part of any cell. Some of this was associated with the persistent high-reflectivity core discussed earlier, but a large part also comprised the extensive overhang to the east and south of the main body of the storm. It is believed, however, that with the adopted analysis scheme most of the heavy precipitation from the storm is accounted for in terms of the central high-reflectivity zone and the cells whose detailed behavior we now discuss.

Standards adopted for analysis were met by 23 cells during the period from 1545 to 1730. Figure 13.9 shows their horizontal tracks and Table 13.1 lists the duration, mean speed and direction of movement, average height, $\bar{h}$, at which the cell's maximum reflectivity resided, and maximum reflectivity factor observed during the cell's lifetime. Table 13.1 also includes an estimate of cell size as determined near its peak intensity from the horizontal area covered by the 50-dBZ_e contour. An equivalent diameter, d, computed from major and minor axes of usually irregular areas, is tabulated, representing the diameter of a circle having approximatly the same area as that of the cell. Most size estimates for weaker cells were taken at or near cloud base altitude because they usually had no earlier history with reflectivity ≥ 50 dBZ_e. For

larger cells size determination, and often tracking continuity as well, were complicated by the fact that they had usually merged with already intense echo by the time they were reaching maximum intensity. In such cases a closed 50-dBZ_e contour defining a unique cell seldom existed, so that size determination had to be estimated from irregularities in the shape of the 50-dBZ_e contour. Given these inherent uncertainties the diameter estimates in Table 13.1 are meaningful within about 1 km.

In Fig. 13.9 and Table 13.1 and subsequent analyses, individual cells are identified by the letter C and are numerically subscripted according to their order of appearance. Complete histories for cells C_1 through C_3, constituting the earliest storm components, were impossible to define because the radar was programmed to scan only the lower portions of the storm (below 7 or 8 km) before 1550. Also, termination of radar coverage due to a power failure at 1730 resulted in incomplete history for cells forming after about 1715. However, the generally decaying character evident in Fig. 13.7 after 1720 and the nature of available coverage up to 1730 indicate that few, if any, of these would have met analysis criteria.

Average characteristics of the overall sample of cells are given at the bottom of Table 13.1. Mean cell lifetime is close to 20 min, with a range of 12 to 36 min. This range is comparable to that given in *The Thunderstorm* (Byers and Braham, 1949) for the mature stage of dynamical cell development and, although radar criteria for cell definition undoubtedly differ from case to case, numerous earlier investigations of High Plains multicellular hailstorms (Dennis et al., 1970; Renick, 1971, Chalon et al., 1976; and Fankhauser and Mohr, 1977) also report typical cell durations on the order of 20 min.

The speed of cell movement varies from 10 to 20 m s^{-1}, with the mean being 15 m s^{-1}. All cells have a predominantly south-to-north trajectory, with a mean direction for the sample as a whole from the south-southwest. Fankhauser and Mohr (1977) compared the average direction of cell motion on a large number of hail days in northeastern Colorado with the mean environmental wind from the surface to the altitude of first echo formation and found a nearly one-to-one correlation. If the winds in Fig. 13.4 are integrated to a height where in-cloud temperature is $-20\,°C$ (≈ 8 km), the resulting vector is 192° and 14 m s^{-1}, in quite close agreement with mean cell motion. The implication is that horizontal momentum of air in both the subcloud and cloud-bearing layers combines to determine general cell movement.

Relatively uniform cell lifetimes and tracks and the tendency for them to form at a common distance south of the main body of the storm are primary factors influencing the storm's geographical stationarity. From tracks in Fig. 13.9 it can be seen that nearly all of the cells form in a region that lies 10 to 40 km east and 0 to 10 km south of the radar, while Fig. 13.7 shows that the main low-level echo resides in the area from 10 to 20 km north and 20 to 50 km east. Thus, the prime formation zone lies some 10 to 20 km south of the storm and, with predominantly northward motion at an average speed of 15 m s⁻¹ and duration of ≈ 20 min, cells having typical path lengths of 15 to 20 km continually replenish the echo in approximately the same location.

The net result is that propagation, determined by the preferred frequency, location, and distance of new echo growth relative to the parent echo, is almost exactly canceled by mean cell motion, and the storm remains essentially stationary for a period of more than an hour. This of course

leads to locally heavy precipitation amounts, which, due to recurrent cell passage, exceed total accumulations of 100 mm (≈ 4 in) at some raingages located within a dense network designated by dashed boundaries in Fig. 13.7. Some gages, for example, receive precipitation from as many as five individual cells (see tracks in Fig. 13.9).

This behavior differs from the classic model of multicellular storms (Browning and Ludlam, 1960; Marwitz, 1972b; and Chisholm and Renick, 1972), wherein periodic formation of discrete new cell growth on some favored side (usually the right forward flank) contributes a significant propagational component to the storm's overall displacement. In the present case, although significant new growth is concentrated mainly on the southern flank, there appears to be no preferred sequential growth pattern relative to the storm's center. For example, after the early development of cells C_4 through C_7 on the southwestern side of the storm, C_8 appears on the south and C_9 and C_{10} form in the southeastern sector. Subsequent development then shifts back to the southern and southwestern flanks.

Some deviation from mean characteristics is exhibited by cell tracks in Fig. 13.9 and by data in Table 13.1. Subsets can be formed from the total sample to reveal discriminating behaviorial features that seem to be a function of where individual cells develop and how they move relative to the center of the main echo. Cells C_9 and C_{10}, for example, form in the southeastern sector, and their tracks show strikingly similar cyclonic curvature. Another group (C_4, C_6, C_8, C_{11}, C_{15}, C_{16}, and C_{23} through C_{26}) develops within or near shelf cloud extending well to the south of the storm's existing high-intensity core, and most of these follow a northward trajectory. All others (C_7, C_{13}, C_{14}, and C_{17} through C_{22}) appear in the flanking cloud line illustrated in Fig. 13.6b and move in a

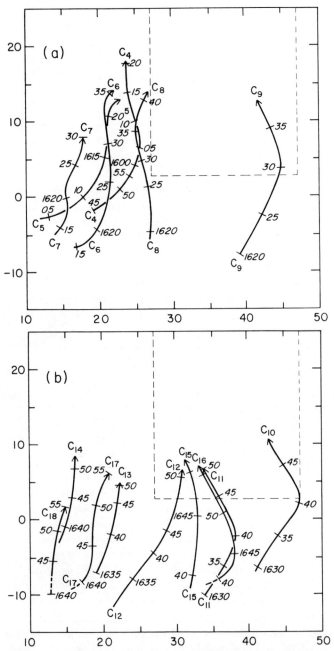

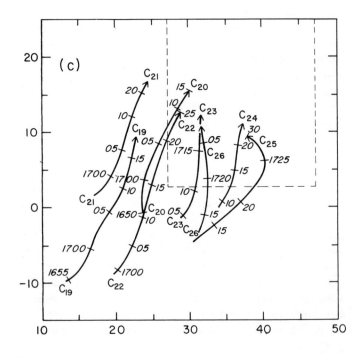

Figure 13.9 Horizontal tracks of radar reflectivity maxima representing lifetimes of cells forming during the periods (a) 1545-1625, (b) 1630-1645, and (c) 1655-1720. Cells labeled as in Fig. 13.7.

TABLE 13.1
Cell characteristics

Cell	Lifetime Begin	Lifetime End	Duration (min)	Mean motion (deg)	Mean motion (m s^{-1})	Diameter (km)	Area (km²)	Mean height (km)	Maximum Z_e (dBZ$_e$)
C_4	1545	1621	36	193	9.8	11.7	108	9.0	70.8
C_5	1603	1623	20	215	17.4	7.2	41	7.3	64.7
C_6	1613	1636	23	194	16.8	6.7	35	7.4	62.6
C_7	1614	1629	15	198	14.3	4.3	15	6.7	53.3
C_8	1619	1641	22	180	13.4	6.0	28	8.2	68.0
C_9	1620	1639	19	187	19.1	5.2	21	6.9	59.9
C_{10}	1629	1649	20	187	13.2	3.2	8	6.2	53.2
C_{11}	1630	1651	21	180	13.5	7.6	45	7.7	66.0
C_{12}	1631	1651	20	209	16.2	6.5	33	6.1	52.5
C_{13}	1635	1650	15	193	14.3	3.5	10	6.9	56.5
C_{14}	1636	1653	17	192	12.9	1.0	1	6.5	53.8
C_{15}	1636	1653	17	178	19.7	9.2	66	8.1	68.3
C_{16}	1638	1656	18	168	15.5	7.6	45	8.3	67.8
C_{17}	1636	1654	18	196	15.9	2.2	4	6.7	53.1
C_{18}	1640	1655	15	197	12.6	1.8	3	5.6	53.5
C_{19}	1654	1718	24	208	14.6	1.1	1	5.8	51.7
C_{20}	1655	1714	19	203	18.1	3.4	9	5.7	52.7
C_{21}	1657	1714	17	208	15.0	2.7	6	6.0	55.2
C_{22}	1700	1725	25	201	15.9	3.6	10	5.9	52.5
C_{23}	1704	1719	15	190	17.3	5.8	26	6.7	56.4
C_{24}	1707	1724	17	198	12.0	7.4	43	6.8	62.6
C_{25}	1711	1729	18	210	16.7	5.6	25	7.4	57.9
C_{26}	1714	1726	12	182	18.4	3.5	10	6.4	55.2
N = 23		Means:	19	194	15.0	5.1	26	6.9	58.6

predominantly north-northeasterly direction. This tendency for cells that form in common positions relative to the storm's center to move along similar paths suggests that some fairly steady larger-scale circulation prevails with respect to the storm as a whole. This premise will be investigated further in Chapter 14. The full ensemble of cell tracks exhibits a net convergence toward the southwestern sector of the precipitation network, accounting for the storm's geographical stationarity at location $(x,y) = (40,13)$ km.

Figure 13.10 shows the evolution of cell pairs selected to represent typical features of cells forming in the three fairly distinct sectors. Histories are illustrated by the position of the maximum reflectivity factor in vertical y-z planes, shown as a function of time at twice the available scan interval. The time series indicate the way all cells were tracked in the basic analysis scheme, i.e., the selected vertical sections represent the cells' full life cycles and the y-z profiles are chosen to intersect the cells' maximum reflectivity factors at given times.

Cells C_9 and C_{10} are unique in that they develop well ahead of the surface outflow boundary shown in Fig. 13.7. Both have similar lifetimes (≈ 20 min) and achieve a maximum reflectivity factor of < 60 dBZ$_e$ while descending through cloud base altitude, where the temperature is only slightly greater than 0 °C. Hence, melting of hail may have contributed to an enhanced reflectivity factor. Both form near an altitude of 8 km and reside in a temperature regime of -5 °C

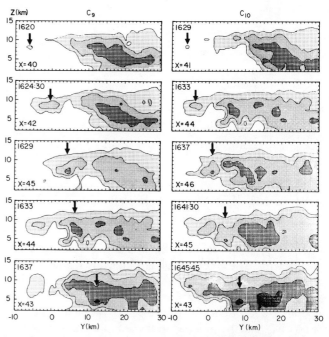

Figure 13.10a Vertical radar reflectivity cross sections showing the evolution of cells C_9 and C_{10}, both of which form in the storm's southeastern sector and follow similar cyclonically curved paths, as indicated in Fig. 13.9. At each time shown, the y-z planes are chosen to intersect the cell's core of maximum reflectivity, designated by heavy arrows. Reflectivity contours are presented as in Fig. 13.7.

to $-15°C$ (see Fig. 13.4) for 12 to 15 min while traveling rapidly northward toward the eastern sector of the storm. The impact of cell C_9 on the horizontal radar reflectivity pattern can be seen in Fig. 13.7f at the 2.5 and 5.0 km levels.

Histories of C_7 and C_{13}, given in Fig. 13.10b, are representative of cells forming to the southwest of the main storm. Again, striking similarities are exhibited with respect to the altitude of development, residence time aloft, and maximum intensity achieved. Compared with C_9 and C_{10}, however, lifetime is shorter (15 min) and intensity is somewhat less, with the maximum Z_e again occurring somewhere near cloud base altitude. As indicated by data in Table 13.1, not all of the cells that developed to the southwest were short-lived, with C_{19} and C_{22} being notable exceptions. Developments that took place in this sector tended, however, to be weaker in general than those that formed elsewhere. Vertical profiles in Fig. 13.10b show that cells of this type are never a part of the intense storm core, and their primary influence is seen in Fig. 13.7 as a westward extension of echo at low altitudes evident at all times after about 1620.

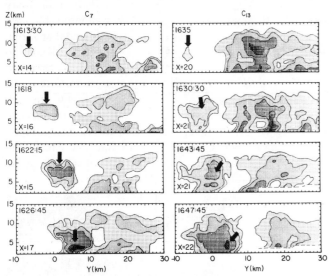

Figure 13.10b Same as Fig. 13.10a, except for cells C_7 and C_{13}, which are representative of cells that form in the storm's southwestern sector and follow northeasterly trajectories, as indicated in Fig. 13.9. See also Figs. 13.7e and g at 2.5 km.

Those cells that form more or less directly south of the storm have the dominant influence on the shape and position of the storm's more intense echo pattern. Life histories of two of these, C_8 and C_{11}, are given in Fig. 13.10c. Tracks begin some 15 to 20 km south of the existing core of high reflectivity factor. These cells dwell at somewhat higher altitudes and colder temperatures while traveling northward over what will be shown later to be the main inflow and primary updraft zone. Cells of this type all achieve intensities ≥ 60 dBZ_e, and, by moving northward into preexisting echo, play the primary role in maintaining and sometimes rearranging the configuration and location of the persistent high-reflectivity zone. For example, collapse of the vaulted structure occurring between 1625 and 1630 (Figs. 13.8 and

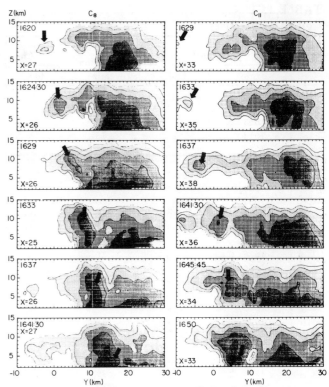

Figure 13.10c Same as Fig. 13.10a, except for cells C_8 and C_{11}, characterizing major developments that form south of the main storm and have the dominant influence on storm evolution. Reference to the 2.5-km CAPPI's in Figs. 13.7f and g indicates that both of the cells eventually constitute a part of the storm's zone of highest reflectivity.

15.18) is a consequence of the intrusion of C_8 into the weak echo region. Similarly, the perceptible eastward shift in the location of the more intense low-level echo pattern occurring between 1630 and 1650 (Fig. 13.7) is caused mostly by descent of cells C_{11}, C_{15}, and C_{16}.

Insight into the relationship between the maximum reflectivity factor ($Z_{e_{max}}$) achieved by cells and the other characteristics listed in Table 13.1 is provided by the scatter diagrams in Fig. 13.11. Cell size is compared with $Z_{e_{max}}$ in Fig. 13.11a, and the scatter displays a distinct tendency for the more intense cells to be the largest. Furthermore, those cells identified earlier as having favorable tracks relative to the primary updraft zone (filled circles) are shown to constitute the largest and most intense cells in the sample. These undoubtedly account for the largest areal extent of heavy precipitation and hail at the ground. A demonstrated tendency for the smaller cells also to be the weakest suggests that finer-scale radar echo features of the kind tracked by Barge and Bergwall (1976) have little significant impact on the storm's overall radar reflectivity structure or its relevant precipitation production in the present case.

Correlation between the average height, $\bar{h}$, at which cells reside during their lifetimes and the maximum reflectivity they ultimately attain is illustrated in Fig. 13.11b. It can be seen that the most intense cells dwell at distinctly higher mean altitudes than do those of lesser intensity. This is also

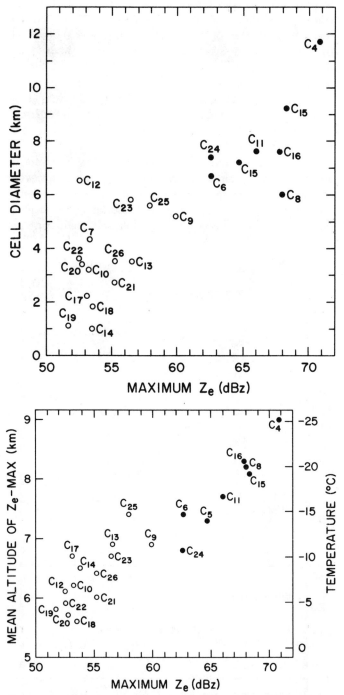

Figure 13.11 Scatter diagrams showing the relationship of the maximum reflectivity (Ze_max) achieved by cells to (a) cell diameter (km) near the time of peak intensity, and (b) the altitude of maximum reflectivity (km) averaged over the cell's lifetime. Filled circles designate cells having $Z_{e_{max}} \geq 60$ dBZe.

data in Table 13.1 show that two of the more intense (C_{15} and C_{24}) have comparatively short lifetimes.

Periodic or cyclic patterns of new cell development, documented in the studies of multicellular storms cited earlier, do not appear to be the primary mode of propagation in the present case. Rather, new growth in the form of discrete cells tends to occur in series. Data in Table 13.1 indicate a gap of around 10 min between the appearance of cells C_5 and C_6 and again between C_9 and C_{10}. Later there is a nearly 15-min lapse between the formation of C_{18} and C_{19}. Following each of these lulls, clusters of new cells appear in fairly rapid succession, and the development of each new series seems to coincide in more than a casual way with the time when echo from earlier cells was cascading toward the surface.

Chronology of cell development is shown graphically in Fig. 13.12, where time of cell initiation is given by both the abscissa and the ordinate, while the abscissa identifies total lifetime as well. Histories of cells achieving maximum intensity ≥ 60 dBZe are designated as bold lines. Recalling that a cell's history was terminated when its maximum reflectivity descended through cloud base altitude, it is of interest to note that C_4 and C_5 collapse almost simultaneously and that within 7 or 8 min the bracketed series C_{10} through C_{18} begins to appear in rapid succession. In the same context, it is assumed that C_1 through C_3, for which there is incomplete radar coverage, had some influence on the formation of C_6 through C_8. Similarly, the decline of C_6 through C_9 seems to associate in a loose way with the appearance of the closely grouped series beginning with C_{19} some 12 to 15 min later. Finally, the latest cells, C_{24} through C_{26}, appear after the dissipation of the major components, C_{11}, C_{15}, and C_{16}.

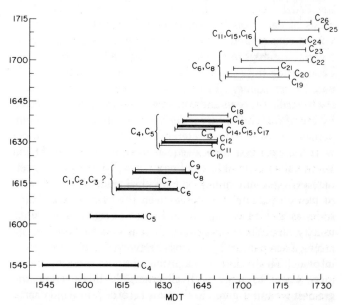

Figure 13.12 Chronology of cellular development where both the abscissa and the ordinate give the time of cell initiation, while line segments parallel to the abscissa define cell lifetime as well. Histories of cells achieving intensity ≥ 60 dBZe are designated as bold lines. Braces show loose relationship between decay of cells on the left and the initiation of those to the right.

evident in Fig. 13.10 and is presumably the result of stronger supporting updrafts. A plot similar to those in Fig. 13.11, relating cell duration to maximum intensity, failed to reveal any obvious trend. It was already mentioned that two of the weaker cells, C_{19} and C_{22}, were among the longest-lasting, and

A possible implication is that downdraft associated with the collapse of each wave of cells, particularly the more intense ones, accelerates outflow, spreading southward beneath the storm so as to enhance convergence at the interface with nearly uniform ambient southeasterly flow. The interlude between decay of earlier cells and successive eruption of each new series is generally consistent with the time required for air to travel from the point of cell decay to the outflow boundary. The extent to which this premise is confirmed by surface mesonetwork analysis and aircraft data is investigated in the following sections.

13.4 SURFACE MESOSCALE ANALYSES

The distribution of the 46 continuously recording mesometeorological sites is given in Fig. 13.1. Instruments comprising the NCAR Portable Automated Mesonetwork (PAM) were located at positions designated by triangles. The PAM system, designed to provide central recording of meteorological data transmitted automatically from remote sampling stations, received its first field test during the 1976 NHRE field program (Brock and Govind, 1977). At all other sites (black dots) data were recorded on conventional drum-mounted and clock-driven strip charts, which required daily service for changing charts and performing quality control calibrations. Basic meteorological variables measured at installations of both types included air temperature, station pressure, relative humidity (wet-bulb temperature in the case of the PAM system), and horizontal wind direction and speed.

Once suitable software was developed, data retrieval from the PAM magnetic tape archive was straightforward. Synthesis of data from the conventional strip chart records involved considerably more effort, however, requiring several painstaking steps to ensure data quality and consistency. The basic analog traces, originally recorded in curvilinear coordinates, were converted to digital form for computer processing to facilitate access and handling. Data were then corrected by use of calibration curves derived from the daily field comparisons with standard reference instruments. After these were incorporated, analog sequences were regenerated from 1-min values plotted in rectilinear form. To minimize uncertainties in absolute timing on individual records it is desirable to plot all parameters from a given site at a common time scale, as is done in Fig. 13.13. By use of significant events, usually identifiable on all records, it is possible to eliminate timing discrepancies by forcing the event of interest to occur in phase. This is done by adjusting absolute time on one or more traces to agree with all others. In this exercise the greatest weight is given to the time recorded on wind charts, since these possess the greatest accuracy (on the order of ±1 min). The timing accuracy achieved for the integrated data is about ±5 min.

An example of time-adjusted and calibrated digital data

reconstructed in analog format is given in Fig. 13.13a for a conventional station situated near the storm. Associated conservative properties of the surface air (θ, θ_e, and q) calculated from the basic pressure and thermodynamic variables, are represented by the dashed curves in Fig. 13.13c. Similar records from a nearby PAM site, also directly affected by the storm, are shown for comparison in Figs. 13.13b and c (solid curves). The power failure at the Grover field headquarters, which terminated radar coverage, also resulted in the loss of PAM data after 1730.

Vertical dashed lines in Fig. 13.13 identify the time of arrival of outflow air at the two selected stations. This event is marked by a sharp temperature drop, a surge in relative humidity and a wind shift typically associated with a wind speed minimum. The wind speed lull and wind shift are

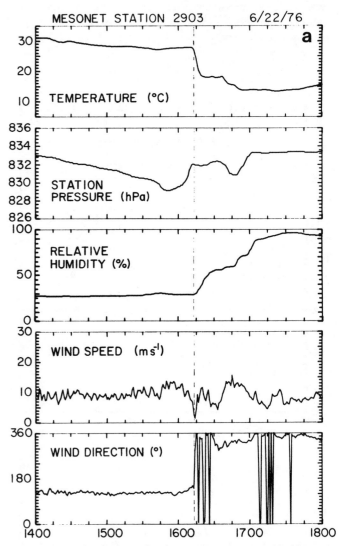

Figure 13.13 Analog records of indicated surface parameters measured by (a) a conventional mesonetwork installation located 29 km east and 3 km north of the Grover field headquarters, and (b) a Portable Automated Mesonetwork station situated 27 km east and 13 km north of the Grover site. Conservative thermodynamic variables computed from data on the records in (a) and (b) are given in (c). Vertical dashed lines identify the time that cool outflowing air arrives at each site. Solid station circles in Fig. 13.14 show site positions relative to the storm.

followed by increasing wind speed of a gusty nature, which in this case reaches peak velocity at both sites some 30 min after the wind shift. These are well-known characteristics of transitions associated with the arrival of thunderstorm outflow at the surface (see, e.g., Humphreys, 1914; Byers and Braham, 1949; and Charba, 1974), but until the advent of the simultaneous sampling capability of the PAM system the precise sequential relationship between the various relevant parameters has been somewhat uncertain. The problem has to do with independent timing mechanisms for recording individual parameters, all of which have poor absolute timing accuracy. The fact that temperature, relative humidity, and wind events all occur simultaneously at the PAM site lends credibility to the manner in which phase adjustments were applied in correcting timing errors in the conventional data.

For the purpose of investigating synoptic changes in relevant surface parameters, data are extracted from each station's time series and plotted in horizontal array, as illustrated for one selected time in Fig. 13.14. Maps of this type,

combining data from both observing systems, are produced at 5-min intervals.

In Fig. 13.14 the environmental conditions shown south and east of the surface outflow boundary are fairly representative of the entire active storm period under discussion. Cool outflowing air, evident as northerly flow over the northwestern portions of the grid, spread continually southward and eastward from the location of the nearly stationary storm, eventually affecting all of the stations in the network by 1845. Isochrones of the position of the outflow boundary were determined at 5-min intervals from a plot of the times of wind shift and temperature break at all sites. Positions of the surface inflow-outflow interface shown in Figs. 13.5, 13.7, 13.14, 13.15, and 13.19 are based on this analysis. The axis of a unique line of radar echo ($Z_e = -5$ to 15 dBZ_e) observed on low-level scans between 1600 and 1700 is also shown as a heavy dashed line in Figs. 13.7b through h. Targets responsible for this radar return, located generally 2 to 4 km behind the leading edge of the outflow, are undefined, but may be caused by such things as refractive index gradient, airborne insects, or surface debris raised by gusty winds produced by downdraft surges. Dust clouds rising from the surface are frequently observed from research aircraft flying in the inflow sector, and the airborne observers' notes indicate that such was the case on this date. This feature, then, most likely corresponds to what is frequently referred to as the "gust front," but in this case it is not coincident with the leading edge of the outflow boundary.

Liberal use was made of each station's time series data in preparing the analyses of thermodynamic variables presented in Fig. 13.14. From a conventional time-to-space transformation (see, e.g., Fujita, 1963; and Foote and Fankhauser, 1973),

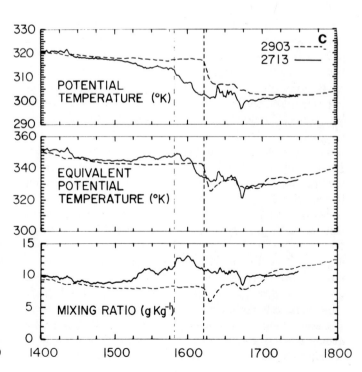

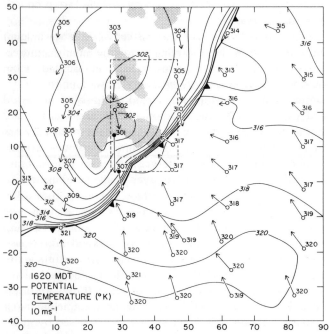

Figure 13.14a Potential temperature (°K) field near the surface at 1620. Heavy barbed line shows the interface between ambient southeasterly airflow and the outflow from precipitation regions, represented by stippled PPI radar reflectivity patterns ($Z_e \geq 35$ dBZ$_e$) near cloud base altitude (≈ 4 km). Border is labeled in kilometers from Grover field site (0,0). Circled X shows location of descent sounding by Queen Air N304D from cloud base altitude to near the surface.

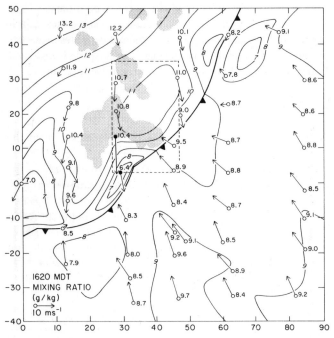

Figure 13.14b Same as Fig. 13.14a, except contours show the field of mixing ratio (g kg^{-1}).

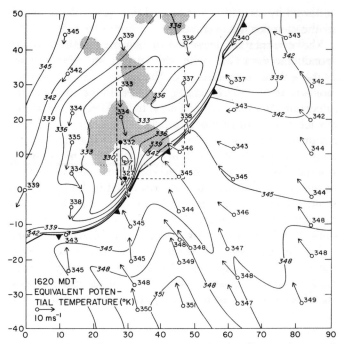

Figure 13.14c Same as Fig. 13.14a, except contours show the field of equivalent potential temperature (°K).

ment using a single criterion such as overall storm motion. In the present case, because the storm itself was stationary, it was necessary to distribute the data spatially, according to time and to the station's location with respect to the storm. In the case of stations lying close to the inflow-outflow interface, displacement was aligned normal to the advancing outflow boundary. For stations located well behind the outflow boundary off-time data were spatially distributed according to the mean cell motion. In the environment southeast of the storm, time series were displaced along the direction of the local wind by use of a 30-min average of the local wind velocity to determine the spatial offset from each station's location.

Incorporation of time-to-space conversion enabled definition of fine-scale features smaller than those resolvable from the given station density alone. For example, the sharp drop in θ seen in Fig. 13.13c, following the wind shift, is illustrated in Fig. 13.14a as a zone of tight contour packing behind the outflow boundary. The gradient is largest immediately to the south and east of the primary storm, as represented by radar echo ≥ 35 dBZ$_e$ in the altitude range of 6 to 8 km. Two pools of cold air delineated by 302 K contours are in good spatial agreement with the locations of the main storm and the next largest echo, seen in a declining stage over the northern sector of the surface precipitation network. The field of θ is comparatively flat in the environment ahead (southeast) of the outflow boundary, decreasing from ≈ 320 K along the southern edge of the network to as low as 313 K in the northeast. While some of this gradient from south to north may reflect large-scale conditions, it is primarily due to shading caused by anvil cloud blown downwind from the main body

off-time reports within ± 15 min of map time (1620) were displaced along spatial increments determined by the local speed of significant perturbations. Normally it is possible to determine a representative time-to-space displacement incre-

of the storm and by cooling beneath shelf cloud in the southeast sector of the storm.

The horizontal distribution of surface mixing ratio, q, corresponding to the θ field in Fig. 13.14a, is given in Fig. 13.14b. In the ambient air southeast of the storm, moisture content ranges from 8 to nearly 10 g kg^{-1}, with 8.5 g kg^{-1} being a fairly typical value. Regions of somewhat drier air are found to the left and right of the generally confluent southeasterly airflow. Moisture content in excess of environmental values exists over the area to the north, west, and northwest of the storm's center and is presumably a result of the evaporation of fallen precipitation. A small area of distinctly drier air is located south of the main precipitation zone immediately to the rear of the outflow boundary.

Fields of θ and q combine to produce the θ_e pattern shown in Fig. 13.14c. Two fairly distinct tongues of potentially warm and moist air are seen in the inflow, one advecting on mostly southerly wind south of the storm and the other approaching from the southeast relative to the storm's center. Air in these sectors a few kilometers ahead of the outflow boundary is characterized by θ_e in the range of 345 to 348 K. Anomalously dry air noted in the mixing ratio field south of the main precipitation zone and behind the inflow/outflow interface associates with a small region of low θ_e in the same relative position. The minimum value of ≈ 327 K is comparable to the coolest and driest air noted earlier on the representative sounding at an altitude of 7 km or so. Evidently some of this air had its origin aloft and was carried downward in downdraft circulation to the indicated position south of the storm. Foote and Fankhauser (1973) found a similar relationship between the locations of the potentially coldest surface air in the outflow and the main precipitation zone.

The circled x located at $(x,y) = (32, -15)$ km in Fig. 13.14a marks the location of the spiral descent sounding used to modify the subcloud portion of the Potter sounding mentioned in Section 13.2. Comparison of the uniform subcloud parameters shown in Fig. 13.4 with nearby surface conditions in Fig. 13.14 shows close agreement, verifying that the layers near the surface in this sector represent the source for air entering the storm's circulation through cloud base on its southern flank. Further details of the thermodynamic structure at cloud base altitudes will be elaborated in the next section.

After completing the descent sounding, the aircraft assumed an east-northeasterly heading, flying at low altitude essentially cross-wind toward the northeastern border of Fig. 13.14. Horizontal variations in θ, θ_e, and q measured along this leg corroborate the horizontal fields shown in Fig. 13.14. In particular, the aircraft data confirm the existence of a narrow tongue of low θ_e extending southeastward from the cusp in the outflow boundary, as shown in Fig. 13.14c, and also verify the decrease in θ from southwest to northeast illustrated in Fig. 13.14a. Presence of lower θ and drier air ahead of the outflow boundary along its northern extent provides explanation for the general decline in convective activity over the northern portions of the grid, since both would tend to decrease the thermal instability of the ambient air in that region. Thus, cooling caused by anvil shadow, as mentioned above, may be an element of self-destruction from the standpoint of available potential instability.

The evolution of surface kinematic features for the period 1600 through 1700 is represented by wind vector and divergence fields in Fig. 13.15. Vector fields were obtained at 5-min intervals by use of a modified objective analysis scheme (Achtemeier et al., 1978) of the type developed by Barnes (1973). The approach applies classic harmonic theory in the design of weighting functions for interpolating between irregularly spaced data points in both the time and space domains. Options include: (1) the number of data points to be considered in the spatial interpolation at each grid location, (2) the number of time steps ($\Delta t = 5$ min in the present case) to be displaced through time-to-space transformation, (3) the criteria for displacement of off-time data, and (4) adjustable parameters controlling the shape of exponential time/space weighting functions.

Evaluation of the wind field at a selected time, based on a careful subjective analysis of streamline, isotach, and isogon patterns, was the basis for choosing appropriate options to be applied in the objective analysis. By invoking inherent kinematical properties of horizontal flow and by conversion of time series data to the spatial domain (as was done in the case of the thermodynamic analysis), it was possible to obtain spatially continuous flow fields with considerably greater detail than would result from consideration of spatially distributed data alone. The validity of such an approach, whether applied subjectively or objectively, is, however, highly dependent upon the temporal stationarity of the flow field under consideration.

Off-time data within ± 15 min of map time were incorporated in the present analysis, and displacement from the point of observation followed the same lines used in the analysis of the thermodynamic variables. An example of the data density following time-to-space transformation is given on the right hand side of Fig. 13.15a. Results of the objective interpolations are represented by the vector fields shown at 20-min intervals on the left hand side of Fig. 13.15. For illustrative purposes vectors are plotted at 4-km intervals, but horizontal wind components were actually obtained at 2-km grid spacing, and these constitute the input for the calculation of the divergence fields also given in Fig. 13.15.

As indicated by the mean conditions given in Table 13.1, cells had an average diameter of about 5 km when they were reaching their greatest radar intensity. This usually coincided with their early period of descent toward the surface from aloft, and probably also represents approximately the time when outflow from their downdrafts is first detectable at the ground. Even though the effects of downdrafts from individual cells were detected in the time series data from specific sites beneath the storm (see, e.g., the small-scale fluctuations in Fig. 13.13b), it is unlikely that these would be reliably preserved in the horizontal flow fields, given the initial data density and smoothing imposed during objective interpolation. Thus, little significance can be attached to the relative positions of the divergence maxima and the primary low-level radar return (≥ 35 dBZ$_e$) represented by the light stippling in Fig. 13.15.

Emphasis here is on the elongated band of convergence centered along the inflow-outflow interface where opposing flows converge toward a wind speed minimum. While "bubbles" of convergence within this zone may also be artifacts of the analysis scheme, the overall scale of this feature is sufficiently large to be treated consistently well by the analysis technique. The grid in Fig. 13.15 is slipped southward with

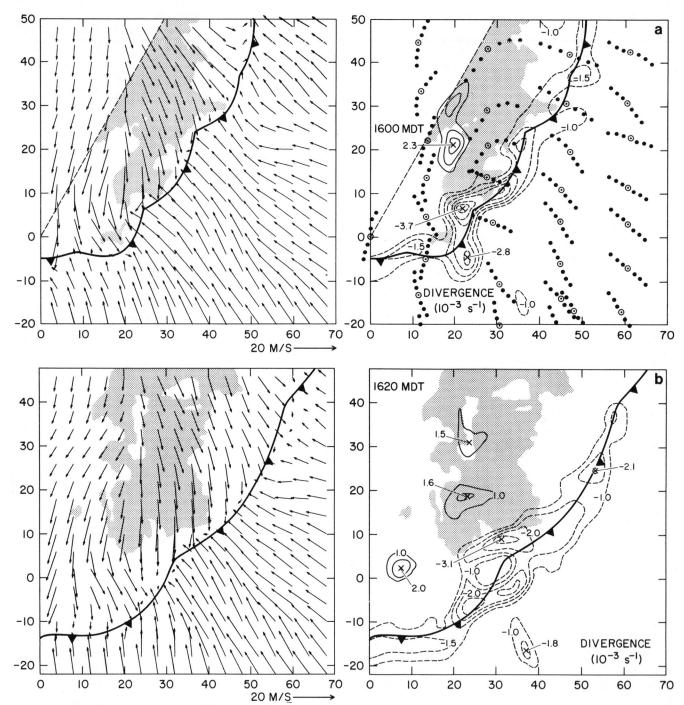

Figure 13.15 Objectively interpolated wind vectors (left) and associated fields of horizontal mass divergence (right) for (a) 1600, (b) 1620, (c) 1640, and (d) 1700. Heavy barbed line again shows leading edge of advancing cold air outflow. Note that figure boundaries are displaced southward with time to keep low-level precipitation echoes (stippled areas) and the zone of maximum convergence (dashed contours) well within the grid. The series of dots distributed about station locations at 1600 in Fig. 13.15a demonstrates how off-time data were incorporated in the spatial interpolations. Dashed line in Fig. 13.15a denotes one limit of the radar scan at that time. Echoes shown represent reflectivities ≥ 35 dBZ$_e$, observed at an antenna elevation angle of $0.5°$.

time to keep this zone within the boundaries of a rectangular area that also accommodates most of the significant low-level precipitation echo.

One noteworthy feature in the divergence fields before 1640 is the tendency for the strongest convergence to extend some distance to the rear of the outflow boundary. At 1600 a maximum of 3.7×10^{-3} s^{-1} is seen a few kilometers west of the cusp on the outflow boundary, and at 1620 a band of strong convergence is found oriented east to west at a position just to the south of the low-level precipitation echo.

Intermediate output showed that such a maximum persisted within the cold outflow air throughout the period 1600 to 1630, and that its position was always in close agreement with the location of the weak echo vault discussed in Section 13.3. If the vaulted region of the storm does indicate a persistent broad and upward-accelerating buoyant updraft zone, then a logical dynamical consequence at the surface would be an associated non-hydrostatic pressure deficit (see, e.g., Yau, 1979). The surface pressure trace at station 2903, given in Fig. 13.13a, shows that a marked mesodepression affected

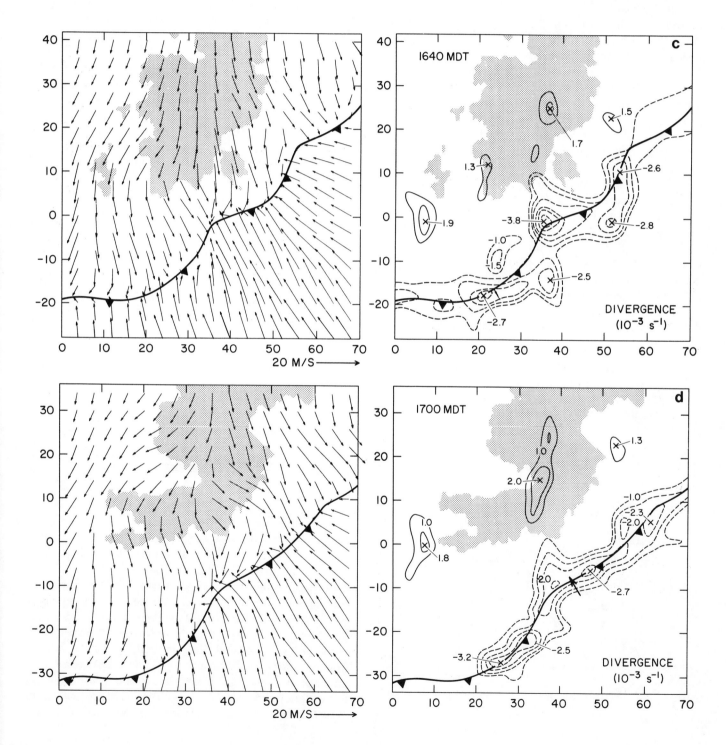

that site between 1545 and 1615. This site lay closest to the features under discussion and no other network station recorded a perturbation of similar magnitude. The evidence, therefore, suggests that the convergence maximum and observed surface pressure deficit are directly related to an organized updraft in the vicinity of the vault aloft (illustrated most clearly at 7.5 km in Fig. 13.7d). Of further possible significance is the fact that after 1630, when the three-dimensional radar echo loses its distinctly vaulted structure, convergence maxima are no longer found to the rear of the outflow boundary.

Another distinctive region of convergence is the one found along the southern boundary of the grid at 1600, which steadily grows in intensity as the outflow advances toward it. This maximum is a result of the marked confluence between strong southeasterly winds to its right and more southerly flow on its left, as evident in the vector fields south of the storm. It is difficult to say whether this confluence is the result of accelerations associated with the mesodepression mentioned above, or whether it might be influenced by the substantial upward slope of the terrain in this locale (Fig. 13.1). The isolated area of convergence does, however, provide an explanation for the development of cells C_9 and C_{10}, which were anomalous in that they formed a considerable distance ahead (southeast) of the surface position of the outflow boundary. Comparison of their points of origin (given in Fig. 13.9), with the location of this secondary convergence maximum shows that the cells first appear only a few kilometers to its north.

All other cells were first detected near or to the north of the primary convergence zone, as illustrated in Fig. 13.16 by

the position of their points of origin, plotted relative to the location of the surface outflow boundary at the time of appearance. Filled circles identify those cells, typified by histories shown in Fig. 13.10c and emphasized in Fig. 13.12, which ultimately reached the greatest intensity and had greatest impact on storm evolution. With the exception of one (C_{24}), all of these first appear within a few kilometers of the surface outflow boundary, suggesting that boundary layer convergence was a dominant factor contributing to their growth into major storm components.

A more general relationship between the evolution of surface convergence and storm intensity is demonstrated in Fig. 13.17. The solid curve gives the transition in air mass flux convergence resulting from the areal integration of all of the surface convergence $\geq 1.0 \times 10^{-3}$ s^{-1} (the area enclosed by the outer dashed curves in Fig. 13.15). Surface flux convergence increases steadily to a peak around 1630, followed by a gradual decrease to 1730 when full network analysis was no longer possible. The average of the 2-h period ($\approx 10 \times 10^6$ kg s^{-1} mb^{-1}) is nearly double the magnitude found by Foote and Fankhauser (1973) in a similar analysis of a traveling hailstorm in a strongly sheared environment.

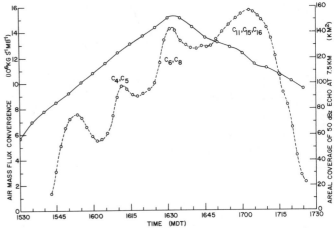

Figure 13.17 Surface air mass flux convergence (units: 10^6 kg s^{-1} mb^{-1}) within area shown in Fig. 13.15, plotted as a function of time (solid curve; scale on the left). Areal coverage (km²) of the 50-dBZ$_e$ radar reflectivity contour at 7.5-km altitude (dashed curve) is shown for comparison (scale on the right). Contributions toward peaks in high-intensity radar echo coverage made by major cells are indicated by labeling.

Areal coverage of radar echo enclosed by the 50-dBZ$_e$ contour at 7.5 km altitude is also given as a function of time in Fig. 13.17 and is intended as a measure of overall storm intensity. The radar data, having finer-scale time and space resolution, exhibits a series of wave-like perturbations superimposed upon a general increase in storm intensity, which appears to lag some 30 min behind the trends in surface flux convergence, as might reasonably be expected. A fairly strong cause and effect relationship can be inferred from the fact that boundary layer flux convergence maximizes prior to the peak storm intensity and that storm decay proceeds rather rapidly following the onset of gradual decline in surface flux convergence. Storm dissipation, occurring at a

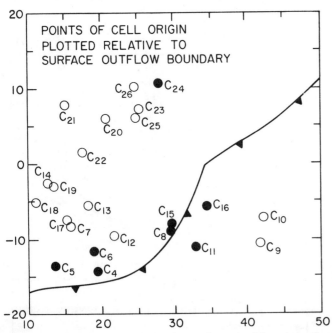

Figure 13.16 Points of cell origin plotted relative to the position of the surface outflow boundary at 1630. Filled circles designate cells with $Z_{emax} \geq 60$ dBZ$_e$. Distance is in kilometers from Grover (0,0).

comparatively faster rate than the decrease in convergent flux, could be due in part to the greater separation between the storm and the outflow boundary indicated in Fig. 13.15 during the later stages.

Each of the maxima in high-intensity radar echo coverage can be attributed in a fairly reasonable way to the times when the more intense cells, emphasized in Fig. 13.12, are reaching their maximum intensity near the center of the storm. This confirms their importance in maintaining storm intensity. Conversely, as indicated in Fig. 13.16, cells constituting later developments (C_{19}-C_{26}) all form a considerable distance away from the primary surface convergence zone, but, according to their characteristics given in Table 13.1 and trends exhibited in Fig. 13.17, few of these contribute significantly to the storm's overall intensity and/or longevity. The failure of these and earlier cells formed along the southwest flank to become major storm components is presumably due to the lack of adequate boundary layer support and less favorable thermodynamic conditions. Thus, it appears that only those cells forming in a favored position relative to the strong boundary layer convergence and main inflow contribute significantly to overall storm structure, and, as the outflow boundary becomes further removed from the main body of the storm, convergence necessary for the development of major new growth diminishes and the storm begins to decay. A decrease in convergence at the inflow-outflow interface is to be expected if the ambient southeasterly flow remains nearly constant, the surface area covered by the outflow air increases with time, and the rate at which the storm feeds mass into the subcloud layer is steady or decreases with time. From the available information all of these conditions appear to have been met.

One final issue deserves discussion in the context of the surface airflow. The vertical structure of the horizontal wind, given in Fig. 13.4 as being representative of the environment, shows that airflow at all levels has a south to north component. In Section 13.3 it was demonstrated that the mean motion of cells was in good agreement with the average wind in a layer from the surface to an altitude of 8 km or so. Given these conditions the question arises as to the source of the strong northerly momentum exhibited in regions of storm outflow at all times after about 1530. The explanation clearly does not lie in a simple model of vertical transport of horizontal momentum, and requires that significant non-hydrostatic forces come into play. Data available here do not permit an adequate treatment, but the issue is discussed further in Chapter 14.

13.5 SUBCLOUD AIRCRAFT MEASUREMENTS

Investigation of air motion and thermodynamic structure in the inflow near cloud base altitude began shortly after 1600 and continued until about 1700. Flight track segments

flown during this period by three Queen Air aircraft (N10UW, N306D, and N304D) are plotted in Fig. 13.18, with radar echo contours and the position of the surface outflow boundary included for reference. Flight legs covering horizontal distances of 20 to 30 km typically had 5-min durations and, as indicated by horizontal wind vectors plotted at 15-s intervals, were usually flown on cross-wind headings.

All three aircraft were equipped to measure static pressure, air temperature, dew point temperature, and horizontal wind; however, instrument malfunctions developed on two of the aircraft. The inertial navigational system on N304D failed just prior to 1600, resulting in the loss of its three-dimensional wind measurement capability and accounting for its departure after obtaining the descent sounding discussed earlier. On N10UW the dynamic pressure measurement was lost because of an electrical failure, but an ingenious technique incorporating different recovery factors for two types of temperature sensors (Biter and Anderson, 1979) made it possible to retrieve estimates of the aircraft's true air speed. These were sufficiently accurate to compute all of the meteorological variables, including horizontal wind.

For both N10UW and N306D the vertical air velocity, w, was calculated from equations involving aircraft attitude and rate of climb (see, e.g., Lenschow, 1976). In the case of N306D, vertical motion of the aircraft is obtained from an inertial platform, and Kelly and Lenschow (1978) show that its use in computation of vertical air motion can resolve frequencies as high as 0.3 Hz. This is equivalent to a minimum resolvable horizontal space scale of ≈ 250 m. Resolvable scale lengths using rate of climb from a variometer, as is done in vertical motion calculations for N10UW (Cooper, 1978), are somewhat larger (≈ 1 km).

13.5.1 Structure of the Inflow Near Cloud Base Altitude

Comparison of wind vectors obtained at flight altitude with those at the surface (Figs. 13.14 and 13.15) shows that south-southeasterly flow of 10 to 15 m s^{-1} at low levels veers to a generally southerly direction in ascent to cloud base. This agrees with the wind distribution with height given in the sounding in Fig. 13.4. Confluence found at the surface is also evident at times in the winds near cloud base, particularly at points where the tracks intersect the surface position of the outflow boundary. Bolder portions along the flight tracks designate segments where the strongest updrafts were encountered by the aircraft flying near cloud base. It can be seen that these also tend to occur near or above the leading edge of the outflow at the surface.

Figure 13.19 presents analog records of vertical velocity, static pressure, and thermodynamic parameters corresponding to selected track segments plotted in Fig. 13.18. Reference lines at $\theta_e = 344$ K, $q = 8$ g kg^{-1}, $\theta = 319$ K, and

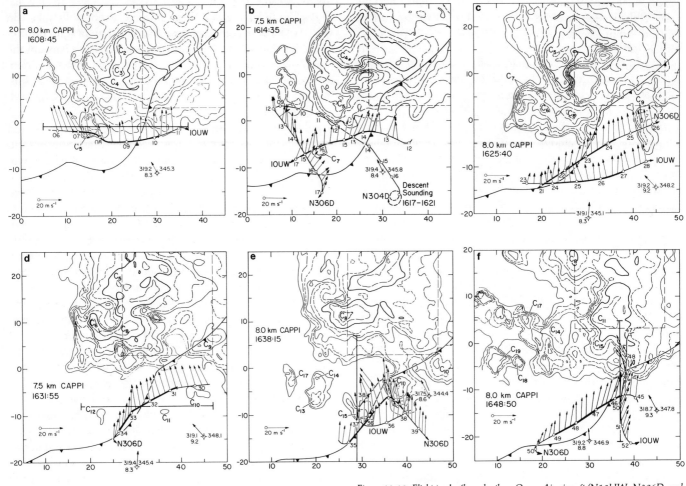

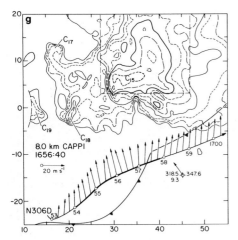

Figure 13.18 Flight tracks flown by three Queen Air aircraft (N10UW, N306D, and N304D, as indicated) in the inflow region near cloud base altitude (≈ 4 km) between 1600 and 1700. Track segments, typically of 5-min duration, are labeled in minutes after 1600 and show horizontal wind vectors (scale at lower left) at 15-s intervals. Heavier portions of the track designate regions of significant vertical velocity. Flight legs are centered in time at about (a) 1608:45, (b) 1614:35, (c) 1625:40, (d) 1631:55, (e) 1638:15, (f) 1648:50, and (g) 1656:40. These times pertain to CAPPI radar echoes shown for reference at altitudes of 7.5 or 8.0 km. Radar reflectivity contours are given at 5-dB intervals for $Z_e \geq 20$ dBZ$_e$, and reflectivity maxima are identified wherever they refer to discrete cells, as discussed in Section 13.3. Straight line segments correspond to vertical radar cross sections given in Fig. 13.19. Heavy barbed line shows position of outflow boundary, and representative ambient inflow conditions from selected surface mesonetwork stations are shown at each time.

$w = 0$ are held constant on all of the time plots, and regions of significant departure are shaded. Marwitz (1972a) has identified the characteristic thermodynamic structure in the strongest updrafts at cloud base altitudes as having maxima of q and θ_e inversely correlated with relative minima in θ. Such a relationship is clearly illustrated on many of the profiles in Fig. 13.19 (see, e.g., Figs. 13.19b and f). As mentioned

previously, the descent sounding in the primary inflow zone at the time and location given in Fig. 13.18b verifies that air with properties observed in the strongest updrafts ascends essentially unmixed from layers close to the surface. This is further substantiated by comparing flight level data with surface observations plotted for reference ahead (southeast) of the outflow boundary in Fig. 13.18. In a convectively unstable environment where θ increases and q decreases with height, as indicated by the Sterling sounding in Fig. 13.4, the result is that virtual temperature near cloud base is typically 1 °C or 2 °C colder than the surrounding air at the same level (Grandia and Marwitz, 1975).

Although the classical updraft structure is observed at

times, significant variations with respect to flight altitude, time, and location relative to radar echoes and the surface outflow boundary are also evident in Fig. 13.19. CAPPI's shown at 7.5 or 8.0 km in Fig. 13.18 were chosen to illustrate the locations of new cells on the storm's southern flank, and it can be seen that in a number of instances the aircraft pass beneath cells in their early stages of development. Whenever tracks coincide with cell positions, vertical radar echo profiles are shown with the corresponding analog data in Fig. 13.19 at a scale that matches, as nearly as possible, the space equivalence of the analog records. Typical aircraft ground speed was 85 m s^{-1}, so that 1 min of flight corresponds to a horizontal distance of ≈ 5 km. Those times on the updraft profiles when flight tracks intersect the surface outflow boundary are also indicated in Fig. 13.19.

Queen Air N10UW made the first pass through the inflow

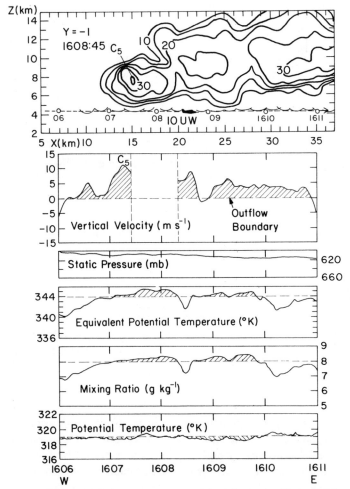

Figure 13.19a Vertical velocity, static pressure, and thermodynamic variables (as labeled) from aircraft (N10UW) measurements along flight track given in Fig. 13.18a. Upward motion and thermodynamic conditions matching those near the surface are shaded. The aircraft track in the vertical plane and a radar echo cross section aligned along the straight line segment in Fig. 13.18a are given in the upper panel. Horizontal distance and time scales are equated by use of the aircraft ground speed. Radar reflectivity contours given at 5-dB intervals (≥ 10 dBZ$_e$). The time when the aircraft passed over the surface position of the outflow boundary is identified on the vertical velocity record, and subscripted cell nomenclature identifies radar reflectivity and vertical velocity maxima, which are presumably related.

sector on an easterly heading between 1606 and 1611 while flying quite near cloud base altitude. Its track is given in Fig. 13.18a with corresponding analog data and echo profiles shown in Fig. 13.19a. The aircraft passed directly beneath the rapidly growing cell C_5 and flew quite close to the southern extent of the overall echo mass at 8 km. The vertical velocity profile shows a general region of ascent some 25 km in width, and the strongest upward motion (≈ 10 m s^{-1}) corresponds quite well with the position of cell C_5. Gaps in vertical velocity data are due to computational constraints associated with changes in aircraft attitude. The mixing ratio profile in Fig. 13.19a identifies drier air on both lateral boundaries of the inflow zone, but upward motion tends to persist beyond the limits that would be implied from thermodynamic data alone. Maximum θ_e and q are 345 K and 8.3 g kg^{-1}, respectively, in good agreement with reference surface data in Fig. 13.18a; but θ is actually somewhat colder at cloud base, indicating the possibility of superadiabatic conditions near the surface.

While N10UW reversed heading and made a westbound traverse (Fig. 13.18b), N306D and N304D flew initial passes on southeasterly headings. Corresponding profiles for N10UW and N306D are given in Figs. 13.19b and c, respectively. Conditions observed by N10UW, flying a few kilometers farther south than on the previous pass, are similar to those observed on the eastbound leg. The primary difference is the somewhat higher q and θ_e, indicating an increase in the moisture content in the inflow with time. This is consistent with the horizontal gradients of q at the surface given in Fig. 13.14b. Again, there seems to be a reasonable correspondence between the strongest region of ascent and the developing cells, C_6 and C_7, aloft.

Since the track of N306D is mostly from north to south, the vertical echo profile shown with analog data in Fig. 13.19c is given in the y-plane. Here perturbations in the thermodynamic parameters serve to identify encounters with outflow air and afford a means for estimating the shape of the outflow boundary in the vertical plane. Low θ after 1612 and near 1614 identifies rain-cooled outflow. Connecting the event at 1614 with the position of the surface outflow boundary gives the simplest possible interpretation of slope, as drawn in Fig. 13.19c. Vertical velocity of about 5 m s^{-1} over a 5-km extent correlates well with both cell C_7 and the steepest incline in the outflow boundary. A representative slope for the airstream at flight altitude is given at ≈ 1615:30 by the resultant of the vertical velocity and the horizontal wind component, v, in the y-z plane shown. It seems clear that air measured here represents the source feeding cell C_6.

The primary updraft zone indicated on the track of N304D was determined from variations in static parameters, observers' notes, and aircraft attitude, and is shown to extend some 10 km rearward from the surface position of the outflow boundary. This agrees with observations from N306D flying on a similar heading but somewhat farther to the southwest of the storm. Dry air (≥ 7 g kg^{-1}) observed by

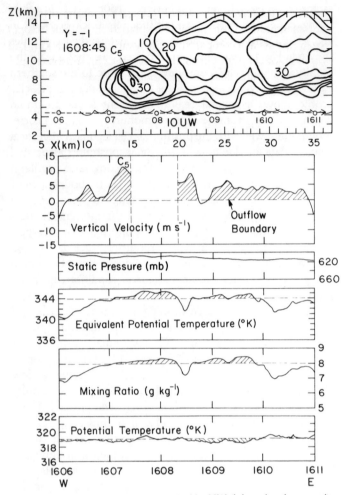

Figure 13.19b As in Fig. 13.19a, except for N10UW flight track and east-west line segment given in Fig. 13.18b. Cloud outline shown schematically.

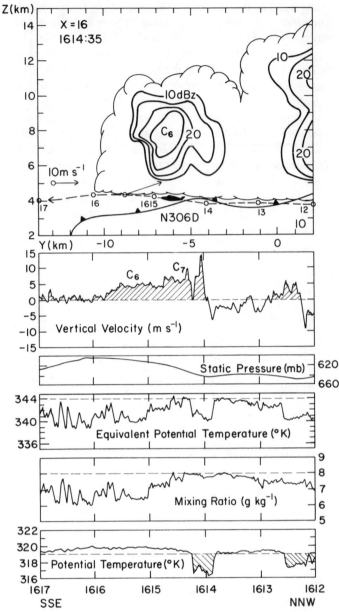

Figure 13.19c As in Fig. 13.19a, except for N306D flight track and north-south line segment given in Fig. 13.18b. Representative inflow vector shown at 1615:30. Barbed line designates outflow boundary. Cloud outline shown schematically.

After 1620, N306D and N10UW joined for nearly simultaneous eastbound legs (Fig. 13.18c) with N306D flying closest to the main precipitation echo aloft and N10UW positioned some 5 km farther to the south. In traversing the inflow both aircraft observed an increase in horizontal flow from a common value of 10 m s^{-1} found near the western end of the tracks. A maximum of 22 m s^{-1} was encountered by N306D just prior to 1624, while at a distance of 8 km upstream N10UW recorded a peak velocity of 16 m s^{-1}. From application of a simplified equation of horizontal motion, which neglects frictional drag, the observed acceleration of 6 m s^{-2} in 8 km associates with a negative non-hydrostatic pressure perturbation of somewhat less than 1 mb. It is interesting to note that this is observed near the surface pressure deficit discussed in the previous section and at a time when the three-dimensional radar echo structure was in its most steady phase (refer to Figs. 13.7d and 13.8). Although the observed surface perturbation is larger, there is no assurance that maximum values were obtained either at flight altitude or at the surface. A non-hydrostatic pressure gradient

in the vertical, thought to be essential for compensating negative parcel buoyancy at cloud base (see, e.g., Grandia and Marwitz, 1975), probably did exist in the inflow sector to the vaulted regions of the storm, but the data are inadequate to verify this.

Vertical velocity profiles (Figs. 13.19d and e) again show the zone of significant ascent to be on the order of 20 to 25 km in lateral extent. The only direct encounter with cells aloft occurs near the eastern end of the leg flown by N306D, where it passes beneath C$_9$. A somewhat puzzling result is that vertical velocity observed by N10UW farther from the storm exceeds that found by N306D in a region where

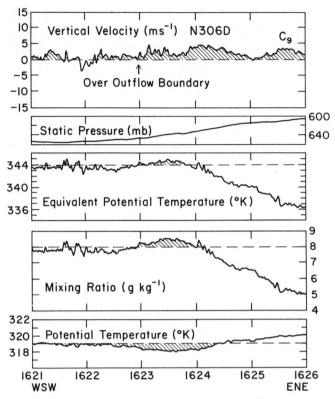

Figure 13.19d As in Fig. 13.19a, except for N306D flight track given in Fig. 13.18c.

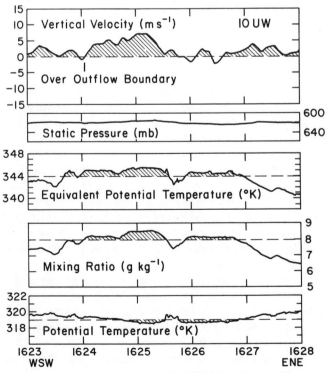

Figure 13.19e As in Fig. 13.19a, except for N10UW flight track given in Fig. 13.18c.

thermodynamic variables would indicate updrafts of significant magnitude. One possible explanation is that N306D began its pass at a comparatively low altitude (≈ 3.5 km) and was in gradual ascent during the traverse. This must have put the aircraft quite close to the sloping inflow/outflow interface, as illustrated in Fig. 13.19c, and below the main branch of the updraft. The "choppy" character of the vertical motion profile between 1622 and 1624 lends some support to this.

After passing to the east of the surface outflow boundary both aircraft observe a gradual decrease in mixing ratio and an increase in θ, with values near the ends of the legs approaching environmental conditions, as represented in the Sterling sounding in Fig. 13.4. Although weak updraft is observed in this zone, the upward motion is apparently not sufficiently strong or sustained to transport air upward from the lowest part of the subcloud layer, as is the case in the main inflow branch.

Between 1630 and 1640 the two aircraft flew in the inflow region immediately south of the storm (Figs. 13.18d and e). On a southwest heading (Fig. 13.18d) N306D passed near three new cells (C_{10}, C_{11}, and C_{12}), and encountered the most sustained upward motion of the day while flying over the surface position of the outflow boundary. Again relative maxima in w can be associated with individual cells, but the strongest and broadest updraft, just prior to 1633, has no concomitant echo aloft in Fig. 13.19f. From data given in Table 13.1 it can be seen, however, that first echo from cell C_{15} appears within 3 min after the updrafts shown in Fig. 13.19f are observed. On a reverse heading (Fig. 13.18e) an equally strong updraft is observed, and Fig. 13.19g illustrates a close relationship between the zone of strongest ascent and the rapidly intensifying echo from C_{15}.

As indicated by the track of N306D in Fig. 13.18e, only that portion of the analog record between 1635 and 1637 corresponds well with the vertical cross section in Fig. 13.19g. During this period, however, the close relationship among cell C_{15}, the surface position of the outflow boundary, and the location of major updraft near cloud base is well illustrated by the vertical y-plane oriented along an x-coordinate that intersects both the weak echo region and the high-reflectivity core of the storm. A representative inflow slope is again given at 1635:30 by the resultant of w and the v component of the horizontal wind. The mean velocity listed for cell C_{15} in Table 13.1 indicates that its motion is quite similar to the horizontal air velocity observed at cloud base (≈ 20 m s^{-1}). Thus, relative to the moving cell the updraft at cloud base altitude is essentially vertical. Air in the core of the cell shown near 8 km in Fig. 13.19g, rising at a rate of 10 m s^{-1} or greater (since w probably increases with height), must therefore have entered the cloud base sometime earlier and at a location some distance south of its indicated position. From the available evidence extrapolation backward in time and southward in space places the origin of the cell in a favorable updraft regime centered over the surface convergence maximum.

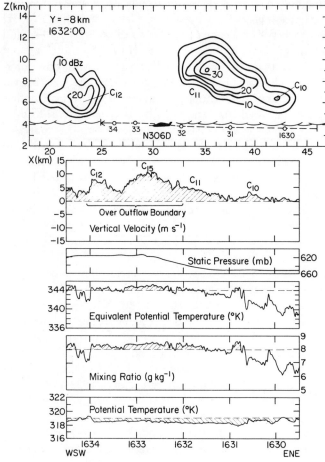

Figure 13.19f As in Fig. 13.19a, except for N306D flight track and east-west line segment given in Fig. 18d.

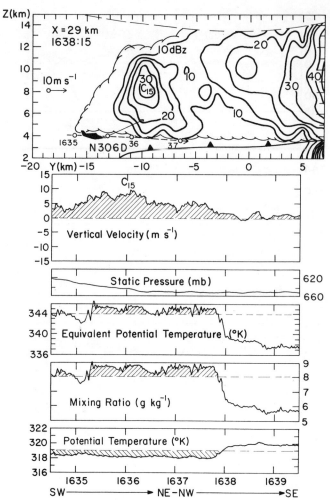

Figure 13.19g As in Fig. 13.19a, except for N306D flight track and north-south line segment in Fig. 13.18e. Representative inflow vector given at 1635:30. Barbed line designates outflow boundary. Cloud outline shown schematically.

Inflow profiles for the tracks plotted in Fig. 13.18f are given in Figs. 13.19h and i. N306D obtained data on a crosswind leg at an altitude somewhat below cloud base, so that N10UW could complete an excursion into the weak echo region and subsequently exit the storm on a southerly heading. While the updraft width and magnitude observed by N306D (Fig. 13.19h) are similar to the conditions observed on other crosswind traverses, considerably drier air results in much lower and more erratic θ_e. Presumably, the lower flight altitude put the aircraft closer to the sloping outflow boundary, where drier and potentially cooler air (indicated in surface analyses in Fig. 13.14b and c) was being mixed with inflow air having the typical characteristics observed during other passes nearer cloud base altitude.

While it was the aim of N10UW to penetrate the weak echo region located at $x = 36$ km and $y = -4$ km in Fig. 13.18f, severe turbulence and heavy precipitation, apparently associated with C_{16}, necessitated a change in the heading, resulting in the outbound leg shown. The coldest θ, shown in Fig. 13.19i just prior to 1647:30, identifies the encounter with the strongest downdraft. Thereafter, profiles in Fig. 13.19i show the distribution of vertical velocity and

thermodynamic variables in the inflow as the aircraft flies along the upwind leg, away from the storm. Except for an inexplicable minimum in w shortly before 1649, the observed updraft profile displays a gradual decrease from 8 to 10 m s⁻¹ above the sloping outflow boundary to no perceptible upward motion some 5 to 10 km ahead of the surface outflow position. Thermodynamic parameters, on the other hand, show a rather sharp demarcation between air characteristic of the strongest updrafts and that of the surroundings, the transition occurring in a horizontal distance of 2 km or so. Temperature and moisture gradients observed at other times (by N306D between 1637 and 1639 and by N10UW between 1639 and 1641), but on similar headings and at the same position relative to the outflow boundary, are shown superimposed upon the N10UW data in Fig. 13.19i. These data exhibit the same sharp decrease in moisture content and associated increase in θ as the distance from the cloud increases. This is evidence that the primary updraft zone does not extend far to the south and east of the surface outflow

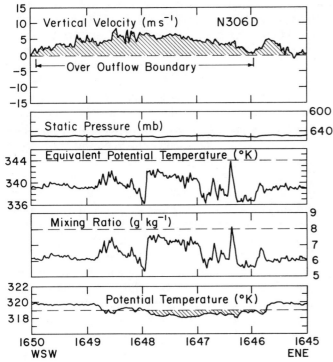

Figure 13.19h As in Fig. 13.19a, except for N306D flight track given in Fig. 13.18f.

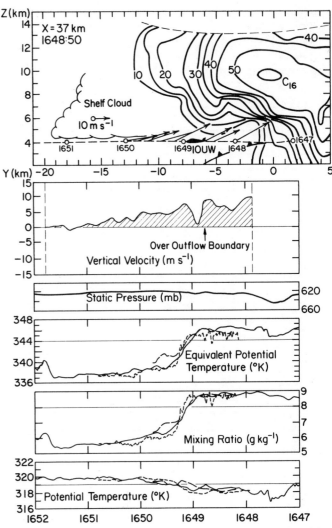

Figure 13.19i As in Fig. 13.19a, except for N10UW flight track and north-south line segment given in Fig. 13.18f. Slope of inflow given by vectors between 1648 and 1650. Barbed line designates outflow boundary. Cloud outline shown schematically. Thermodynamic data from other flight legs but in a similar position relative to the storm are shown for comparison as long and short dashed curves.

boundary. Representative airflow vectors in the plane of Fig. 13.19i again show a reasonable relationship between maximum w and cell C_{16}.

The final complete traverse through the inflow sector is represented by the track of N306D in Fig. 13.18g and corresponding analog data appear in Fig. 13.19j. The low mixing ratio observed between 1653 and 1654 identifies that region of drier air encountered on earlier traverses to the southwest of the storm (Figs. 13.19a, b, and c). Erratic w and θ_e from 1654 to near 1656 again suggest that the aircraft was flying near the top of the outflow while in gradual ascent. As on earlier passes, the smoothest and broadest updraft, although diminished somewhat in comparison with Figs. 13.19f and g, is encountered in the vicinity of the surface outflow boundary.

13.5.2 Structure of the Outflow Boundary Near the Surface

Observations of thermodynamic and airflow structure across the inflow-outflow interface near the surface were obtained during inbound and outbound penetrations of the outflow by N10UW shortly after 1700 in the relative location identified by the heavy arrow in Fig. 13.15d. Analysis of the relevant data is given in Fig. 13.20. Radar "skinpaint" defined the position of the aircraft in horizontal space 4 s after it had encountered a sharp θ gradient at the interface, and the boundary between inflow and outflow had passed a

PAM site (4593) lying almost directly beneath the aircraft track 8 min prior to the time of the aircraft event. Using a uniform southeastward displacement of 6.4 m s^{-1} (determined from isochrones as discussed in the previous section) both the aircraft and PAM time series were converted to a common relative space domain. Wind vectors and θ spaced ≈ 2 km apart represent the maximum 1-min time resolution in the basic surface observations. At flight altitude 1-s time resolution is illustrated by the line of small circles, with representative θ given at approximately 5-s intervals. Isotherms are drawn at intervals of 2 K. At the interface aloft a decrease of 5 K was observed in the 1-s data. A somewhat larger decrease is shown in the 1-min data at the surface, but whether the gradient as drawn represents a realistic magnitude is not known because of the poorer spatial resolution. A superadiabatic lapse rate near the surface, mentioned

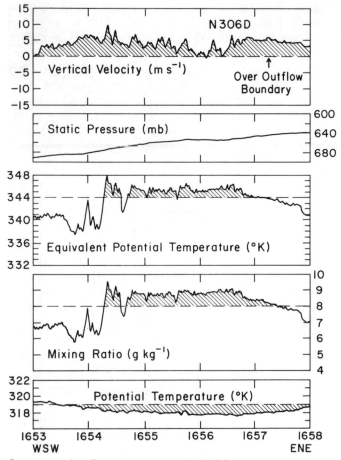

Figure 13.19j As in Fig. 13.19a, except for N306D flight track given in Fig. 13.18g.

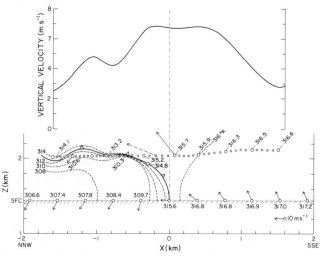

Figure 13.20 Vertical cross section (bottom) showing shape of the outflow boundary (heavy barbed line), based on measurements at the surface from a PAM site and aloft (≈ 600 m AGL) from Queen Air N10UW. Associated vertical velocity at flight altitude is given at the top. Potential temperature isotherms (dashed contours) are drawn at intervals of 2 K, and solid vectors represent horizontal wind plotted such that north is up and east is to the right. The dashed vector at flight altitude gives representative slope of airflow in the vertical plane. The time of the observations corresponds closely to that in Fig. 13.15d, and the position and orientation of the cross section are indicated by the arrow on the right of that figure.

earlier in the comparison of surface and cloud base conditions, is confirmed by the decrease of θ with height on the warm air side.

A nearly 1-to-1 slope is indicated for the interface from the surface to flight altitude (≈ 600 m AGL), with the event aloft occurring ≈ 600 m to the rear (northwest) of the surface position. Further to the rear the boundary appears to become nearly horizontal at or near flight altitude, where oscillations in θ of 2 to 4 K suggest a wavelike structure, which, in similar analyses by Charba (1974) and Goff (1976), was attributed to successive downdraft surges.

Of greatest interest here is the vertical velocity maximum observed at flight level on the warm side of the interface and centered essentially over the surface position. The slope of the ascending air immediately ahead of the outflow is given by the resultant of the horizontal and vertical components and is shown as a dashed vector. It can be seen that its orientation in the vertical plane agrees well with the slope defined by aircraft penetrations at two different altitudes. Convergence in the layer below flight level required to produce the observed w would be an order of magnitude greater than that resolved in the objective analysis of surface wind data (Section 13.4). However, this is physically consistent, since, as indicated by surface wind vectors, the directional shift accounting for a large part of the convergence occurs over an equivalent space interval of 200 to 400 m, while minimum grid spacing in the mesoscale analysis was 2 km. Another note of interest in Fig. 13.20 is that the horizontal flow doubles in magnitude in ascent from the surface to flight altitude.

Two factors that could contribute to vertical velocity in the inflow sector are the ascent of air flowing over rapidly increasing terrain and the vertical impetus associated with the strongest horizontal convergence at the inflow-outflow interface. It is of some interest to compare these. Topographic contours in Fig. 13.1 show that in the region immediately to the south of the storm's location, the terrain rises some 125 m in a horizontal distance of about 5 km (e.g., between sites 3088 and 3096). An air current flowing steadily toward this increasing topographic gradient at a rate of 15 m s^{-1} would be lifted at a rate of ≈ 0.4 m s^{-1}. If maximum surface convergence values indicated in Fig. 13.15 (≈ 4 × 10^{-3} s^{-1}) were applied through a layer of 125-m thickness, the vertical velocity at the top of that layer would be 0.5 m s^{-1}. Thus, the possible contributions from the two mechanisms are comparable in magnitude, and indeed they are complementary and additive when the outflow boundary lies immediately downwind (north) of the highest point in the escarpment. It is interesting to note that the time when the most favorable relationship between the two exists coincides with that period in the storm's development where vaulted radar echo structures are presumably sustained by the broadest and strongest updrafts.

13.6 PRECIPITATION EFFICIENCY

A time-averaged rainfall rate at the surface was derived for the storm as a whole from the analysis of timed accumulations measured by a dense network of weighing gages located within boundaries identified on earlier figures. The approach followed a technique suggested by Brandes (1975) in which accumulation rates from all gages affected by the storm between 1625 and 1700 (the period of heaviest rainfall within the network) were positioned relative to contours of radar reflectivity factor representing low-level echo configurations near the middle of the time interval. Data composited by use of this qualitative relationship between reflectivity factor and rain rate were then analyzed to obtain the analysis given in Fig. 13.21. When this field was integrated by use of a planimeter the result was a mean precipitation output rate of 7.6 kt s⁻¹.

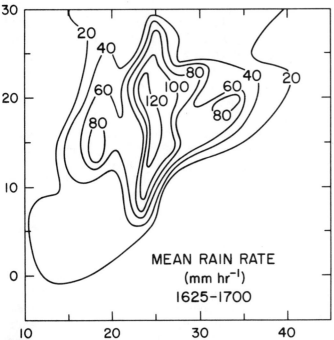

Figure 13.21 Contours of rainfall rate at the surface (mm h⁻¹) representing an average condition for the period 1625 to 1700. Boundaries are labeled in kilometers from Grover.

As outlined by Foote and Fankhauser (1973), airflow and moisture content at the surface and at flight altitude can be used to estimate the efficiency of the storm in converting ingested water vapor into precipitation at the ground. If, as was done in the case of airmass convergence discussed in Section 13.4, all of the surface water vapor flux convergence, $\nabla \cdot q\vec{V}$, is integrated over the domain in Fig. 13.15 and averaged for the period 1600-1700, the result is 11.1×10^4 kg s⁻¹ mb⁻¹. A similar calculation based on the expression

$$m_p = \frac{1}{g} \int_{\Delta L} q v_n dL$$

applied along eight cross-wind traverses in Fig. 13.18 (mean length $\Sigma \Delta L = 20$ km), gives an average value of 2.5×10^4 kg s⁻¹ mb⁻¹. While legs flown at cloud base altitude probably do not traverse all of the relevant inflow, the convergence zone included at the surface undoubtedly accounts for more flux than is actually entering the circulation of the principal storm. One approach to obtaining a measure for the entire depth of the subcloud layer is simply to average the two estimates. This gives a flux rate of 6.8×10^4 kg s⁻¹ mb⁻¹. Upon application of this value throughout the layer from the surface to cloud base ($\Delta p = 235$ mb) the rate of water vapor inflow becomes 16.0 kt s⁻¹.

An independent assessment is obtained by using the general region of cloud base ascent delineated in Fig. 13.22. Water vapor flux through cloud base is given by $F_w = \varrho q w A$, where air density $\varrho = 0.75$ kg m⁻³, and the area covered by significant updraft is taken as $A = 300 \times 10^6$ m². Setting $q = 8.3$ g kg⁻¹ and $w = 5$ m s⁻¹ (reasonable values according to profiles in Fig. 13.19) gives a water vapor inflow rate of 9.3 kt s⁻¹; somewhat less than the estimate obtained from horizontal flux.

The ratio of precipitation deposition at the ground to the rate of water vapor inflow obtained with the two different flux calculations gives a precipitation efficiency ranging from 50% to 85%. Whether either extreme is a true representation is doubtful, considering the crudity of the flux calculations, but it can be said with some confidence that the storm was certainly more efficient than some earlier cases analyzed in a similar manner. Foote and Fankhauser (1973), for example, obtained an efficiency of only about 15% in the case of a highly sheared quasi-steady hailstorm, while Chalon et al. (1976) report a value of 40% for a multicellular storm similar in some respects to the one dealt with here. Marwitz (1972c) has demonstrated an inverse relationship between precipitation efficiency and shear of the horizontal wind through the cloud bearing layer, and explains the correlation in terms of a shear-related erosion or detrainment process. The cloud-layer shear quoted in Section 13.2 is not small and would therefore, according to Marwitz's criteria, relate to somewhat lower efficiency than is implied by the results above. High efficiency in the present case may be explained by noting that larger and more intense cells, accounting for the bulk of the heavy precipitation, were effectively shielded by the weaker growth occurring on the west and southwest flanks of the storm, as illustrated in Figs. 13.9 and 13.16.

13.7 SUMMARY

Figure 13.22 represents an attempt to synthesize observations presented in Sections 13.3 through 13.5, and provides a basis for summarizing results. Generalized conditions as shown apply best to a period near 1630, when, as discussed in Section 13.3, radar echo structure was changing from a

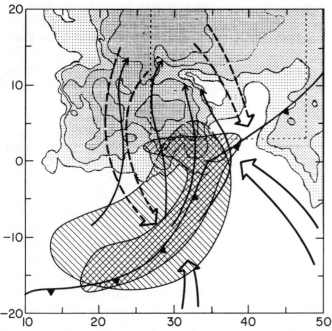

Figure 13.22 Schematic drawing synthesizing observations from radar, the surface mesonetwork, and aircraft platforms. Gray shading with increasing density designates 10-dB steps in radar reflectivity from 20 dBZ_e, taken for reference from Fig. 13.7e at 7.5 km. Heavy barbed line shows interface between ambient south-southeasterly airflow (broad solid arrows) and branches of outflow (broad dashed arrows) originating from beneath zones of highest reflectivity. Cross-hatching oriented from lower left to upper right identifies region of surface air mass convergence $\geq 2 \times 10^{-3} s^{-1}$. Hatching extending from lower right to upper left is the area of significant vertical velocity ($w \geq 1$ or 2 m s^{-1}) near cloud base altitude and beneath shelf cloud on the southern flank of the storm. Streamlines represent horizontal tracks of the more intense radar cells, which dominate the storm's evolution. Border gives distance in kilometers east and north of the Grover field site, and conditions as shown relate best to a period around 1630.

quasi-steady configuration to one dominated by cellular evolution.

The hatching that slopes from lower left to upper right and is centered on the outflow boundary identifies the zone of maximum surface convergence represented here by magnitudes $\geq 2 \times 10^{-3} s^{-1}$. Analyses in Section 13.3 show that the strongest convergence is always found near or along the boundary between opposing flows in the environment and outflow produced by downdrafts, and that maximum values are typically found near or to the southwest of the cusp in the outflow boundary. An extension of surface convergence westward from the cusp was prominent only during the storm's quasi-steady phase. Presumably, this is dynamically linked with broad organized updraft, which was associated with a weak echo vault present at the same time.

It should be emphasized that the outflow boundary did not always occupy the same position relative to the primary echo as illustrated in Fig. 13.22. Rather, as shown in Fig. 13.7, it progressed steadily southeastward away from the main precipitation zone at a rate of about 6 m s⁻¹, in apparent response to additional mass being fed into the subcloud layer by the successive downdrafts from individual decaying cells. As the size of the area covered by the outflow increased the strength of the surface convergence diminished, and this ap-

parently accounts for the storm's eventual dissipation.

Regions of significant vertical motion ($\geq \approx 2$ m s⁻¹) at cloud base altitude, identified in Fig. 13.18 by the bolder segments of the aircraft tracks, were plotted relative to the surface outflow boundary, as shown in Fig. 13.22. An envelope enclosing the limits of the updraft encountered on the individual traverses defines the zone of primary ascent beneath the shelf cloud on the storm's southern flank (illustrated by hatching that slopes from lower right to upper left). Except for the dashed boundary nearest the storm the limits are well defined by the variety of tracks flown on numerous headings and at varying distances from the storm. It can be seen that the area of primary updraft centers over the zone of maximum surface convergence, extending ≈ 5 km ahead and some 8 to 10 km to the rear of the outflow boundary. The other horizontal dimension, based primarily on the longest crosswind traverses in Fig. 13.18, is on the order of 25 km.

Two distinct scales of ascent are evident in the vertical velocity profiles given in Fig. 13.19. One has dimensions corresponding to the broad zone identified here, where w increases gradually from near zero at the edges to maxima of 5 to 6 m s⁻¹ near the core. Superimposed upon this larger-scale ascent are local maxima of 10 m s⁻¹ or less having scale lengths on the order of 2 to 5 km. In most cases these latter perturbations are correlated with radar reflectivity maxima of similar scale forming aloft, and this finer-scale updraft structure undoubtedly represents the dynamical counterpart of cells as defined in terms of radar reflectivity factor. Horizontal tracks of four major cells, typified by histories in Fig. 13.10c, are plotted in Fig. 13.22. The starting points of the tracks demonstrate that they all originate as radar entities within the region of primary cloud base updraft and in most cases not far from the zone of strongest surface convergence. Considering the juxtaposition of the surface convergence maximum and the large-scale updraft it is not unreasonable to ascribe the broad region of ascent to boundary layer forcing. Mechanisms controlling the scale of the smaller vertical velocity perturbations responsible for the cellular development are less clear, however.

Using an empirical relation between mixed boundary layer depth and the horizontal dimension of thermals of the type given by Lenschow and Stephens (1980), one would anticipate vertical velocity perturbations with a mean scale length of between 400 and 500 m for the present boundary layer conditions. Indeed, fluctuations of this size are quite evident in the N306D w profiles, but these do not correlate well with the dimensions of newly formed echo cells. Furthermore, since those updraft maxima occurring at the scale of the cells do not always exhibit in-phase temperature and moisture anomalies they are probably not thermally induced. Instead they are probably a consequence of the same mechanical forcing responsible for the broader-scale updraft zone. Considering the gusty nature of surface winds in the outflow and pulsations from periodic decay of individual cells the existence of small-scale localized convergence maxima is quite likely, but they were not detectable with the resolution available in the surface mesonetwork data.

Thermodynamic conditions on inflow traverses occasionally exhibit characteristics in regions of strongest ascent that identify the lowest part of the subcloud layer as the only possible source of updraft air. Frequently, however, upward motion was sustained beyond the limits where, based on classical models, temperature and moisture profiles would place the edges of significant ascent. This may be explained in terms of a mixing process wherein thermodynamic properties near the edges of the mechanically driven updraft are eroded toward the environmental state. Evidence for such a mechanism is given in a number of profiles in Fig. 13.19.

The shape of the inflow-outflow interface in the vertical and the depth and extent of the outflow air relative to the storm are illustrated by y-z planes given in Figs. 13.19c, g, and i, and by the vertical cross section in Fig. 13.20. In all instances the surface boundary position is well known from isochronic analysis of the "temperature break" and wind shift, and occasional aircraft encounters with the outflow aloft serve to identify slope and depth. As demonstrated in the reference profiles, shape and depth are clearly a function of location relative to the storm and, as mentioned before, probably also of time. In the sector southwest of the storm the outflow layer is shallow and, except near the surface, it has a very gradual upward slope toward the storm. Where the surface boundary is close to the storm, the outflow depth is greatest and the slope of its leading edge is steepest.

Synthesis of all the available data leads to the delineation of a zone of maximum rate of ascent, within the inflow sector, that centers over the surface outflow boundary and that probably has a nearly vertical axis up to cloud base altitudes. Horizontal velocity in this zone increases from 8 or 10 m s^{-1} near the ground to ≥ 20 m s^{-1} at cloud base. Consistent with this is a rapid increase in w with height. Where defined in vertical cross section, the slope of the inflow is usually parallel to the underlying outflow boundary. Cells first detected over this preferred region tend to move toward the storm with the horizontal velocity of the inflow air at cloud base (see also discussion in Chapter 14), and associated vertical velocity attributed to net boundary layer convergence provides for their growth and sustenance. Detailed radar echo histories given in Chapter 15 demonstrate that radar reflectivity growth rate is greatest in the early stages, when cells reside over the strong subcloud convergence zone. Forming first at an altitude of 8 km or so, some cells actually descend somewhat as they move northward away from the boundary layer support and then later ascend again as they approach the storm's core, where, according to analyses in Chapter 14, a general region of updraft is maintained.

As illustrated in Fig. 13.22, the zone of maximum convergence usually extended from a cusp in the inflow-outflow interface southwestward along the outflow boundary. However, significant convergence was also found at times along that portion of the outflow boundary extending northeastward from the cusp. The reason why little or no cellular development was observed in the eastern sector of the storm has to do primarily with the fact that inflow there was cooler and, therefore, less thermodynamically unstable. Cooling of the environment in that sector is attributed to shading of the surface by anvil "blow off" and by shelf cloud extending southward from the main body of the storm.

Radar echo history can be divided into two fairly distinct phases. One is the 30-min period when three-dimensional echo structure persisted in a relatively unchanged position and shape, characterized by a principal core of high radar reflectivity factor and little cellular evolution. This phase is followed, indeed interrupted, by a series of cells that tend to form in waves and clearly dominate the latter stages of storm evolution. Arguments presented by Browning (1977), based largely upon the analyses of Browning and Foote (1976) and Browning et al. (1976), suggest that internal airflow and associated hail formation mechanisms might be significantly different in the two phases. Highest radar reflectivity observed in the central precipitation zone differed very little, however, between the two stages, and this fact, along with other evidence in Chapter 14, raises the question whether any real physical significance can be attached to the dissimilar radar echo structure from the standpoint of hail production at the ground. Foote (1978) points out that attempts to classify storms rigorously and objectively according to type often fail under the scrutiny of high-resolution radar data. Furthermore, Fankhauser and Mohr (1977) note that on a given day convective development might evolve through any number of specified categories. Such seems to be the case here, and the important questions of whether internal flow patterns associated with the distinctive radar echo configurations differ significantly will be addressed in Chapter 14. Distinguishable hail growth processes are discussed further in Chapter 15.

References

Achtemeier, G. L., P. H. Hildebrand, P. T. Schickedanz, B. Ackerman, S. A. Changnon, Jr., and R. G. Semonin, 1978: Illinois precipitation enhancement program (Phase 1) and design and evaluation techniques for High Plains Cooperative Program. Final Rep., Contract 14-06-D-7197, Atmospheric Science Section, Illinois State Water Survey, Champaign, Ill., 313 pp.

Barge, B. L., and F. Bergwall, 1976: Fine scale structure of convective storms associated with hail production. Atmos. Sci. Sec. Rep. 76-2, Alberta Research Council, Edmonton, Alberta, 43 pp.

Barnes, S. L., 1973: Mesoscale objective map analysis using weighted time series observations. NOAA Tech. Memo. ERL NSSL-62, National Oceanic and Atmospheric Administration, Boulder, Colo., 60 pp. (NTIS COM-73-10781).

Beebe, R. G., and F. C. Bates, 1955: A mechanism for assisting in the release of convective instability. *Mon. Weather Rev.* 3, 1-10.

Biter, C. J., and J. L. Anderson, 1979: Estimating aircraft true air speed using temperature from two different probes. *J. Aircraft* 16, 893-894.

-----, and M. E. Solak, 1979: On-top seeding for hail suppression: an NHRE operational feasibility study. National Center for Atmospheric Research, Boulder, Colo., unpublished manuscript.

Brandes, E. A., 1975: Optimizing rainfall estimates with the aid of radar. *J. Appl. Meteorol.* 14, 1339-1345.

Breed, D. W., 1978: Case studies of convective storms: Case study 1. 2 June 1976: first echo case. NCAR Tech. Note (NCAR/TN130 + STR), National Center for Atmospheric Research, Boulder, Colo., 49 pp.

Brock, F. V., and P. K. Govind, 1977: Portable automated mesonet in operation. *J. Appl. Meteorol.* 16, 299-310.

Browning, K. A., 1964: Airflow and precipitation trajectories within severe local storms which travel to the right of the winds. *J. Atmos. Sci.* 21, 634-639.

-----, 1977: The structure and mechanisms of hailstorms. *Meteorol. Monogr.* 16 (38), 1-43.

-----, and G. B. Foote, 1976: Airflow and hail growth in supercell storms and some implications for hail suppression. *Q. J. R. Meteorol. Soc.* 102, 499-533.

-----, and F. H. Ludlam, 1960: Radar analysis of a hailstorm. Tech. Note No. 5, Dept. of Meteorology, Imperial College, London, England, 106 pp.

-----, J. C. Fankhauser, J-P. Chalon, P. J. Eccles, R. G. Strauch, F. H.Merrem, D. J. Musil, E. L. May, and W. R. Sand, 1976: Structure of an evolving hailstorm: Part V. Synthesis and implications for hail growth and hail suppression. *Mon. Weather Rev.* 104, 603-610.

Byers, H. R., and R. R. Braham, 1949: *The Thunderstorm.* U.S. Govt. Printing Office, Washington, D. C., 287 pp.

Chalon, J-P., J. C. Fankhauser, and P. J. Eccles, 1976: Structure of an evolving hailstorm: Part I. General characteristics and cellular structure. *Mon. Weather Rev.* 104, 564-575.

Charba, J., 1974: Application of gravity current model to analysis of squall-line gust front.*Mon. Weather Rev.* 102, 140-156.

Chisholm, A. J., 1973: Alberta hailstorms: Part I. Radar studies and airflow models. *Meteorol. Monogr.* 14 (36), 1-36.

-----, and J. H. Renick, 1972: The kinematics of multicell and supercell Alberta hailstorms. Hail Studies Rep. No. 72-2, Alberta Hail Studies 1972, Research Council of Alberta, Edmonton, Alberta, 24-31.

Cooper, W. A., 1978: Cloud physics investigations by the University of Wyoming in HIPLEX 1977. Rep. No. AS119, U.S.B.R. Grant 7-07-93-V0001, Dept. of Atmospheric Science, Univ. of Wyoming, Laramie, Wyo., 69-75.

Dennis, A. S., C. A. Schock, and A. Koscielski, 1970: Characteristics of hailstorms of western South Dakota. *J. Appl. Meteorol.* 9, 127-135.

Fankhauser, J. C., and C. G. Mohr, 1977: Some correlations between various sounding parameters and hailstorm characteristics in northeast Colorado. Prepr. 10th Conf. on Severe Local Storms, Omaha, Nebr., Oct. 18-21, 1977, Am. Meteorol. Soc., Boston, Mass., 218-225.

Foote, G. B., 1978: Response to "The structure and mechanisms of hailstorms." *Meteorol. Monogr.* 16 (38), 45-47.

-----, and J. C. Fankhauser, 1973: Airflow and moisture budget beneath a northeast Colorado hailstorm. *J. Appl. Meteorol.* 12, 1330-1353.

Fujita, T., 1963: Analytical mesometeorology: review. *Meteorol. Monogr.* 5 (27), 77-125.

Goff, R. C., 1976: Vertical structure of thunderstorm outflows. *Mon. Weather Rev.* 104, 1429-1440.

Grandia, K. L., and J. D. Marwitz, 1975: Observational investigations of entrainment within the weak echo region. *Mon. Weather Rev.* 103, 227-234.

Humphreys, W. J., 1914: The thunderstorm and its phenomena. *Mon. Weather Rev.* 42, 348-380.

Kelly, T. J., and D. H. Lenschow, 1978: Thunderstorm updraft velocity measurements from aircraft. Prepr. 4th Symp. on Meteorological Observations and Instrumentation, Denver, Colo., Apr. 10-14, 1978, Am. Meteorol. Soc., Boston, Mass., 474-478.

Lenschow, D. H., 1976: Estimating updraft velocity from an airplane response. *Mon. Weather Rev.* 104, 618-627.

-----, and P. L. Stephens, 1980: The role of thermals in the convective boundary layer. *Boundary-Layer Meteorol.* 19, 509-532.

Marwitz, J. D., 1972a: The structure and motion of severe hailstorms: Part I. Supercell storms. *J. Appl. Meteorol.* 11, 180-188.

-----, 1972b: The structure and motion of severe hailstorms: Part II. Multi-cell storms. *J. Appl. Meteorol.* 11, 180-188.

-----, 1972c: Precipitation efficiency of thunderstorms on the High Plains. Prepr. 3rd Conf. on Weather Modification, Rapid City, S. Dak., June 26-29, 1972, Am. Meteorol. Soc., Boston, Mass., 245-247.

Mohr, C. G., and R. L. Vaughan, 1979: An economical procedure for Cartesian interpolation and display of reflectivity factor data in three dimensional space. *J. Appl. Meteorol.* 18, 661-670.

Nelson, S. P., and R. R. Braham, Jr., 1975: Detailed observational study of a weak echo region. *Pure Appl. Geophys.* 113, 735-746.

Newton, C. W., 1966: Circulations in large sheared cumulonimbus. *Tellus* 18, 699-713.

Renick, J. H., 1971: Radar reflectivity profiles of individual cells in a persistent multicellular Alberta hailstorm. Prepr. 7th Conf. on Severe Local Storms, Kansas City, Mo., Oct. 5-7, 1971, Am. Meteorol. Soc., Boston, Mass. 63-70.

Reynolds, D. W., 1980: Observations of damaging hailstorms from geosynchronous satellite digital data. *Mon. Weather Rev.* 108, 337-348.

Yau, M. K., 1979: Perturbation pressure and cumulus convection. *J. Atmos. Sci.* 36, 690-694.

CHAPTER 14
The 22 June 1976 Case Study: Structure and Evolution of Internal Airflow

L. Jay Miller, F. Ian Harris, and James C. Fankhauser

14.1 INTRODUCTION

Recently, several multiple-Doppler radar studies of convective storms have been reported. These studies have included observations of severe storms in Oklahoma (e.g., Brandes, 1977a and b; Ray, 1976; Ray et al., 1975, 1978; G. M. Heymsfield, 1978), sea breeze storms in Florida (Lhermitte and Gilet, 1975), and high plains convective storms in Colorado (e.g., Miller, 1975; Kropfli and Miller, 1976; Heymsfield et al., 1980). All of these investigators except Ray et al. (1978) and Heymsfield et al. (1980), who used three radars, have used measurements from only two Doppler radars. Ray et al. (1978) dealt more with the analysis technique than with the meteorology of convective storms. Heymsfield et al. (1980), on the other hand, used radar-derived wind fields to help explain hail formation and growth processes within a multicellular hailstorm.

We use measurements from three Doppler radars to reveal the kinematic structure during and after the nearly steady phase of an intense rain and hail storm. Although the storm was multicellular in nature, we feel that observed airflow in its steady phase was very similar to flow in a broad class of storms called supercells (e.g., Browning and Ludlam, 1962; Marwitz, 1972; Fankhauser, 1971; Chisholm, 1973; Nelson and Braham, 1975; Browning and Foote, 1976; Lemon and Doswell, 1979). Such storms are characterized by a single, large reflectivity cell that remains more or less steady in appearance and exhibits a weak echo vault on its right flank and a high reflectivity core that extends to the ground and borders the vault (Browning, 1977). The vault indicates the presence of a large, strong updraft that is thought to be important in the production of large-sized hail.

Besides this supercell-like echo configuration, the 22 June storm also consisted of several smaller cells that moved around and into the vault (Chapter 13). Smaller cells were probable sources of embryos for hail formation within the main updraft. One of these, cell C_8, was followed throughout most of its lifetime as it moved into and obliterated the vault by filling it with precipitation. The names of cells and storms used in Chapter 13 are retained.

14.2 THE NATURE OF THE DATA

Three pulsed Doppler radars, one from NCAR's Field Observing Facility (FOF) and two from NOAA's Wave Propagation Laboratory (WPL), were operated on 22 June.

Several volume scans of the southernmost storms within a line of radar echoes were made; however, we are concerned mainly with scans taken in the time periods 1620-1625, 1625-1630, and 1635-1640 (hereafter referred to as 1620, 1625, and 1635 scans, respectively). The 1620 scan was made while Storm K was nearly motionless and still had the vaulted structure of a supercell. Remaining scans were made while a large cell, C_8, was moving through the vaulted area. Storm K's position one hour before the 1620 volume scan is shown in Fig. 14.1, along with the Doppler and surveillance radar

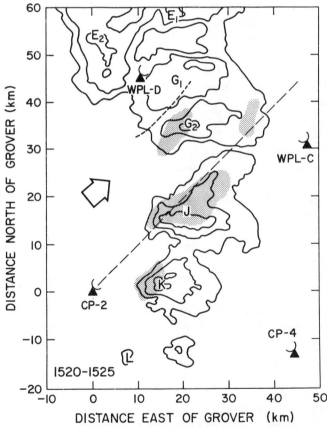

Figure 14.1 Positions of the three Doppler radars (CP-4, WPL-C, and WPL-D), the surveillance radar (CP-2), and the low-level radar echoes from the line of storms designated E_1, E_2, G_1, G_2, J, K, and L. Isopleths of radar reflectivity are shown at 10-dB intervals starting at 15 dBZ$_e$. Echo in the region south of the long-dashed line extending northeast from CP-2 is from the 1525 elevation scan at 0.5°, whereas the echo north of this line is from the 1520 scan and has been displaced northward at the speed of Storm K. Areas of horizontal divergence $\geq 1 \times 10^{-3}$ s^{-1} deduced from radial gradient of radial velocity from WPL-D are stippled. The short-dashed line located southeast of WPL-D and between Storms G_1 and G_2 represents the boundary between northwesterly flow (to the northwest of this line) and southeasterly flow (to the southeast). The broad arrow west of the line of storms shows the relative direction of ambient airflow in the layer of lowest equivalent potential temperature ($\theta_e \leq 330K$).

locations. The Doppler radars were WPL-C and WPL-D, operated at 3.22-cm wavelength, and CP-4, operated at 5.45-cm wavelength. The surveillance radar was CP-2, operated at 10-cm wavelength.

Returned signals from each Doppler radar measurement volume were analyzed for average return power and mean (reflectivity-weighted average) precipitation-particle radial velocity. Sampling locations were separated by 300 to 800 m in range and $1°$ to $1.5°$ in angle, depending upon the radar. Fields of average return power for each radar were converted to conventional equivalent radar reflectivity factor, or simply reflectivity. The synthesis of multiple-Doppler radar measurements is detailed in the appendix to this chapter. Briefly, three radial velocity estimates, one from each radar, were combined after linear interpolation to common grid points to give orthogonal components of the three-dimensional vector particle velocity. This vector field consisted of the two horizontal components of air motion and the vertical component of precipitation motion (wind speed plus still-air precipitation fall velocity). Only the horizontal components of air motion have been retained in the analysis presented here.

As described in the appendix, vertical airflow was obtained from integration of the mass continuity equation downward from the first occurrence of an estimate of horizontal wind divergence. At each (x,y) location near the top of the echo (at 20 dBZ_e) the value of vertical mass flux (density times vertical air motion, eq. A14.9) was arbitrarily assigned as the product of vertical grid spacing ($\Delta z = 500$ m) and density-weighted divergence. Therefore, divergent flow at the top of the storm was always associated with upward motion. In downward integration the influence of the boundary condition varies as $w_T \exp(-\alpha H)$ for the exponentially-varying air density used here, where H is the distance below the storm top and w_T is the upper boundary value. Both w_T and H are functions of (x,y) location, and we are integrating from different heights along the upper surface containing divergence values. Nearly all integration paths were initialized above 12 km, so that at the middle levels, say 7 km with $\alpha \approx 0.1$ km^{-1}, 60% of the boundary value was still present in the integration for vertical motion,

$$w(x,y,z) = w_T(x,y)e^{-\alpha H} +$$
$$e^{\alpha z}\int_z^{z_T} e^{-\alpha\xi}D(x,y,\xi)d\xi,$$
(14.1)

where z_T is the height of first occurrence of divergence D and $H = z_T - z$. Less than 45% of the boundary value remained below cloud base (4 km).

Boundary values of w_T were typically near 5 m s^{-1}, so that magnitudes of w greater than 3 m s^{-1} at middle levels were a consequence of net divergence between the storm top and middle levels. It is, however, impossible to know accurately what error exists in vertical motion estimates, since this requires knowing the true boundary conditions. The largest particle size associated with the 20-dBZ_e reflectivity near echo top was about 2 mm, which was computed from Auer's

(1972) graupel and hail size distribution. It falls at about 3 m s^{-1} (A. J. Heymsfield, 1978) at 13 km altitude. This fallspeed was nearly balanced by upward air motion, since little vertical movement of echo top was observed, the top remaining between 14 and 15 km for at least 30 min (see Fig. 13.8). Therefore, we feel that no more than a 5 m s^{-1} error in the upper boundary condition could have been made in the scheme that was used. Except near storm top, the resulting error in our estimates of vertical air velocity probably did not exceed 2 to 5 m s^{-1}.

In order to assess fully random error propagation in the analysis method, we replaced all input radial velocity estimates from all radars with a random value taken from a uniform distribution between plus and minus 1 m s^{-1}. We then subjected these noise data to the same analysis method used for signal data to obtain estimates of random errors for output quantities. At grid locations where only two radial velocities existed, we used them to approximate horizontal motion whenever we could safely assume that the vertical motion contribution was less than 10% of the total radial velocity. Spatial variations of the output quantities u, v, and w were computed throughout the analysis domain. For input standard deviations* in radial velocities of about 0.6 m s^{-1}, the resultant standard deviations on output were $\sigma(u) = 0.4$ m s^{-1}, $\sigma(v) = 0.3$ m s^{-1}, and $\sigma(w) = 0.2$ m s^{-1}. Maximum magnitudes of u, v, and w were, respectively, 2.8 m s^{-1}, 2.7 m s^{-1}, and 1.5 m s^{-1}. These values can be used to bound the errors in derived quantities since errors in the input radial velocities were certainly less than 1 m s^{-1} in magnitude. Errors in w were less than those in horizontal components because divergence was smoothed before the integration was done.

Evaluation of error propagation also gave us a way to estimate which scale sizes (one half of a wavelength) were present in the synthesized fields. Inspection of the distances across local maxima or minima in u, v, and w indicated that scale sizes smaller than about 3 to 5 km were effectively removed in the analysis process. We are, therefore, confident that features in the wind fields larger than this size were reliably replicated. This was sufficient for most of the cellular structure to be evident, since the average diameter of major reflectivity cells analyzed in Chapter 13 was 5.1 km. Smaller cells were not resolved in the Doppler analysis, and nothing can be said about their kinematic structure from data presented here.

14.3 DEVELOPMENT LEADING TO THE STATIONARY, NEARLY STEADY PHASE OF STORM K

As discussed in Chapter 13, the 22 June convective system was characterized by a north-south line of radar echoes with

---

*Standard deviations refer to spatial fluctuations from the mean value within the entire volume of data.

as many as four to eight storms present at any time. Most of the storms formed on the south end of the line, but a few formed on the west side. Although separate storms moved quickly northward, the line as a whole moved slowly eastward at less than 10 km h^{-1}. A large component of storm motion along rather than normal to the line is common for (squall) line storms (e.g., Brunk, 1953; Fujita and Brown, 1958; Boucher and Wexler, 1961; Newton, 1963; Newton and Fankhauser, 1964).

By 1500, Storm K, the subject of this study, had formed about 10 km south of the older Storm J, and by 1515 Storm L had also formed yet farther south. All three storms were moving from about 210°, with K slowly overtaking J. At 1525 the line of radar echoes was situated as shown in Fig. 14.1. Since the part of the line north of the long-dashed line was not covered at 1525, low-level scans from the CP-2 radar taken at 1520 and 1525 were combined to produce this figure. Radar echoes from the 1520 scan were displaced northward at the average storm speed to produce the portion of the figure north of the long-dashed line emanating from the position of the CP-2 radar. The remainder of the figure is from the 1525 scan.

14.3.1 The Surface Outflow

The analysis in Chapter 13 suggests very strongly that the evolution of Storm K past 1525 was importantly modulated by the development of surface outflow from the line of storms and from Storm K itself. The following discussion explores that relationship further, using Doppler radar data to provide more information. At 1525, divergent flow at the surface, indicated by positive radial gradient of the radial velocities measured with the WPL-D radar,[*] showed the location of developing downdrafts in Storms G$_2$, J, and K. These divergent flow regions, stippled in Fig. 14.1, were located close to the highest reflectivity, but displaced to the west toward the side from which air with low equivalent potential temperature (θ_e) was entering these storms aloft.

Winds measured by a rawinsonde released at 1450 from Potter, Nebraska, located 75 km east and 35 km north of Grover, represent the undisturbed environmental flow east of the line. The hodograph in Fig. 14.2 shows considerable veering of the environmental winds (solid line) from southeasterly near the surface to southwesterly aloft. Mean winds within the storm (dashed line), found by taking the areal average of Doppler-derived horizontal flow at each height at 1620, were similar to the environmental winds, except below cloud base. There the Doppler wind measurements were dominated by outflow, since radars could not detect the inflow that contained no precipitation particles. There was about 5 to 8 m s^{-1} of relative flow

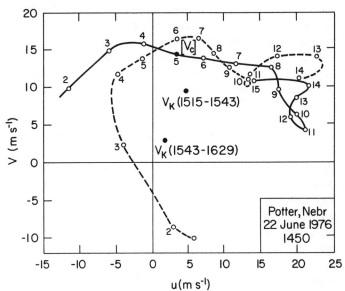

Figure 14.2 Hodographs of environmental (solid) and in-cloud (dashed) winds. Ambient winds were taken from the 1450 sounding released at Potter, Nebraska, and the in-cloud winds were determined from the areal averages at each level of Doppler-derived horizontal flow at 1620-1625 during the nearly stationary phase of Storm K. The echo motion for Storm K in the periods 1515-1543 and 1543-1629 and average cell motion [V$_c$] are shown as bold dots. The lower levels of the in-cloud winds were dominated by outflow, since much of the inflow was invisible to the radars.

between storm and environment at nearly all heights.

The velocity of Storm K in its early stage, 1515 to 1543, is shown by the appropriately labeled V$_k$ in Fig. 14.2, and represents the motion of the other storms during this time period as well. Thus the early low-level flow into these storms was easterly, whereas upper-level flow was westerly. Environmental air within the layer of lowest equivalent potential temperature ($\theta_e \leq 330$ K, height 6.2 to 7.2 km, see Fig. 13.4) was overtaking these storms from the southwest at nearly 7 m s^{-1} (broad arrow in Fig. 14.1). Downdrafts are likely to develop when precipitation falls into such middle tropospheric air because its low wet-bulb potential temperature is capable of producing cold, descending currents. Kamburova and Ludlam (1966) argued that the strongest downdrafts should occur in regions where rainfall is most intense, since here evaporative chilling and mass loading would complement each other in the production of negative buoyancy. The negative buoyancy produced by cooling air 1 °C is roughly equivalent to that from a rainwater mixing ratio of 4 g kg^{-1}.

Even though at 1525 there was divergence in the flow beneath Storms G$_2$, J, and K, most of the air still retained the southeasterly momentum of the low-level environmental flow. Surface layer winds 10 km southeast of radar D, however, had shifted around to northwesterly at the short-dashed line in Fig. 14.1, according to radial velocity measurements from that radar. Northwest of this line air motion near the ground consisted entirely of outflow from older storms such as E$_1$, E$_2$, and G$_1$. This same pattern was found in the surface flow presented in Chapter 13 for the later time

---

*Since most of the horizontal flow was northwesterly or southeasterly, it was nearly parallel to the radial direction from this radar. The radial gradient of the measured velocity, therefore, represents most of the divergence of horizontal air motion.

periods when Storm K was most intense, that is, southeasterly flow southeast of the gust front and northwesterly flow northwest of the front.

The movement of this outflow boundary between 1525 and 1645 could be followed by observing the zero radial velocity lines in both of the WPL radar measurements. Successive positions of this line, which we will call the gust front, are shown in Fig. 14.3. The areas inside the closed contour lines near Storm K's high-reflectivity region (which is cross-hatched) at 1552 contained outflow air with northwesterly momentum, whereas the surrounding area consisted of inflow air with southeasterly momentum. The gust front was also visible in the CP-2 radar data as a line echo emerging from the main radar echo mass associated with Storm L at 1601. An extension of this line echo to the north at 1610 coincided very well with the zero velocity line detected by radar C at 1610. For comparison, three gust front positions determined by analysis of surface mesonetwork winds are also shown. There is very good agreement between network- and radar-determined positions of the gust front, except along the section directly south of radar C. This discrepancy is probably a result of the slope in the outflow boundary between the surface and the height of Doppler radar data, about 500 m above the ground.

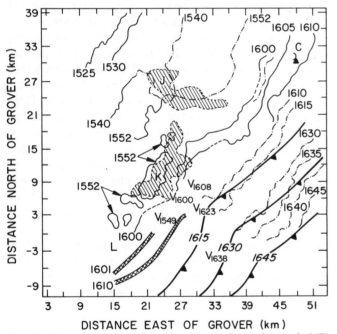

Figure 14.3 *Successive positions of the surface gust front as determined from radar WPL-D (solid line), radar WPL-C (dot-dashed line) and surface mesonetwork (heavy, barbed line). The broad, double cross-hatched lines are the positions of a low-level thin line echo at times 1601 and 1610. Low-level high-reflectivity (Z > 50 dBZ$_e$) regions at time 1552 for Storms J and K are cross-hatched. Middle-level echo vault positions are marked by a V at the various times. The solid lines at time 1552 enclose areas in the surface layer that had northwesterly momentum (surrounded by areas with southeasterly momentum).*

Simple vertical transport of horizontal momentum by the downdraft cannot explain the fact that most of the momentum in the surface outflow layer was northerly, since no such

flow existed anywhere in the environment. Significant horizontal pressure gradient forces evidently must have acted beneath the downdrafts to accelerate air away from the storm and against the low-level environmental winds. One might expect such a flow to resemble source flow within a uniform stream from the southeast; however, this simple flow regime may have been disrupted by the orographic ridge line west of Storm K's 1552 position (see Fig. 13.1), such that downdraft air drained preferentially downslope, creating primarily northerly to northwesterly winds in the outflow. There were no data to the northwest of the older storms, so we cannot say whether these storms developed a more symmetric outflow toward the northwest, as might be expected since they were farther from the ridge line.

Most of the time the gust front moved southeastward at 6 to 8 m s^{-1}, occasionally moving discontinuously as new downdrafts deposited cold air into the surface layer ahead of the established outflow layer. This discrete propagation of the outflow boundary is clearly illustrated by the developing outflow from Storm K at 1552, shown in Fig. 14.3. Shortly after this time Storm K's radar echo took on the more steady, vaulted structure characteristic of supercells. This vault persisted near coordinates (27,7) km until about 1630, when new, more transient weak echo regions were found farther southeast (see Chapter 13). In the nearly steady phase from about 1555 to 1625, the radar echo vault was located northwest (downstream relative to inflow) of the cusp in the gust front formed by older outflow, particularly from Storm J, merging with more recent outflow from Storm K. The Doppler radar analysis to be presented in Section 14.4 indicates that updraft speeds in excess of 10 m s^{-1} were occurring at 1620 over a broad region around the vault, with speeds of 35 m s^{-1} centered on the vault at 7 km.

Radial velocities measured from WPL-D at 1600 were used to deduce the variation in depth of the outflow layer, shown in Figs. 14.4 and 14.5. Three depths of surface outflow are indicated in Fig. 14.4 by three densities of stippling. Figure 14.5 illustrates how the depths were derived from the radial velocity data. Much of the outflow was confined to a rather shallow layer less than 750 m thick, particularly beneath Storms J, K, and L (shown as cross-hatched regions in Fig. 14.4). The outflow layer deepened to more than 1250 m mainly to the west of these storms. Deep outflow remained near (18,9) km, just west of Storm K's core reflectivity, and a strong downdraft was found there in the Doppler-derived wind fields. Two other regions of deep outflow that were associated with Storms M and G$_2$ (north of the figure) were found, respectively, near (9,33) km and (36,39) km.

Radial velocities from radar WPL-D are presented in Fig. 14.5 in a vertical section along the storm's inflow direction, the line marked NW-SE in Fig. 14.4. The contour interval is 5 m s^{-1}, with the zero velocity shown as a solid line and the traces of the original elevation scans shown as short-dashed lines sloping upward from the left side of the figure. The southeast end of this section is situated just north of the

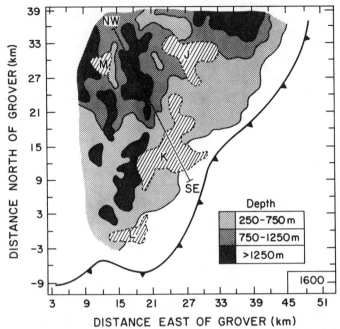

Figure 14.4 *Topography of the outflow pool at 1600 as determined from WPL-D radial velocity measurements. The heavy barbed line is the surface position of the gust front. Outflow over the areas that have low-, medium-, and high-density stippling is between 250 m and 750 m, between 750 m and 1250 m, and more than 1250 m deep, respectively. The solid line marked NW-SE is the position of the vertical section shown in Fig. 14.5. Cores of low-level reflectivity for Storms J, K, L, and M are depicted by cross-hatched areas.*

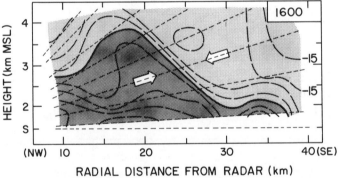

Figure 14.5 *Vertical section of radial velocity from WPL-D at 1600. The position of this section is the line marked NW-SE in Fig. 14.4. Velocities away from and toward the radar are heavily and lightly stippled, respectively. The heavy solid line is the zero radial velocity and the contour interval is 5 m s⁻¹. Short-dashed lines sloping upward toward the right are positions of the original elevation angles that were scanned. The vertical scale is exaggerated 5:1 over the horizontal scale. Broad arrows represent radial flow direction in this plane. The radar weak echo vault was situated above the nose of the outflow.*

center of the vault at 7 km. There was little flow normal to this plane in the sub-cloud layer, so that radial flow closely represented air motion within the outflow (heavily stippled) and inflow (lightly stippled) regions. A shallow layer about 500 m deep within the outflow behind (northwest of) the surface position of the gust front had velocities as high as

12 m s^{-1} away from the radar. Farther northwest the velocities first decreased in magnitude and then increased again to values in excess of 20 m s^{-1} beneath the deepest portion of the outflow. This deeper outflow was well behind (north-northwest of) Storm K and between the high-reflectivity regions of Storms J and M. Air ahead (southeast) of and above the sloping interface was approaching the storms at speeds of 15 m s^{-1} or greater, which is in good agreement with horizontal winds measured by aircraft flying above the gust front in the inflow air (see Chapter 13).

Horizontal flow in this vertical section resembles flow within laboratory density currents (e.g., Keulegan, 1958; Middleton, 1966; Simpson, 1969), and it is also similar to flow near other gust fronts described, for example, by Charba (1974) and Goff (1976). Using a non-hydrostatic, two-dimensional numerical model, Mitchell and Hovermale (1977) produced a similar flow pattern during the persistent phase of outflow from a cold downdraft. They found that, with surface drag in the model, two elevated wind maxima developed—one just behind the nose of the outflow within the elevated head and the other closer to the downdraft source. Both maxima were about the same speed, but Mitchell and Hovermale had no environmental winds in their model to oppose the spreading outflow. In our case low-level ambient flow opposite to the outflow may have mixed into the current and reduced the maximum speed in the nose.

Some idea of the horizontal convergence between inflow and outflow can be obtained by examining the radial gradient of the radial velocity presented in Fig. 14.5. Convergence values of 8 to 10 × 10^{-3} s^{-1} were found near the surface and along the sloping interface between inflow and outflow. Convergence of more than 1 × 10^{-3} s^{-1} was concentrated in a 10-km band centered on this interface. This result is similar to that found in the analysis of surface winds presented in Fig. 13.15. Above the surface position of the zero velocity (37.5-km radial distance in Fig. 14.5) radial gradient was nearly nonexistent above 3 km height. Thus, an approximate average convergence of 4 to 5 × 10^{-3} s^{-1} was present through a 1.5-km thick layer. Layer convergence of this magnitude would have resulted in updraft speeds of 6 to 7.5 m s^{-1} at the top of the layer. This agrees with updraft speeds of 5 to 10 m s^{-1} measured by aircraft as they flew over the surface gust front position (see Chapter 13).

14.3.2 Radar Echo Evolution

We now turn to a discussion of the echo evolution as Storm K became stationary and nearly steady. A sequence of horizontal sections of CP-2 radar reflectivity data at a height of 5.5 km is shown in Fig. 14.6. These sections were created by use of Mohr and Vaughan's (1979) scheme to interpolate data from the sloping, scan surfaces onto constant-height surfaces. The time period shown, 1530 to 1640, covered most of the mature phase of Storm K as well as some of the history of

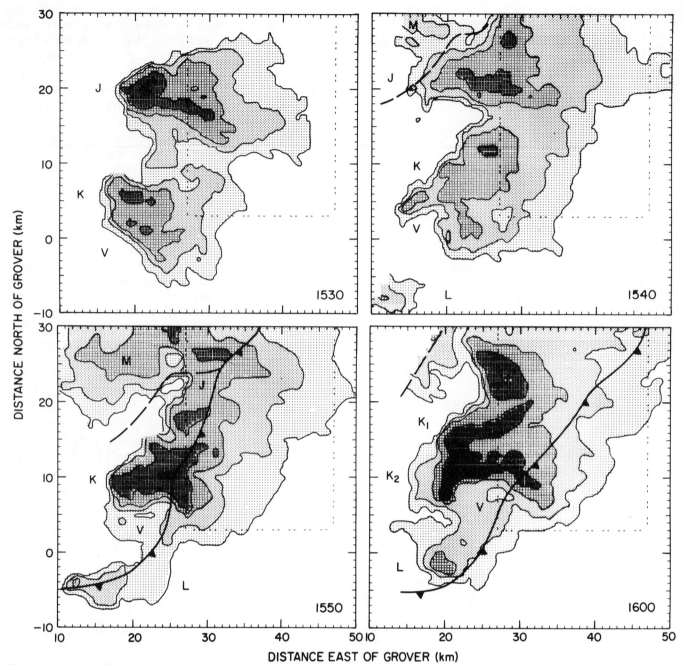

Figure 14.6a Successive echo configurations of the line of storms for the period 1530 to 1600 at 5.5 km taken from CP-2 measurements. Graduated shading represents 10-dB intervals in reflectivity starting at 20 dBZ$_e$. Heavy barbed line shows the surface position of the gust front as determined from analysis of surface winds. The long-dashed line in the up- per left-hand corner (1540 and 1550 panels) is the surface position of the gust front boundary determined by the Doppler radar radial velocities. Letters alongside (or within) the core reflectivities designate the various storms within the line. A letter V shows the horizontal position of the weak echo vault, usually found between 7 km and 9 km.

Storms J, L, and M. All of these storms except M formed along the north-south line, as previously discussed.

The location of a vault in the radar echo is depicted on each frame except the last by the letter V, and its motion throughout the period is shown at 1620 by the solid arrow. As the gust front moved southeastward beneath the storm, the vault moved at 7 m s^{-1} northeastward, parallel to the front. Initially the vault was southwest of the high-reflectivity region, but by 1600 it was southeast, on the low-level inflow side of the storm. The storm configuration from about 1600 to 1630 was similar to the supercell structure summarized by Browning (1977). This now-familiar radar echo structure is associated with storms that develop in a strongly sheared, veering wind environment and attain an intense, quasi-steady phase that produces severe weather, including large hail and strong winds at the ground.

Heavy-dashed lines at 1540 and 1550 show the position (as found in Doppler radial velocities) of the outflow boundary

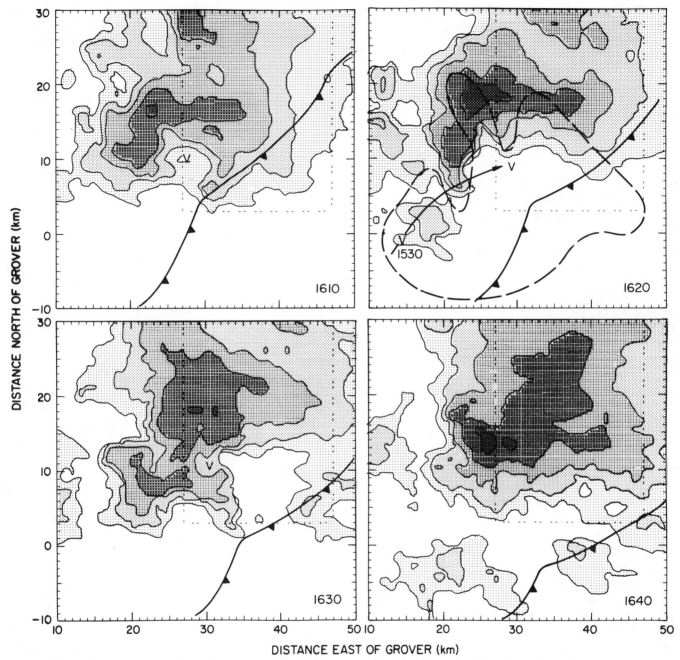

Figure 14.6b Same as Fig. 14.6a, except for the period 1610 to 1640. The cloud base updraft region (w > 5 m s⁻¹) is delineated by a long-dashed line at 1620. The solid, curved line shows the movement of the vault over the period 1530 to 1620.

from older storms to the northwest, while the solid, barbed line is the gust front position as determined by the analysis of surface mesonetwork wind data (see Chapter 13). The gust front position at 1550 is somewhat tentative, since the Doppler radar data presented in Fig. 14.3 showed that this portion of the outflow was just beginning to develop as Storm K's downdraft deposited significant amounts of cool air into the surface layer. By 1620 a large region of upward-moving air ($w > 2$ m s⁻¹) existed at cloud base on the south-southeast side of the radar echo, encircled by the dashed line. This updraft area was defined by the results from Doppler radar data

analysis as well as aircraft measurements at cloud base (Fig. 13.22).

In the early period, especially before 1600, Storm K was directly exposed to low-θ_e air between 6 and 7 km, coming from the west. With the exception of the time when a small cell was present just north of the vault at 1540, the gradient of reflectivity remained rather steep on the southwest side of the main echo mass. This steep gradient, coupled with a vaulted overhang near 8 km, indicated the presence of updraft on this side of the storm (Marwitz et al., 1972). tion produced in the updraft was carried northeastward to fall out downstream of the updraft. Relative locations of updraft and core reflectivity stayed the same until after 1550.

Middle-level environmental air overtook the storm and flowed around the northwest side of the updraft. This air was entrained into the storm and cooled by the evaporation of rainwater and cloud material, thereby establishing a low-level downdraft on the west side of Storm K (Figs. 14.3 and 14.4). As this downdraft reached the surface, the result was discrete propagation of the outflow boundary from the long-dashed line to the solid, barbed-line position shown at time 1550. Outflow therefore undercut the existing updraft southwest of Storm K's main echo mass. At this time the relative positions of vault (updraft) and core reflectivity changed, with the vault now located more to the south-southeast side of the highest reflectivity. This change in appearance of the echo was a result of heavy precipitation falling out on the southwest side of the core echo.

Vigorous uplifting over a broad area along the gust front brought an increased supply of low-level, potentially unstable air into the main updraft circulation. However, it is not very clear why the updraft did not remain intense and nearly steady as the gust front moved southeastward. The prolific development of new cells south and southeast of Storm K after 1630 clearly indicated that a broad area of uplifting near the gust front still existed, but now the updraft was no longer as intense and was concentrated more at smaller-scale sizes than before, as evidenced by Doppler-derived winds to be presented in Section 14.4.

An interesting aspect of storm interaction was evident in the echo structure between Storms J and K at 1530 and Storms K and L at 1550. At each of these times there was clearly a stream of precipitation emanating from the north side of the southern storm into the updraft of the northern storm. Implications of this natural seeding of an updraft by precipitation-sized embryos will be explored more fully in Chapter 15. Browning (1965) observed a similar feature in the radar echo between two storms in Oklahoma and used this to infer a great deal about the updraft structure within the storm being seeded. In our storm, cell C_4, which appeared as a local reflectivity maximum in the ceiling of the vault, was apparently a result of rapid growth of small precipitation particles that were injected from Storm L into the updraft of Storm K. The inverted U-shape in the reflectivity at 1610 resulted as precipitation that was grown around the updraft core fell through this level.

There is no evidence that Storm L's updraft intensified and replaced the existing updraft associated with Storm K, as might be suggested by reference to Browning's (1977) concept of "daughter clouds." Browning refers to the developing cells of an organized multicell storm as daughter clouds when they grow and, in succession, become the mature storm. The term used during Project Hailswath (Goyer et al., 1966) was "feeder cloud," implying merger with the mature storm, a different concept. In the present case Storm L is clearly feeding Storm K and, indeed, was quickly ingested by K, so that L

was no longer distinguishable after about 1605. Shortly thereafter there was a significant increase in the volume of reflectivity greater than 60 dBZ_e, as seen at 1620, when the cross-sectional area of 60 dBZ_e was larger than the one at 1610.

By 1630 the vault was nearly filled by precipitation from cell C_8 as it moved northward into the nearly stationary updraft. Air lifted out of the boundary layer by the gust front could no longer penetrate this far northwestward, but instead was rising through cloud base to the southeast of the vault at 1630. Shortly after this time there was another surge in the volume of reflectivity greater than 60 dBZ_e, followed by the prolific generation of many new cells to the south and southwest of Storm K, particularly at 1640. This second surge in high reflectivity resulted from additional growth and fallout of the many precipitation particles deposited from cell C_8 into the vaulted area. The appearance of a vault in the radar echo prior to this time meant that few precipitation-sized particles had been present to deplete the liquid water. Precipitation embryos from cell C_8, therefore, could continue growing in an environment with high values of liquid water content.

14.4 STRUCTURE AND EVOLUTION OF THE AIR MOTION AND REFLECTIVITY FIELDS AS REVEALED BY DOPPLER RADAR MEASUREMENTS

Radial velocity data from each of the radars were interpolated onto a grid with 750-m horizontal spacing and 500-m vertical spacing. These data were then used to derive the three components of air motion. The horizontal grid spacing was comparable to the original radar sampling increment, but the vertical grid spacing was roughly one-half to three-quarters of this increment. Wind vectors will be shown on a coarser grid yet, and represent nearly independent measurements. Figures 14.7-14.11 show time sequences from the 1620, 1625, and 1635 scans of horizontal sections of reflectivity (top panels) and horizontal vector winds with overlaid contours of vertical motion (bottom panels) at different heights.

In each figure reflectivity is contoured at 10-dB intervals with 20 dBZ_e shown as a solid line. The 10-dBZ_e echo overhang at 9 km is shown as cross-hatching in the 4-km panels (Figs. 14.8a, b, and c). Vertical air motion is contoured at 7.5 m s^{-1} intervals, with a solid line denoting zero velocity. In addition, regions of updraft speed greater than 7.5 m s^{-1} are heavily stippled, and regions of downdraft speed greater than 7.5 m s^{-1} are lightly stippled. Light stippling was extended to zero vertical motion in the 2-km reflectivity panels. Vector lengths are indicated by a scaled vector of length 30 m s^{-1} in the lower left-hand corner of the vector plots. Surface gust front positions are shown by a solid, barbed line in each 2-km panel (Figs. 14.7a, b, and c). Low-level inflow toward the storm and upper-level environmental flow are represented schematically by broad arrows on the 2- and 13-km panels, respectively. Streamlines are also drawn over the reflectivity fields at 13 km (Figs. 14.11a, b, and c). Flow is presented relative to the ground, and insofar as the storm was moving only very slowly northward, these flow fields also closely represent motion relative to the storm.

Although the 1620 scan was made close to the end of the quasi-steady phase of Storm K, we feel that it still represents flow throughout this phase because of similarities between the echo structure at 1620 and that of earlier times. Radar echo had a steady appearance, except for new cells that developed on the south side, from about 1555 to 1625, which was long enough for air to move through the storm. The 1625 and 1635 scans show the history of cell C_8 as it moved into the vault region.

14.4.1 Storm Features in the Surface Layer (2 km)

Figures 14.7a, b, and c show the reflectivity and air motion for 1620, 1625, and 1635 at a height of 500 m above the ground. Air moved from the southeast at speeds of 25 to 35 m s^{-1} over the storm outflow and entered the main updraft cells along the southeast side of the high-reflectivity region. A fairly abrupt change in wind direction and speed occurred across the inflow-outflow boundary. Even in the relatively smooth flow fields presented here, convergence values were as large as 8×10^{-3} s^{-1} at a grid scale of 750 m. Inflow was accelerated to about 15 m s^{-1} faster than the undisturbed environmental flow, which implies that a 1- to 2-mb pressure deficit must have existed beneath the updraft core (Browning, 1977). Outflow speeds averaged about 15 to 20 m s^{-1}, with peak values of 30 m s^{-1}.

Near the gust front, inflow and outflow combined to give cyclonic vorticity behind (northwest of) the front, whereas only weak anticyclonic vorticity existed in the small portion of inflow detected by the Doppler radars. Weak cyclonic vorticity was also found within the western half of the outflow with a north-south band of anticyclonic vorticity near the storm center. Typical values of vorticity were about 2 to 4×10^{-3} s^{-1} with some local maxima as large as 19×10^{-3} s^{-1}. Outflow was mainly cyclonic in the storm's western half and anticyclonic in the eastern half.

Downdrafts were located mostly in the higher-reflectivity regions and at this level had peak speeds of 10 to 14 m s^{-1}. Strong updrafts that protruded well into the precipitation behind the gust front position are thought to be artifacts of integrating downward to obtain vertical motion. A broad area of low-level convergence extended to the northwest of the front, but the true magnitude of convergence was probably much greater than observed. This convergence zone was also evident in the surface wind data presented in Chapter 13. Most of the upward-moving air was found on the southeast side of the storm, with the cores of the updraft cells 8 to

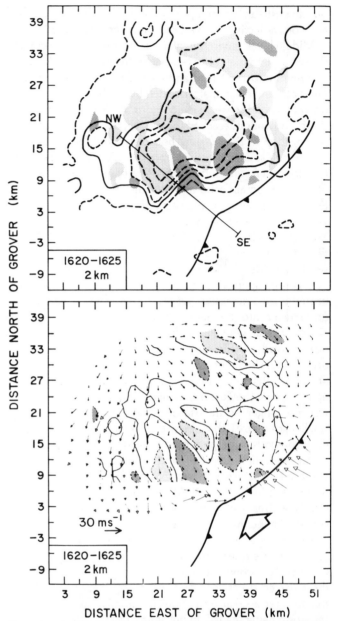

Figure 14.7a Isopleths of reflectivity (top panel) and vertical motion with overlaid horizontal winds (bottom panel) for the period 1620-1625 at a height of 2 km (500 m above the ground). Regions of significant updraft (w > 7.5 m s^{-1}) and downdraft (w < −7.5 m s^{-1}) are heavily and lightly stippled, respectively. Light stippling in the downdraft was extended to zero motion in the top panel. Reflectivity contours are at 10-dB intervals with the 20-dBZ$_e$ isopleth shown as a solid line. Vertical motion contours are at 7.5 m s^{-1} intervals with the zero velocity isopleth shown as a solid line. The surface position of the gust front is shown by a heavy, barbed line, and the ambient inflow direction is depicted by a broad arrow. A vector scaled at 30 m s^{-1} per 3 km is shown near the lower left corner. The solid line marked NW-SE on the reflectivity panel is the position of the vertical section in Fig. 14.13.

10 km behind (northwest of) the front's surface position. At 500 m above the ground, the maximum derived updrafts were about 10 to 12 m s^{-1}, which is probably excessive since the analysis scheme did not allow accurate representation of

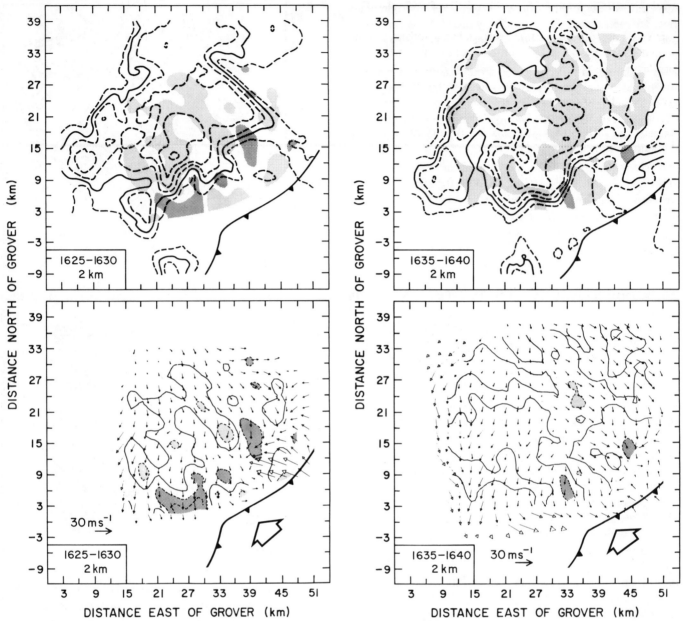

Figure 14.7b Same as Fig. 14.7a, except for period 1625-1630.

Figure 14.7c Same as Fig. 14.7a, except for period 1635-1640.

the magnitude of subcloud convergence. Strong low-level convergence that should have compensated the strong divergence aloft was not detected. Therefore, integration of the mass continuity equation downward from the top led to somewhat excessive updraft speeds near the surface. Actual upward speeds at this height were probably closer to one-half the values presented.

The radar echo consisted of a large area of intense precipitation surrounded by smaller cells on the south and west sides. These reflectivity cells represented the first stages of precipitation from updraft cells that were mechanically

driven as outflow pushed outward to the southwest and south. The area of more than 50-dBZ$_e$ reflectivity remained at about 60 to 85 km^2 throughout the period. The low-level echo structure was more or less steady except for occasional small cells. Strongest horizontal gradients of reflectivity were found on the southeast side of the storm.* A large reflectivity gradient on the low-level inflow side of the storm is typical of many severe storms (e.g., Browning, 1977).

*The gradient in the northeast portion of Fig. 14.7b is an artifact of using 3-cm reflectivity data from the WPL-D radar to fill in to the northeast. These data were heavily attenuated by an intervening cell near (24,35) km.

14.4.2 Storm Features Near Cloud Base (4 km)

Even though outflow near the surface had a northerly component, there was no such wind component above the surface layer (Figs. 14.8a, b, and c). Instead, flow was entirely from the south, with the western half of the storm having predominantly southwesterly flow and the eastern half having mostly southeasterly flow. Direction of flow in the western half of the storm was parallel to ambient airflow in the layer between 4 and 6 km, whereas flow in the eastern half was parallel to ambient flow at 2 km. These two inflow branches combined into confluent flow that was evident in the layer from 4 to 6.5 km, along an axis that was oriented

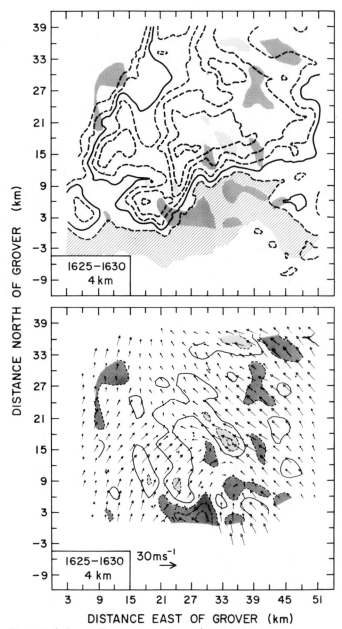

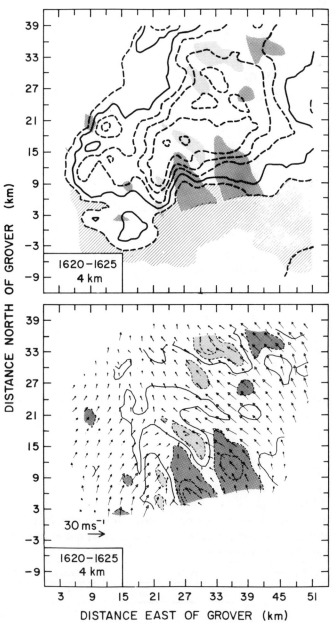

Figure 14.8a As in Fig. 14.7a, except for a height of 4 km. The extent of echo overhang ($Z \geq 10$ dBZ$_e$) at 9 km is cross-hatched.

Figure 14.8b Same as Fig. 14.8a, except for period 1625-1630.

northwest-southeast. Flow was mainly cyclonic in the southeasterly branch and anticyclonic in the southwesterly branch. Speeds were somewhat faster in the southeasterly branch, peak speeds being 25 to 30 m s^{-1}, compared to 20 to 25 m s^{-1} in the southwesterly branch. The area where speeds exceeded 20 m s^{-1} was significantly larger in the southeasterly branch, especially before 1635, when the two areas became nearly equal.

The most intense updrafts were located within the southeasterly inflow branch with only smaller, weaker updrafts and associated reflectivity cells within the southwesterly inflow. At 1620 (Fig. 14.8a) these two branches came together near a notch in reflectivity at (27,9) km, where the

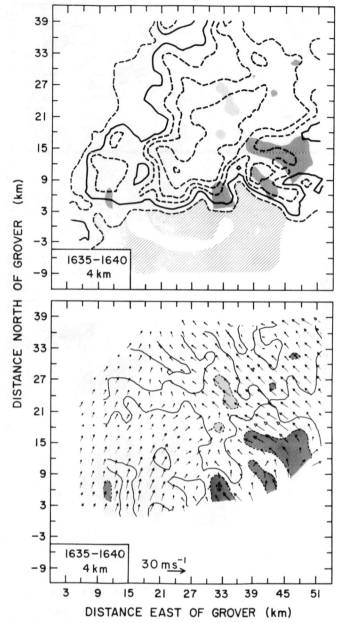

Figure 14.8c *Same as Fig. 14.8a, except for period 1635-1640.*

At this level we see cells C_6 and C_7 (see Chapter 13) moving along the southwesterly inflow stream into the storm's main echo mass. By 1635 (Fig. 14.8c) these two cells were no longer distinguishable from the main echo. Because the bulk of the echo was found downstream of the updraft, it would appear that particles growing in the updraft were transported northward as they rose and then fell out. Particle growth trajectories will be described in detail in Chapter 15. Maximum reflectivity remained at about 60 dBZ$_e$ throughout the period 1620 to 1640.

We attempted to assess the relative importance of the two inflow branches by separating updraft mass flux into portions from two areas, one with southeasterly flow and the other with southwesterly flow. We calculated vertical mass flux using

$$F_M = \rho_A \bar{w} N \delta^2, \qquad (14.2)$$

where ρ_A is air density, w is the areal average vertical speed within updrafts whose speeds exceeded 2 m s^{-1}, N is the number of grid points used in the average, and δ is the horizontal grid spacing (in this case $\rho_A = 0.77$ kg m^{-3} and $\delta = 750$ m). Results of the flux computations at cloud base for the three time periods are presented in Table 14.1. At 1620 most of the influx of air into the storm (78%) was occurring in the southeasterly branch, compared with only 22% in the southwesterly branch. However, at 1635 the southeasterly branch had become less dominant. At this time the total mass flux had also significantly decreased to 2300 kt s^{-1} from the 3200 kt s^{-1} at 1620.

According to the analysis of surface air mass flux convergence presented in Fig. 13.17, flux convergence reached a maximum between 1630 and 1635 that corresponds well with our findings about vertical mass flux in the updraft at cloud base. If we assume that surface convergence was representative of convergence throughout the depth of the subcloud layer, the air mass flux convergence maximum of 15 × 10^6 kg s^{-1} mb^{-1} converts to 3200 kt s^{-1} vertical flux at cloud base, compared with 3700 kt s^{-1} found by Doppler analysis. This is good agreement considering the potential for error in both methods.

maximum updraft speed was 26 m s^{-1}. A secondary updraft maximum was located 9 km to the northeast and had a speed of 18 m s^{-1}. By 1635 (Fig. 14.8c) these two intense updraft cores were no longer evident.

There were downdrafts as strong as 15 m s^{-1} northwest of the updraft cores. These downdrafts were located mostly in the 60-dBZ$_e$ reflectivity region, which was in the wake (northwest) of the main updraft region. Downdraft positions shifted northeastward with time as the confluence zone penetrated farther into the storm. At the earlier times, 1620 and 1625, the core of the echo was located in the confluent flow in the updraft wake. Maximum values of horizontal convergence in the wake were about 3 × 10^{-3} s^{-1}.

TABLE 14.1

Vertical air mass flux partitioned into southwesterly (SW) and southeasterly (SE) branches at 4-km height for the times shown

Time period	Flux (kt s^{-1})			Percentage of total	
	SW	SE	Total	SW	SE
1620-1625	700	2500	3200	22	78
1625-1630	1300	2400	3700	35	65
1635-1640	900	1400	2300	39	61

14.4.3 Storm Features Near Middle Level (7 km)

Reflectivity and motion fields at 7 km are shown in Figs. 14.9a, b, and c for the 1620, 1625, and 1635 scans. In the earliest period there was a well-defined weak echo vault at the intersection of the lines N-S and NW-SE. A vaulted structure in the radar echo can be the result of strong updrafts not allowing larger particles to enter from above and sweeping growing particles rapidly upward before they reach large sizes (Browning and Ludlam, 1962). This vault was surrounded by cells C_6, C_7, and C_8 on the southwest side, the main echo on the north and northeast sides (actually old cells C_4 and C_5), and the stream of precipitation particles emanating from cell C_8 around the southeast side (Fig. 14.9a shows cell locations). Updraft speed within the vault reached a maximum of 38 m s^{-1}, which was fast enough to carry even the largest particles (3-cm-diameter hail was observed at the ground) upward past this level. The only particles that could have remained near this level for any appreciable time must have had fall speeds in excess of about 30 m s^{-1}, or have been larger than about 4 to 5 cm in diameter (Auer, 1972).

Updraft cells along the southeast side of the storm were roughly elliptical, with dimensions of 14 to 18 km for the major axes and 7 to 9 km for the minor axes. The major axis was always oriented northwest-southeast, along the low-level inflow direction. Although updraft speeds decreased with time, cell orientation and dimensions remained about the same throughout the scans.

At 1620 there was another updraft core to the southwest of the vault, associated with reflectivity cells C_6 and C_7 (Fig. 14.9a). At 1625 there was no longer rapid upward flow within the vault. Instead, all that was found in this area was the weaker updraft associated mainly with cell C_8 and to a lesser extent cells C_6 and C_7, all located south and southwest of the vault. Peak speed in this updraft was only 30 m s^{-1}, and by 1635 no updrafts exceeded 20 m s^{-1}. The T-28 aircraft flew through the vault area at about 6 km from 1630 to 1634, and found variable updraft speeds that did not exceed 15 m s^{-1} (see Figs. 15.8 and 15.9). It was not possible to follow all the updraft cells between scans; however, the time history of cell C_8 was reconstructed and will be discussed in Section 14.4.6.

The direction of horizontal flow in the southeast and northwest quadrants of the storm was close to that in the environmental airstream. Speeds, however, were as much as 20 m s^{-1} faster in the storm, especially to the southeast of the updraft core. Evidently strong horizontal pressure gradients accelerated air past the core. Farther to the northwest of the main updraft, in the heart of the storm, there was a region of southeasterly flow. The area of southeasterly momentum shifted to the northwest with height; at 4 km southeasterly momentum was found within the updraft core near (29,5) km, particularly at 1620. This suggests that some updraft air, preserving its horizontal momentum, was mixed into the local environmental airstream and appeared

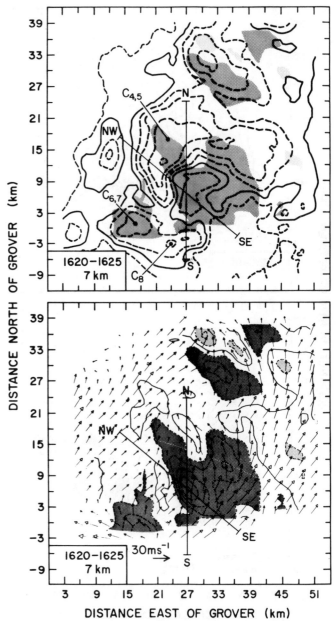

Figure 14.9a As in Fig. 14.7a, except for a height of 7 km. The position of the vertical sections in Fig. 14.12 is the line marked N-S. Subscripted letter C denotes cells referred to in the text.

downstream (relative to the inflow direction) of the updraft core.

The updraft that supported Storm J was located farther to the north near (33,30) km at 1620. As this updraft drifted northeastward a strong downdraft developed between Storms J and K, oriented northwest-southeast with a core of 14 m s^{-1} at (35,25) km and especially evident at 1635 (Fig. 14.9c). Coincident with this downdraft was the saddle in reflectivity separating these two storms.

Throughout this time period small updraft cells pushed up through middle-level air that was flowing into the west side of the storms. The mixing of warm, moist air from these up-

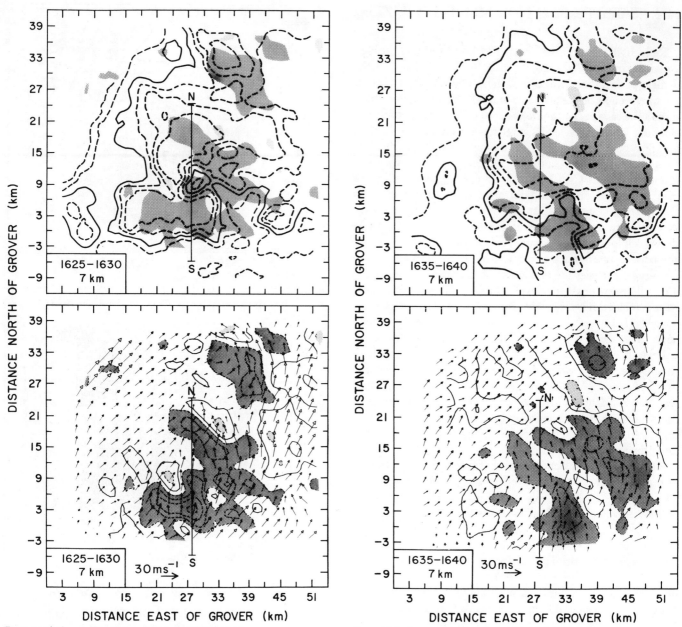

Figure 14.9b Same as Fig. 14.9a, except for period 1625-1630.

Figure 14.9c Same as Fig. 14.9a, except for period 1635-1640.

draft cells with the cool, dry middle-level air decreased the chances of any significant evaporative cooling and consequent downward flow of air. Had there been no updrafts modifying the environment on the west side, middle-level air with low equivalent potential temperature might have undercut the updraft, been cooled, and sunk to the surface, forming a strong, organized downdraft circulation in the upwind sector.

14.4.4 Storm Features Near Cloud Top (10 and 13 km)

Horizontal flow near cloud summit appeared more or less as divergent flow from a source embedded in a uniform stream (see, e.g., Milne-Thomson, 1965). The source consisted of updraft air that decelerated as it penetrated the stable layer at the tropopause. This feature was less promi-

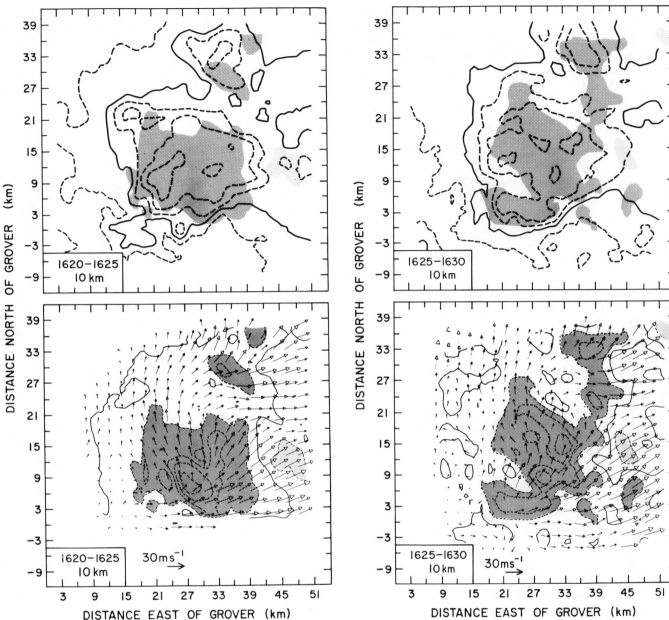

Figure 14.10a As in Fig. 14.7a, except for a height of 10 km.

Figure 14.10b Same as Fig. 14.10a, except for period 1625-1630.

nent at 10 km (Figs. 14.10a, b, and c) than it was at 13 km (Figs. 14.11a, b, and c) throughout the period. Winds were much faster on the downstream side of the updraft than on the upstream side. Outflow into the anvil on the east side of the storm in places exceeded 45 m s⁻¹, which was nearly twice the speed of ambient flow aloft, whereas flow on the upstream side was only a few meters per second, indicating a significant deceleration of the ambient air.

Almost all of the air was moving upward at the 10- and 13-km levels with maximum speeds of 15 to 30 m s⁻¹. At 1620 a broad updraft zone was surrounded by downdraft whose greatest speeds (5 to 12 m s⁻¹) were on the downstream (east) side of the updraft. Much weaker

downward motions, 2 to 5 m s⁻¹, were found on the upstream (west) side of the updraft. In time the updraft summit became more diffuse, with more small-scale features embedded within the larger updraft region. Such small-scale features were especially evident at 1625 and 1635 at 10 km (Figs. 14.10b and c), where there were several local updraft maxima about 5 to 10 km apart.

Maximum reflectivities at 10 km (Fig. 14.10a, b, and c) were about 50 dBZ$_e$, and most of the high-reflectivity region was within updraft. There was no significant temporal change in the area of echo greater than 50 dBZ$_e$, though the location of peak values was quite erratic. The largest values of reflectivity fell off gradually with height to about 40 dBZ$_e$

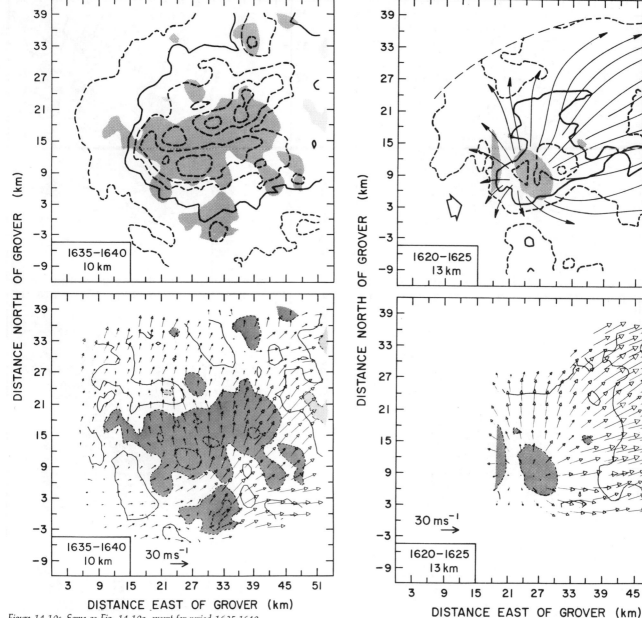

Figure 14.10c Same as Fig. 14.10a, except for period 1635-1640.

Figure 14.11a As in Fig. 14.7a, except for a height of 13 km. Streamlines of internal airflow and ambient wind direction (broad arrow) are shown.

at 13 km (Figs. 14.11a, b, and c). At the 10-km level the 30-dBZ$_e$ isopleth was roughly circular; however, weaker reflectivity isopleths streamed off asymmetrically into the anvil outflow. There were also diffuse areas of weaker reflectivity to the northwest, apparently occurring when precipitation particles were carried in that direction by divergent flow out of the updraft.

The tops of cells C$_6$ and C$_7$ were located southwest of the main storm echo at 1620 (Figs. 14.10a and b). After 1630 these cells were no longer distinguishable, as they had been carried northward and had merged into the echo mass to the northeast. Small-scale, transitory features in reflectivity that were probably a result of small updraft turrets that pushed up

toward storm top and then collapsed were found on the perimeter of the high-reflectivity region. Since these features were typically smaller than 3 to 5 km in diameter, their updrafts were not reproduced in the Doppler wind field analysis. Such features were probably of minor importance in the total production of precipitation because their contribution to the vertical transport of air and water substance was much smaller than that of the flux within the large updraft area.

At 1620 km the center of divergence at 13 km (Fig. 14.11a) was located upwind of most of the highest reflectivities. The areas of maximum divergence and maximum reflectivity were coincident in the later scans (Fig. 14.11b and

50

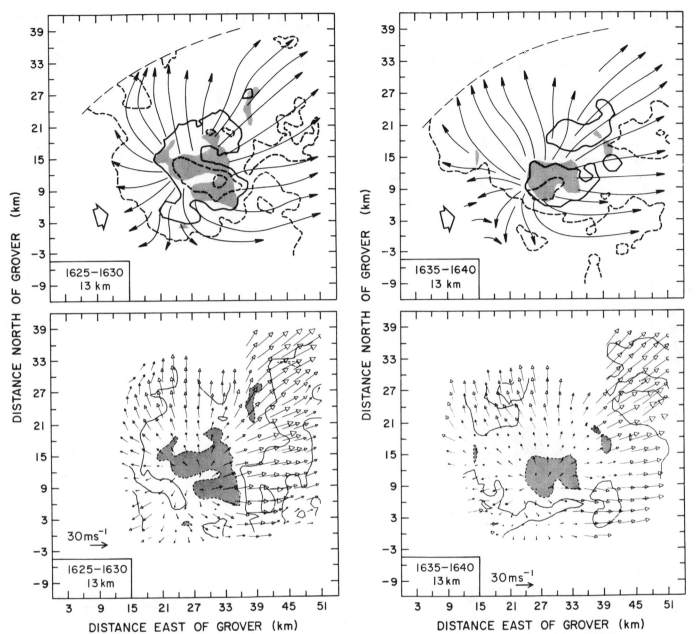

Figure 14.11b Same as Fig. 14.11a, except for period 1625-1630.

Figure 14.11c Same as Fig. 14.11a, except for period 1635-1640.

c). The way boundary conditions were set for integrating downward forced updraft cores to be coincident with divergence maxima. We cannot say with absolute certainty, therefore, where the actual updraft cores were located. However, the locations of the low horizontal speeds, updraft, and core reflectivity were probably nearly coincident at

cloud summit. Ice particles present at 13 km were transported from lower in the cloud rather than grown at these cold temperatures (about $-65\,°C$). Fall speeds of these particles could not have exceeded the 10 to 15 m s^{-1} updraft speeds, so that the biggest particles had diameters of about 8 to 12 mm (A.J. Heymsfield, 1978).

14.4.5 Storm Features in Vertical Sections

Figures 14.12a, b, and c are north-south vertical cross sections, with isopleths of vertical motion (top panels) and two-dimensional airflow (bottom panels) combined with radar reflectivity, for the three scan times. Horizontal positions of these vertical sections are indicated by lines marked N-S in Figs. 14.9a, b, and c. These planes pass through the main updraft core and its associated weak echo vault at 1620 and 1625. By 1635 cell C_8 had completely filled the vault with precipitation so that this vertical section is more through C_8's updraft.

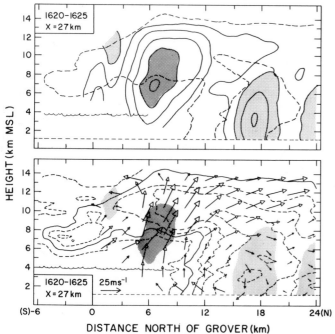

Figure 14.12a Isopleths of vertical motion (top panel) and reflectivity with overlaid vector winds (bottom panel) in a north-south plane at x = 27 km for period 1620-1625. The dashed line (top panel) and solid line (bottom panel) depict the 20-dBZ$_e$ reflectivity isopleth. In addition a dashed line (top panel) delineates the 50-dBZ$_e$ region. The vertical motion and reflectivity fields are contoured at 7.5 m s^{-1} and 10-dB intervals, respectively. The regions of downdraft (w < 0) and updraft (w > 30 m s^{-1}), respectively, are lightly and heavily stippled. A scaled vector of 25 m s^{-1} per 2 km is shown.

Some of the main features of the storm structure are clearly revealed in these vertical sections. The correspondence between positions of the broad updraft region and the reflectivity overhang on the south side of the storm is obvious. In the middle levels, upward-moving air was found over an area about 18 km wide in the north-south direction. Near the updraft core at $y = 6$ km, reflectivity isopleths show an upward bulge in echo structure surrounding the place where vertical motion exceeded 35 m s^{-1} (Fig. 14.12a). The southward extension of overhang past $y = 3$ km is the echo from cell C_8. Downdrafts that had speeds as great as 15 m s^{-1} were located north of the updraft and in the lower half of the storm.

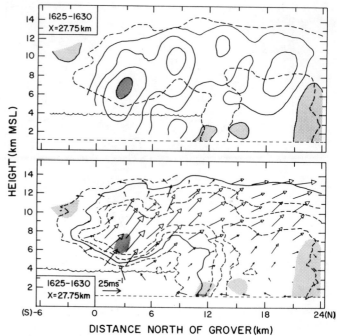

Figure 14.12b Same as Fig. 14.12a, except for period 1625-1630 at x = 27.75 km.

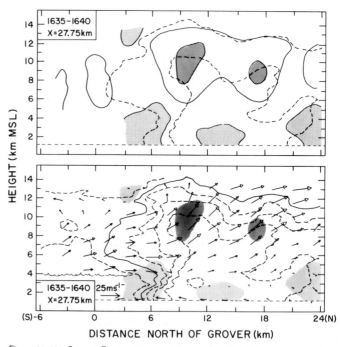

Figure 14.12c Same as Fig. 14.12a, except for period 1635-1640 at x = 27.75 km. The region of updraft greater than 15 m s^{-1} is heavily stippled.

After cell C_8 enlarged and began to fill the updraft vault, the character of the updraft underwent a significant change. There was no longer a single large, intense updraft cell. Instead, the updraft took on a structure characterized by several small cells, as seen, for example, in Fig. 14.12b. Furthermore, the updraft cells were undercut by downdrafts (at least in this plane) by 1635 (Fig. 14.12c), which clearly represented the end of the vaulted phase of this storm.

A vertical section along the inflow direction is shown in Fig. 14.13, with the horizontal location of this section indicated in Fig. 14.9a by the line marked NW-SE. This section was constructed to dramatize the updraft circulation and to illustrate the relative positions of updraft, upper-level echo overhang, precipitation fallout, and outflow behind the gust front. The figure shows a reversal of direction between surface inflow and upper-level outflow within the updraft circulation. Precipitation cascaded out of the updraft core to the northwest, forming an arc of reflectivity greater than 50 dBZ_e that extended to the surface.

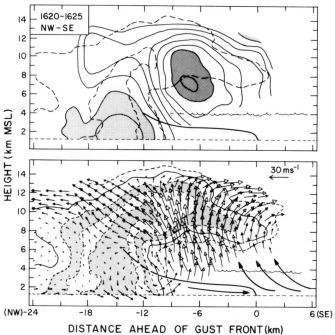

Figure 14.13 As in Fig. 14.12a, except oriented NW-SE along the low-level inflow direction. Vertical motion is contoured at a 5 m s^{-1} interval. The heavy solid line marks the top of the outflow air. Heavy arrows (bottom panel) schematically depict airflow in the plane. Shading of the 30- to 40-dBZ$_e$ and 50- to 60-dBZ$_e$ regions has been added (bottom panel).

A strong downdraft ($w < -9$ m s^{-1}) formed within this heavy precipitation shaft, bringing middle-level air to the surface to flow outward toward the southeast and form a gust front that undercut the updraft position. The weak updraft within the outflow on the southeast edge of the velocity data is thought to be a result of bias in the vertical motion. Strong low-level divergence was not adequately represented here by the analysis scheme so that upward motions were still present upon integration downward. The heavy solid line drawn to show the top of the outflow (northwesterly momentum) is more representative of the surface of zero vertical motion. Aircraft passing between inflow and outflow found no vertical motion at this interface (Chapter 13).

Strictly speaking the streamlines implied in Fig. 14.13 represent trajectories only for steady-state, two-dimensional flow. However, there was significant horizontal flow normal to this plane, especially in the middle levels. Similar simple models of airflow within intense convective storms have been presented by Browning (1963), Kropfli and Miller (1976), and Browning et al. (1976), among others. In fact, there is little difference between the quasi-two-dimensional flow pattern that Browning et al. deduced for a multicellular hailstorm in northeastern Colorado (see their Fig. 14.3) and the flow pattern shown in Fig. 14.13.

14.4.6 Features and Evolution of Cell C₈

Vertical sections oriented north-south and passing through the core of cell C₈ are shown in Fig. 14.14. This updraft cell

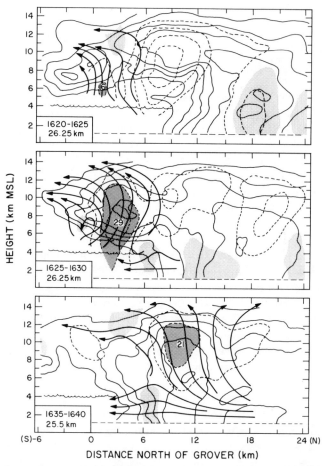

Figure 14.14 Vertical sections oriented N-S through the core of cell C₈ showing reflectivity (solid lines), vertical motion (short-dashed lines), and air motion streamlines (heavy arrows) relative to moving cell. Reflectivity and vertical motion are contoured at 10-dB and 10 m s^{-1} intervals, respectively. The downdraft ($w < 0$) and cell C₈'s updraft ($w > 15$ m s^{-1}) are, respectively, lightly and heavily stippled. The core updraft speeds were 16 m s^{-1} at 1620-1625 (top panel), 29 m s^{-1} at 1625-1630 (middle panel), and 21 m s^{-1} at 1635-1640 (bottom panel). The x positions of the sections are shown in the lower left corners.

and its associated reflectivity could be followed unmistakably in the Doppler-derived wind pattern throughout much of the cell's history. As shown in Chapter 13, this cell was first detected at 1619 as it formed a first radar echo south-southwest of the weak echo vault. It moved northward at 13.4 m s⁻¹, and it was no longer distinguishable as a separate radar echo after 1641. Its track carried it directly into the vault, and by 1625 (middle panel of Fig. 14.14) its updraft was the dominant one. The previous vault updraft was drastically weakened to roughly 15 m s⁻¹, compared with its prior peak of 38 m s⁻¹.

Streamlines of flow relative to this moving cell and within the vertical plane showed that the updraft stream was more erect than seen, for example, in Fig. 14.12, where flow relative to the stationary storm is shown. At 1625 the updraft rapidly attained a peak speed of 29 m s⁻¹, which then decreased to 21 m s⁻¹ at 1635 as the top of the cell impinged upon the tropopause. At this time (bottom panel in Fig. 14.14) there was rapid fallout of precipitation, completely obliterating the vault that had been present.

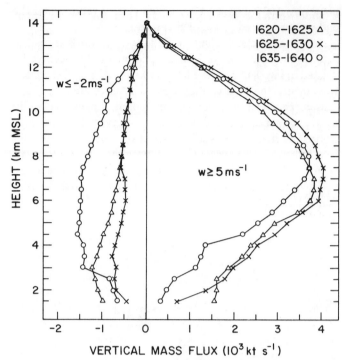

Figure 14.15 *Vertical profiles of vertical mass flux (kt s⁻¹) in the updraft (w ≥ 5 m s⁻¹) on the right and downdraft (w ≤ −2 m s⁻¹) on the left for the three Doppler radar scans.*

14.5 VERTICAL FLUXES OF AIR AND MOISTURE

Profiles of vertical flux of air within the updrafts and downdrafts for the three time periods are shown in Fig. 14.15. Vertical motion was partitioned into significant updrafts ($w > 5$ m s⁻¹) and downdrafts ($w < -2$ m s⁻¹) within the domain of wind data presented in Figs. 14.7-14.11. Average speeds and areas were calculated, and equation (14.2) was used to compute fluxes at each horizontal level. These vertical profiles represent storm-wide fluxes for both Storms J and K and do not apply to single updraft cells.

There was very little change in the updraft flux over the entire period except for a small decrease in flux at all levels below about 7 km at 1635. In Chapter 13 it was reported that the surface flux convergence gradually increased to a maximum value shortly after 1630 and then slowly decreased after that time. This same trend is broadly represented in the flux profiles; however, no significance should be placed on small changes, since uncertainties (perhaps 5% to 20%) in these flux values are comparable to the differences between the profiles.

According to updraft locations determined by aircraft and deduced from the southward echo overhang (see Chapter 13), some of the updraft area was not observed in the Doppler wind fields, especially below about 8 km. Part of the decrease in flux values below this height was a result of this. We estimate that the percent of updraft area missed at 1620, for example, was 10% at 7 km and 20% at 4 km. However, profiles were not adjusted, and values should be assigned this range of uncertainty. The increase in downward flux at 1635, however, is significant, since it represents nearly a threefold increase from the previous two times. As the storms

dissipated, there was an increase in downdraft area that reflected a greater downward flux of air.

We also computed the flux of air through the vault ($w > 5$ m s⁻¹), roughly the area 20 km < x < 43 km, −2 km < y < 22 km at 1620 (see Fig. 14.9a). At middle levels (7 km), we found that 2300 kt s⁻¹ of air was passing through the vault, compared with 3800 kt s⁻¹ for storm-wide flux. Slightly more than 60% of the total upward flux was passing through roughly 50% of the total cross-sectional area of updrafts. Obviously this greater flux within the vault was a result of the larger speeds there: 13.4 m s⁻¹ averaged over the vault, compared with 11.6 m s⁻¹ averaged over the entire storm.

Flux of moisture through cloud base is easily obtained by multiplying air mass flux by a representative value of water vapor mixing ratio. Since the best estimate of the area of updraft beneath the vault not observed by the Doppler radars was at 1620, moisture flux was determined only for this time and portion of the storm. The adjusted air mass flux was 2500 kt s⁻¹ through an area of 330 km², and the moisture flux was 20 kt s⁻¹, based on a subcloud layer mixing ratio of 8 g kg⁻¹ (Chapter 13).

Auer and Marwitz (1968) reported maximum values for air and moisture flux in one of eight northeastern Colorado hailstorms of 660 and 5.8 kt s⁻¹, respectively. In their study of a northeastern Colorado supercell, Foote and Fankhauser (1973) estimated values of 2000 kt s⁻¹ and 13 kt s⁻¹, respectively, for air mass inflow and moisture influx. Fankhauser (1971) reported 28 kt s⁻¹ for the moisture flux into a severe thunderstorm that moved across the Texas Panhandle and into Oklahoma. Foote and Fankhauser (1973) re-examined wind measurements made at a single altitude in the storm reported by Fankhauser, and estimated that subcloud air inflow was 3000 kt s^{-1}. The values of air and moisture flux within the 22 June storm vault clearly lie in the range reported for supercells. Air flux through Storm J's updraft cell near (36,30) km in Fig. 14.9a was 630 kt s^{-1}, which is comparable to Auer and Marwitz's reported value.

14.6 SUMMARY AND CONCLUDING REMARKS

Before 1545 Storm K consisted mostly of a single large-reflectivity cell that was moving parallel to and about 5 m s^{-1} slower than environmental winds in the layer of lowest equivalent potential temperature (heights 6 to 7 km). Core reflectivity was located downstream of the updraft, estimated to be 10 to 12 km in diameter. Cool, dry air ($\theta_e < 330$ K) overtook the storm, was entrained, and formed a downdraft on the west-southwest side of low-level echo. Outflow from this downdraft combined with northwesterly flow out of older storms to the north and northwest, forming a cusp in the gust front. This front moved slowly southeastward and undercut Storm K's updraft.

Shortly thereafter the storm entered a stationary, nearly steady phase, with northward cell motion almost balanced by southward echo propagation. In this phase, from about 1555 to 1625, the radar echo configuration was similar to the one associated with supercells. Characteristics of the echo included a large vault in middle levels that was open toward southeasterly inflow at cloud base and was bounded on the northwest by high reflectivity values and strong reflectivity gradients. Maximum upward speed found in Doppler-derived winds at 1620 was 38 m s^{-1} in the middle of the vault. Nearly three-quarters of the total upward flux of air was through this updraft. The remaining upward flux was through smaller and weaker updrafts on the west side of the storm.

Small intense downdrafts were observed northwest of the vault within the heavy precipitation area. Cool, dry middle-level environmental air, however, could not reach these downdrafts without first mixing with warm, moist updraft air on the west side of the storm. This large-scale mixing process apparently warmed and increased the moisture content of the environmental air, and so prevented an organized downward flow of air into the surface outflow layer from being established. Influx of substantial amounts of dry, middle-level air that can undercut the updraft, be chilled by evaporation of cloud and precipitation material, and sink in a well-organized downdraft stream is often cited as a distinguishing feature of long-lived supercells (e.g., Browning, 1977), but such a downdraft was present only sporadically in this storm. Similar reasoning is used in Chapter 13 to explain the high

precipitation efficiency of this storm.

Air and precipitation trajectories during the steady phase were determined, assuming that the Doppler wind fields at 1620 represented internal flow for 30 min prior to this time. All of these trajectories were fairly simple, so they are not presented in detail. Instead, we choose to discuss trajectories in the framework of horizontal streamlines at cloud base and middle levels (Fig. 14.16). Low-level inflow and upper-level outflow directions, along with cloud base (dashed) and middle-level (solid) streamlines, are shown. The region of heavy precipitation ($Z_e > 50$ dBZ$_e$) in the surface layer is stippled, and isopleths of upward motion at 7 km are presented at 7.5 m s^{-1} intervals, starting at 7.5 m s^{-1}. Air entering the storm from the southeast in the surface layer ascended in the updraft, followed anticyclonically-turning paths and exited into the anvil to the northeast. Air that entered the storm updraft fringes at middle levels slowly rose as it passed through the storm and exited on the north and northeast sides.

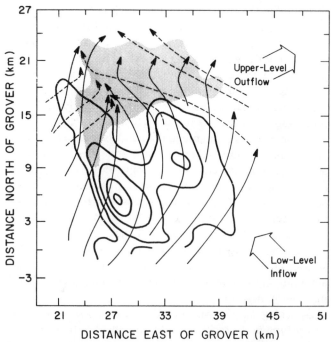

Figure 14.16 Schematic representation of airflow during the nearly steady phase of Storm K. Low-level inflow and upper-level outflow directions are indicated by broad arrows. Light, dashed lines and light, solid lines are streamlines at 4 and 7 km, respectively. Middle-level (7 km) updraft isopleths are shown at intervals of 7.5 m s^{-1}, starting at 7.5 m s^{-1}. Cloud base area of reflectivity greater than 50 dBZ$_e$ is stippled.

Most precipitation particles in Storm K followed fairly simple paths from the south side toward the north side of the storm. Recirculation of precipitation at the scale sizes resolved in the Doppler wind fields was not possible, since there was no southward component of air motion to carry precipitation back into the updraft except in the surface outflow layer and on the southwest side of the storm near cloud top (Figs. 14.11a, b, and c). Particles falling out of the anvil would have been quickly swept back into the storm as they fell on the south and west sides. Such particles could not descend to lower levels in an embryo curtain as suggested by Browning and Foote (1976). Precipitation that fell out of the main updraft stream in middle levels traveled along cyclonically-turning paths to reach the surface northwest of the updraft. Small-sized precipitation that did not fall out of the updraft stream in middle levels was ejected into anvil outflow near the updraft summit. We estimate that these particles were smaller in diameter than about 5 to 10 mm. Larger-sized particles could cross the horizontal breadth of the updraft since they were lifted more slowly because of their larger fall speeds. These particles, however, had to grow from millimeter-sized embryos in order to fall out as centimeter-sized hail.

Browning (1964) suggested that most updraft streamlines should turn cyclonically in rightward-moving severe storms. He presented observed cyclonic rotation of identifiable features in the echo as evidence of a similar curvature of air streamlines. However, in spite of the obvious cyclonic vorticity within the updraft at middle levels in Storm K, air trajectories could not turn cyclonically since the winds within the storm veered with increasing height. Precipitation falling in such a wind regime, however, would follow cyclonic trajectories. Klemp et al. (1981) reached a similar conclusion regarding trajectories within an Oklahoma tornadic storm; that is, that updraft air trajectories turned anticyclonically and precipitation trajectories turned cyclonically.

Besides the large, intense updraft and its related radar echo vault, there were other smaller and weaker updrafts associated with reflectivity cells. Some of these cells moved northward around the vault perimeter in response to the local airstream, somewhat resembling obstacle flow about the main updraft. Shortly after 1620 the strong updraft in the vault decreased, and one large cell that was comparable in size to the vault passed directly into it, ending the nearly steady phase of the storm. This cell deposited copious amounts of precipitation into the vault, completely filling it in a few minutes.

Browning (1977) has recently summarized ideas about storm type and precipitation development within multicell and supercell storms. The evolution of multicell and supercell storm concepts has come, of course, through the Thunderstorm Project (Byers and Braham, 1949) and later investigations of severe storms by Browning and Ludlam (1962), Marwitz (1972), and Chisholm (1973), among others. Browning relied heavily on results from two NHRE cases, the Raymer and Fleming storms, in his discussion of hail formation. In the Raymer storm there was a systematic, periodic development of new cells, and hail was thought to develop and fall out within single updraft cells as they grew and decayed (Browning et al., 1976). In the Fleming storm a steady-state configuration of airflow was deduced and a recycling mechanism was advanced to provide embryos that could grow into large hail as they transited the updraft a second time (Browning and Foote, 1976).

Hail formation in the 22 June storm will be discussed in more detail in Chapter 15. However, it is worthwhile here to try to fit the observed flow field and echo evolution into one or the other of these two categories using evidence from Chapters 13 and 14. The immediate conclusion is that the 22 June storm had features of both storm types at the same time in its history. There were cells, so it was multicellular in one sense by definition, but there was also a larger-scale organization with features often considered diagnostic of supercells, such as a vaulted echo configuration. An important distinction is, however, the impossibility of any organized recirculation of hail embryos. They must have been grown in cells and become injected into the main updraft. Calculated precipitation trajectories indicated that only millimeter-sized particles placed in the updraft on the south side had a chance of falling out as centimeter-sized hail in the core echo. Smaller sizes were ejected into the anvil or traveled across the entire storm to fall out as relatively small-sized hail on the north side of the core echo.

Appendix

Cartesian Wind Components and Precipitation Vertical Motion From Doppler Radar Measurements

A.1 RADIAL VELOCITY MEASURED BY DOPPLER RADAR

If we assume that hydrometeors move with the horizontal winds (u,v) and deviate from the vertical air motion (w) only by their still-air fall velocities $(v_T < 0)$, precipitation particle velocity, projected along the radar radial direction, can be expressed as follows:

$$v_r(\underset{\sim}{r}) = u(\underset{\sim}{r}) \sin a \cos e + v(\underset{\sim}{r}) \cos a \cos e +$$
$$[w(\underset{\sim}{r}) + v_T(\underset{\sim}{r})] \sin e, \qquad (A14.1)$$

where the position vector $\underset{\sim}{r} = r(\sin a \cos e\, \underset{\sim}{i} + \cos a \cos e\, \underset{\sim}{j} + \sin e\, \underset{\sim}{k})$ at range r, azimuth angle a, and elevation angle e. The quantities $\underset{\sim}{i}$, $\underset{\sim}{j}$, and $\underset{\sim}{k}$ are unit vectors in the respective eastward, northward and upward directions. Although the volume of space illuminated by the radar beam contains discrete scatterers, these targets are usually sufficiently close together ($\approx$ 10-20 cm), compared with the dimensions of the pulse volume ($\approx$ 100-500 m), that the relevant fields may be considered as continuous functions of three-dimensional space.

Doppler radar measures a reflectivity-weighted mean radial velocity within the pulse volume (Doviak et al., 1979)

$$\hat{v}_r(\underset{\sim}{R}) = \frac{\iiint v_r(\underset{\sim}{r})\eta(\underset{\sim}{r})I(\underset{\sim}{R},\underset{\sim}{r})d\underset{\sim}{r}}{\iiint \eta(\underset{\sim}{r})I(\underset{\sim}{R},\underset{\sim}{r})d\underset{\sim}{r}}, \qquad (A14.2)$$

where $\eta(r)$ is reflectivity (backscatter cross section per unit volume) and $I(R,r)$ is the radar beam illumination function centered at $R = \tilde{R}(\sin A \cos E \, i + \cos A \cos E \, j + \sin E \, k)$. The center of the radar pulse volume is at range R, azimuth angle A, and elevation angle E. Integration in equation (A14.2) is over the pulse volume roughly defined by the radar pulse length (typically ≈ 150 m) and antenna pattern beamwidths (usually $\approx 1°$ in both the horizontal and vertical directions). If the radar antenna scans while measurements are being taken, the volume viewed is stretched in the direction scanned. In this case, it is more appropriate to refer to the measurement volume, which is larger than the pulse volume, and to include scanning effects in the beam illumination function. At 25 km the volume illuminated is about $2 \times 10^7 \, \text{m}^3$ and may contain as many as 10^9 to 10^{10} particles.

As can be seen from equation (A14.2), individual radial velocities lead to a Doppler velocity spectrum within the measurement volume. The power-weighted average of this spectrum is the mean radial velocity, which is usually obtained with the covariance method (Miller and Rochwarger, 1972; Rummler, 1968; Zrnic, 1977). Average reflectivity, $\hat{\eta}$, in the denominator in equation (A14.2), is proportional to the return power.

To see how the reflectivity-weighted mean radial velocity is related to mean Cartesian components, we can approximate the sines and cosines in equation (A14.1) using Taylor's series expanded about the center point of the measurement volume. Neglecting second and higher order terms, we get

$$\hat{v}_r(R) \approx \hat{u}(R) \sin A \cos E + \hat{v}(R) \cos A \cos E +$$

$$\hat{W}(R) \sin E + \frac{1}{\hat{\eta}(R)} \left\{ \iiint u(r)\eta(r)I(R,r)[a - \right.$$

$$A) \cos A \cos E - (e - E) \sin A \sin E] dr$$

$$- \iiint v(r)\eta(r)I(R,r)[(a - A) \sin A \cos E$$

$$+ (e - E) \cos A \sin E] dr +$$

$$\left. \iiint W(r)\eta(r)I(R,r)(e - E) \cos E \, dr \right\}, \quad (A14.3)$$

where the quantities $\hat{u}, \hat{v},$ and $\hat{W}$ represent reflectivity-weighted averages:

$$\hat{f}(r) = \frac{1}{\hat{\eta}(R)} \iiint f(r)\eta(r)I(R,r)dr, \quad (A14.4)$$

with $f = u, v,$ or W.

The last term on the right-hand side of equation (A14.3) and all higher order terms are negligible only for narrow beamwidth radars, that is, $I(R,r) \approx 0$ for $|a - A|$ and $|e - E| > B$, where B is the beamwidth, typically $1°$. When this condition is not satisfied, the error in neglecting these higher order terms in $(a - A)$ and $(e - E)$ represents bias error in the sense that there is no longer a simple relationship

between mean radial velocity and mean Cartesian components. Further, if all radars do not sense the same reflectivity, have isotropic beam illumination functions, and average over the same spatial volume, the reflectivity-weighting in equation (A14.4) will not be the same for all radars. Therefore, even though point velocities certainly transform as equation (A14.1), mean values may not. The mean radial velocity measured by Doppler radar is

$$\hat{v}_r(R) = \hat{u}(R) \sin A \cos E + \hat{v}(R) \cos A \cos E$$

$$+ \hat{W}(R) \sin E + b(R), \quad (A14.5)$$

where $b(R)$ represents the accumulated spatial-averaging effects, a bias error of unknown magnitude. As seen in equation (A14.3), evaluation of this bias requires that we know the true velocity, reflectivity, and beam illumination functions.

Some of the causes of bias errors were discussed by Doviak and Strauch (1980). The greatest difficulty arises when there are significant gradients of reflectivity over distances comparable to the dimensions of the actual measurement volume. Power is transmitted outside the main lobe into sidelobes. Often these antenna sidelobes pick up highly reflecting regions, such as ground targets and intense rainfall regions, and average those radial velocities ($v_r = 0$ for ground targets) in with return from the main lobe. Though the mean radial velocity is correctly determined in the sense that it is indeed the reflectivity-weighted average over the entire volume of space illuminated, this mean may not be representative of the motions within a small volume near the center of the measurement volume. In fact, the center of reflectivity (analogous to center of mass) relative to the center of the measurement volume is at

$$[r(\hat{\eta})] = \hat{\eta}^{-1} \iiint r\eta(r)I(r)dr, \quad (A14.6)$$

where r is the vector distance from the measurement volume center. The quantity $[r(\hat{\eta})]$ can be interpreted as representing an error in the position R where we assign $\hat{v}_r$ in equation (A14.2). Unfortunately, we cannot determine this error since we do not know the actual point reflectivity function. Weak-reflectivity regions adjacent to stronger reflectivity or ground targets must be eliminated from the measured data set of velocities to remove the regions with potential bias.

In a study of side lobe effects on echo top height estimation, Donaldson (1964) found that tops for reflectivity gradients of about 1 dB km^{-1} were in error by as much as 1.5 km. This result was obtained for a typical radar with a $1°$ beamwidth viewing precipitation at 45 km. The actual location of the region returning the signal was significantly below the position indicated by the mean axis elevation angle and range to the target. These sorts of problems that lead to the bias in equation (A14.5) can be reduced only by improving the characteristics of the radar antenna pattern (Doviak and Strauch, 1980).

A.2 MULTIPLE-DOPPLER RADAR SOLUTIONS

From equation (A14.5), neglecting the bias term, when three noncolinear radial velocities are measured, precipitation particle motion is immediately obtained (Armijo, 1969):

$$\hat{u} = D_3^{-1}[\hat{v}_1(Y_2 Z_3 - Y_3 Z_2) + \hat{v}_2(Y_3 Z_1 - Y_1 Z_3)$$
$$+ \hat{v}_3(Y_1 Z_2 - Y_2 Z_1)],$$

$$\hat{v} + D_3^{-1}[\hat{v}_1(X_3 Z_2 - X_2 Z_3) + \hat{v}_2(X_1 Z_3 - X_3 Z_1)$$
$$+ \hat{v}_3(X_2 Z_1 - X_1 Z_2)], \qquad (A14.7a)$$

$$\hat{W} = D_3^{-1}[\hat{v}_1(X_2 Y_3 - X_3 Y_2) + \hat{v}_2(X_3 Y_1 - X_1 Y_3)$$
$$+ \hat{v}_3(X_1 Y_2 - X_2 Y_1)],$$

with $X_m = \sin A_m \cos E_m$, $Y_m = \cos A_m \cos E_m$, $Z_m = \sin E_m$ and $D_3 = Z_1 (X_2 Y_3 - X_3 Y_2) + Z_2(X_3 Y_1 - X_1 Y_3) + Z_3(X_1 Y_2 - X_2 Y_1)$. Estimates of radial velocities $(\hat{v}_1, \hat{v}_2, \hat{v}_3)$ in equation (A14.7a) are obtained by linear interpolation of data in the spherical (radar) coordinate system to a Cartesian coordinate system common to all the radars. At grid points where only two radial velocity estimates exist the horizontal components are

$$\hat{u} \approx D_2^{-1}(\hat{v}_1 Y_2 - \hat{v}_2 Y_1),$$

$$\hat{v} \approx D_2^{-1}(\hat{v}_2 X_1 - \hat{v}_1 X_2), \qquad (A14.7b)$$

where $D_2 = X_1 Y_2 - X_2 Y_1$. With the magnitude of $\hat{W}$ about 10 m s^{-1} or less, this approximation is usually satisfactory, provided $\sin E_m < 0.1$ for both radars involved, so that the term $\hat{W}(R) \sin E$ in equation (A14.5) can be neglected.

Since vertical air motion and fallspeed are multiplied by the same coefficient in equation (A14.1) or (A14.5), additional radars are insufficient to separate these two quantities; more information must be incorporated into the formulation. We chose not to retain the solution for $\hat{W}$ because of unacceptable errors, but instead to compute $\hat{w}$ from the mass continuity equation. It is assumed that reflectivity-weighted mean values of air motion obey the mass continuity equation in the form:

$$\frac{\partial(\rho \hat{w})}{\partial z} + \rho\left(\frac{\partial \hat{u}}{\partial x} + \frac{\partial \hat{v}}{\partial y}\right) = 0, \qquad (A14.8)$$

with the air density ϱ being assumed to be a function of z alone. The major difficulty in this assumption is probably whether or not the reflectivity-weighted mean values actually obey equation (A14.8). It is likely that the assumption of a locally steady, horizontally homogeneous density field is close enough to the truth that any resulting error is masked by the first assumption. Vertical air motion was computed by integrating equation (A14.8) downward from a z_T boundary

$$\hat{w}(x,y,z) = \rho^{-1}\left[(\rho \hat{w})_{zT} + \int_z^{z_T} \rho\left(\frac{\partial u}{\partial x} + \frac{\partial v}{\partial y}\right) dz'\right], \quad (A14.9)$$

with $\varrho(z) = \varrho_0 e^{-\alpha z}$, where $\alpha \approx 0.1$ km^{-1} and ϱ_0 equals a reference density. The upper boundary condition was assigned

$$(\rho \hat{w})_{zT} = \rho\left(\frac{\partial u}{\partial x} + \frac{\partial v}{\partial y}\right)_{zT} \Delta z, \qquad (A14.10)$$

which assumes that $w = 0$ at $z = z_T + \Delta z$ and density-weighted or mass divergence is constant in this layer of thickness Δz. Insufficient data exist to test the validity of these assumptions so that reasonableness of the results is the only criterion used to measure quality of the data.

We have data only at grid points, so that equation (A14.9) must be represented numerically with difference formulas. Horizontal divergence of the wind is approximated with centered differences, which reliably represent the true divergence value (within 80%) only when the scale size (one-half wavelength) is larger than about three times the grid spacing (Thompson, 1961). These centered-difference estimates of divergence are smoothed to reduce the noise introduced by differentiation of data before they are integrated numerically.

References

Armijo, L., 1969: A theory for the determination of wind and precipitation velocities with Doppler radars. *J. Atmos. Sci.* 26, 570-575.

Auer, A. H. Jr., 1972: Distribution of graupel and hail with size. *Mon. Weather Rev.* 100, 325-328.

——-, and J. D. Marwitz, 1968: Estimates of air and moisture flux into hailstorms on the high plains. *J. Appl. Meteorol.* 1, 196-198.

Boucher, R. J., and R. Wexler, 1961: The motion and predictability of precipitation lines. *J. Meteorol.* 18, 160-171.

Brandes, E. A., 1977a: Flow in severe thunderstorms observed by dual-Doppler radar. *Mon. Weather Rev.* 105, 113-120.

-----, 1977b: Gust front evolution and tornado genesis as viewed by Doppler radar. *J. Appl. Meteorol.* 19, 333-338.

Browning, K. A., 1963: The growth of large hail within a steady updraught. *Q. J. R. Meteorol. Soc.* 89, 490-506.

-----, 1964: Airflow and precipitation trajectories within severe local storms which travel to the right of the winds. *J. Atmos. Sci.* 21, 634-639.

-----, 1965: Some inferences about the updraft within a severe local storm. *J. Atmos. Sci.* 22, 669-677.

-----, 1977: The structure and mechanisms of hailstorms. In *Hail: A Review of Hail Science and Hail Suppression, Meteorol. Monogr.* No. 38, 1-43.

-----, and G. B. Foote, 1976: Airflow and hail growth in supercell storms and some implications for hail suppression. *Q. J. R. Meteorol. Soc.* 102, 499-533.

-----, and F. H. Ludlam, 1962: Airflow in convective storms. *Q. J. R. Meteorol. Soc.* 88, 117-135.

-----, J. C. Fankhauser, J-P. Chalon, P. J. Eccles, R. G. Strauch, F. H. Merrem, D. J. Musil, E. L. May, and W. R. Sand, 1976: Structure of an evolving hailstorm: Part V. Synthesis and implications for hail growth and hail suppression. *Mon. Weather Rev.* 104, 603-610.

Brunk, I. W., 1953: Squall lines. *Bull. Am. Meteorol. Soc.* 34, 1-9.

Byers, H. R. and R. R. Braham, 1949: *The Thunderstorm*, U.S. Government Printing Office, Washington, D.C., 287 pp.

Charba, J., 1974: Application of gravity current model to analysis of squall-line gust front. *Mon. Weather Rev.* 102, 140-156.

Chisholm, A. J., 1973: Alberta hailstorms: Part I. Radar studies and airflow models. *Meteorol. Monogr.* 14, 1-36.

Donaldson, R. J., Jr., 1964: A demonstration of antenna beam errors in radar reflectivity patterns. *J. Appl. Meteorol.* 3, 611-623.

Doviak, and R. G. Strauch, 1980: The multiple Doppler radar workshop, November, 1979: Part III. Single radar data acquisition. *Bull. Am. Meteorol. Soc.* 61, 1178-1183.

-----, D. S. Zrnic, and D. S. Sirmans, 1979: Doppler weather radar. *Proc. IEEE* 67, 1522-1553

Fankhauser, J. C., 1971: Thunderstorm-environment interactions deter mined from aircraft and radar observations. *Mon. Weather Rev.* 99, 171-192.

Foote, G. B., and J. C. Fankhauser, 1973: Airflow and moisture budget beneath a northeast Colorado hailstorm. *J. Appl. Meteorol.* 12, 1330-1353.

Fujita, T. and H. A. Brown, 1958: A study of mesosystems and their radar echoes. *Bull. Am. Meteorol. Soc.* 39, 538-554.

Goff, R. C., 1976: Vertical structure of thunderstorm outflows. *Mon. Weather Rev.* 104, 1429-1440.

Goyer, G. G., W. E. Howell, V. J. Schaefar, R. A. Schleusener, and P. Squires, 1966: Project Hailswath. *Bull. Am. Meteorol. Soc.* 47, 805-809.

Heymsfield, A. J., 1978: The characteristics of graupel particles in north-eastern Colorado cumulus congestus clouds. *J. Atmos. Sci.* 35, 284-295.

-----, A. R. Jameson, and H. W. Frank, 1980: Hail growth mechanisms in a Colorado storm: Part II. Hail formation processes. *J. Atmos. Sci.* 37, 1779-1807.

Heymsfield, G. M., 1978: Kinematic and dynamic aspects of the Harrah tornadic storm analyzed from dual-Doppler radar data. *Mon. Weather. Rev.* 106, 233-254.

Kamburova, P. L., and F. H. Ludlam, 1966: Rainfall evaporation in thunderstorm downdraughts. *Q. J. R. Meteorol. Soc.* 92, 510-518.

Keulegan, G. H., 1958: The motion of saline fronts in still water. Natl. Bur. Stand. Rept. No. 5831, Natl. Bureau of Standards, Boulder, Colo., 28 pp.

Klemp, J B., R. B. Wilhelmson, and P. S. Ray, 1981: Observed and numerically simulated structure of a mature supercell thunderstorm. *J. Atmos. Sci.* 38, 1558-1580.

Kropfli, R. A., and L. J. Miller, 1976: Kinematic structure and flux quantities in a convective storm from dual-Doppler radar observations. *J. Atmos. Sci.* 33, 520-529.

Lemon, L. R., and C. A. Doswell III, 1979: Severe thunderstorm evolution and mesocyclone structure as related to tornadogenesis. *Mon. Weather Rev.* 107, 1184-1197.

Lhermitte, R. M., and M. Gilet, 1975: Dual-Doppler radar observation and study of sea breeze convective storm development. *J. Appl. Meteorol.* 14, 1346-1361.

Marwitz, J. D., 1972: The structure and motion of severe hailstorms: Part I. Supercell storms. *J. Appl. Meteorol.* 11, 166-179.

-----, A. H. Auer, Jr., and D. L. Veal, 1972: Locating the organized updraft on severe thunderstorms. *J. Appl. Meteorol.* 11, 236-238.

Middleton, G. V., 1966: Experiments on density and turbidity currents: Motion of the head. *Can. J. Earth Sci.* 3, 523-546.

Miller, K. S., and M. M. Rochwarger, 1972: A covariance approach to spectral moment estimation. *IEEE Trans. on Inf. Theory*, IT-18, 588-596.

Miller, L. J., 1975: Internal airflow of a convective storm from dual-Doppler radar measurements. *Pure Appl. Geophys.* 113, 765-785.

Milne-Thomson, L. M., 1965: *Theoretical Hydrodynamics.* The Macmillan Co., London, England, 660 pp.

Mitchell, K. E., and J. B. Hovermale, 1977: A numerical investigation of the severe thunderstorm gust front. *Mon. Weather Rev.* 105, 657-675.

Mohr, C. G., and R. L. Vaughan, 1979: An economical procedure for Cartesian interpolation and display of reflectivity factor data in three dimensional space. *J. Appl. Meteorol.* 18, 661-670.

Nelson, S. P., and R. R. Braham, Jr., 1975: Detailed observational study of a weak echo region. *Pure Appl. Geophys.* 113, 735-746.

Newton, C. W., 1963: Dynamics of severe convective storms. In: *Severe Local Storms, Meteor. Monogr.* No. 27, 33-58.

-----, and J. C. Fankhauser, 1964: On the movement of convective storms, with emphasis on size discrimination in relation to water-budget requirements. *J. Appl. Meteorol.* 3, 651-668.

Ray, P. S., 1976: Vorticity and divergence fields within tornadic storms from dual-Doppler observations. *J. Appl. Meteorol.* 15, 879-890.

-----, R. J. Doviak, G. B. Walker, D. Simmans, J. Carter, and B. Bumgarner, 1975: Dual-Doppler observation of a tornadic storm. *J. Appl. Meteorol.* 14, 1521-1530.

-----, K. K. Wagner, K. W. Johnson, J. J. Stephens, W. C. Bumgarner, and E. A. Mueller, 1978: Triple Doppler observations of a convective storm. *J. Appl. Meteorol.* 17, 1201-1212.

Rummler, W. D., 1968: Two-pulse spectral measurements. Tech. Memo. MM-68-4121-15, Bell Telephone Laboratories, Whippany, N. J.

Simpson, J. E., 1969: A comparison between laboratory and atmospheric density currents. *Q. J. R. Meteorol. Soc.* 95, 758-765.

Thompson, P. D., 1961: *Numerical Weather Analysis and Prediction.* The Macmillan Co., New York, N. Y., 170 pp.

Zrnic, D. S., 1977: Spectral moment estimates from correlated pulse pairs. *IEEE Trans. Aero. and Elec. Sys.* AES-13, 344-354.

CHAPTER 15
The 22 June 1976 Case Study: Precipitation Formation

Charles A. Knight, L. Jay Miller, Nancy C. Knight, and Daniel Breed

15.1 INTRODUCTION

Chapters 13 and 14 have provided description and discussions of the 22 June 1976 storm's interactions with its environment, its reflectivity structure and development, and its internal airflow as derived from measurements from three Doppler radars. This concluding part comprises a report of observations of the microphysical aspects of the storm and a discussion of the formation of rain and hail within it. Data to be presented here that were not used in Chapters 13 and 14 include measurements during cloud penetrations by the University of Wyoming Queen Air (N10UW), the South Dakota School of Mines and Technology armored T-28, and the NCAR/NOAA instrumented sailplane; precipitation measurements at the ground, including timed, quenched hailstone collections; and radar reflectivity data at lower reflectivity levels than were used in previous chapters.

The picture that has emerged is of a big, severe storm with two rather distinct scales: that of the overall storm, with a rather steady reflectivity and airflow pattern on a horizontal scale of 20 km or so, and that of individual "cells," defined semi-objectively in Chapter 13 in terms of reflectivity, on a horizontal scale of about 5 km. In its mature phase, this storm lasted over 1 h and was nearly stationary, while the cells formed, moved into it, and either merged with or replaced the high-reflectivity core. Each cell lasted about 20 to 30 min. The storm exhibited a fairly well-defined, bounded weak echo region for a period of about 30 min. It had features of both multicell and supercell storms as summarized by Browning (1977).

Chapter 14 includes a qualitative discussion of trajectories of growing hail. An objective of this chapter is to try to use the data to test present ideas about hail formation in multicell and supercell storms. Other objectives are to say something about the region(s) of embryo formation and, if possible, about hail suppression possibilities.

An enormous amount of data was recorded on this storm, and while not all of it will be presented, it is felt that it is worthwhile to try to be complete in summarizing the main aspects of the data, rather than selecting only those portions that are felt to be relevant to the conclusions. Thus, while some data selection has been made in terms of relevance to the discussion, we attempt to separate the observations from the discussion and conclusions as completely as we can.

15.2 AIRCRAFT DATA

The research objectives of the 1976 NHRE field season

were twofold: to study the first precipitation formation in newly forming cumulus ("first echo studies") and the formation of hail in severe storms ("mature storm studies"). It was thought likely that hail embryos formed in earlier stages of convection—perhaps in new turrets on the flanks of storms*—and thus both stages had to be studied to evaluate hail suppression possibilities. The sailplane was the primary cloud physics platform in the first echo studies; the T-28 in the mature storm studies. Both aircraft were to be supported when possible by N10UW penetrating the clouds usually at or below the −5°C level, and in the inflow below cloud base, where the NCAR Queen Airs also operated. The 22 June 1976 storm complex was the object of both types of study, performed in sequence since tracking and control capabilities, radar support, and safety precluded operating the T-28 and the sailplane simultaneously. The first echo study will be summarized first.

15.2.1 The First Echo Study

This study was hampered because the aircraft penetrations were only some 10 km from the radar, and the echo tops could not be scanned adequately while maintaining adequate time resolution and horizontal coverage of the storm. The study was reported in detail by Breed (1978), from which the following is a brief summary.

The sailplane penetrations took place from 1520 to 1530 and from 1535 to 1550 local time (MDT) at the west end of the cell marked L in Figs. 13.5 and 14.6a. The same cell is evident in Figs. 13.7a and b (1550 and 1600), but it was not one of those analyzed in Chapter 13 because it did not quite meet the reflectivity criteria. Furthermore, it dissipated before it would have moved into the main core of the storm. However, as noted in Chapter 14, the reflectivity pattern indicates strongly that precipitation particles that originated in Storm L did enter Storm K for a period of time around 1600 (see Fig. 14.6a).

Figures 15.1 and 15.2 (reproduced from Breed, 1978) show the sailplane and N10UW tracks with respect to the radar echo and the basic data from the sailplane for the first penetration. The second sailplane penetration, 10 min later, found less active updrafts (the radar echo top was also no longer rising) and is not presented here. N10UW made seven

---

*This is not an original idea to NHRE, but is an outgrowth of the feeder cell concept (Dennis et al., 1970) and evidently was first suggested as a basis for a hail suppression technique by Summers et al. (1972).

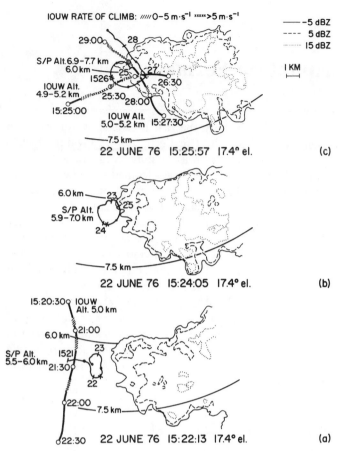

IOUW RATE OF CLIMB: ///// 0-5 m·s⁻¹ ××××× >5 m·s⁻¹

——— −5 dBZ
− − − 5 dBZ
········· 15 dBZ

l KM

29:00   28
S/P Alt. 6.9-7.7 km
6.0 km
1526
IOUW Alt.   26:30
4.9-5.2 km   25:30 28:00
15:25:00
                IOUW Alt.
                5.0-5.2 km 15:27:30
——— 7.5 km
22 JUNE 76 15:25:57 17.4° el.   (c)

6.0 km   23 25
S/P Alt.
5.9-7.0 km   24
——— 7.5 km
22 JUNE 76 15:24:05 17.4° el.   (b)

15:20:30 IOUW
          Alt. 5.0 km
6.0 km   21:00
S/P Alt. 1521 23
5.5-6.0 km 21:30
          22
22:00
——— 7.5 km
22:30 22 JUNE 76 15:22:13 17.4° el.   (a)

Figure 15.1 Radar PPI's with the sailplane track (solid line) and N10UW partial track (heavy solid line) plotted and altitude ranges labeled. Sailplane position is marked every 30 s, labeled every minute, and the X gives its location at the time of the PPI. N10UW position is marked and labeled every 30 s, and hatched and cross-hatched when rate of climb is 0-5 m s⁻¹ and >5m s⁻¹, respectively. Reflectivity contour intervals are 10 dB apart beginning at −5 dBZ$_e$, and altitude arcs at 6.0 and 7.5 km MSL show the height on the PPI surface.

cloud penetrations beneath and to the west of the west end of the echo, at 5.0 to 5.6 km MSL, between 0 °C and −5 °C, three of which are shown in Fig. 15.1. It encounterd regions of substantial updraft, as indicated qualitatively by its rate of climb. Liquid water content measurements were not obtained from N10UW. The PMS 2-D imaging probe revealed some graupel to 2 or 3 mm diameter, and a few hexagonal crystals, but no detectable water drops of precipitation size.

As is shown in Figs. 15.1 and 15.2, the sailplane climbed to nearly 8 km MSL (−18 °C) in an updraft commonly stronger than 10 m s⁻¹, encountering liquid water contents often between 2 and 3 g m⁻³, often reaching adiabatic concentrations (Heymsfield et al., 1978). The updraft used by the sailplane could have been smaller than those of cells closer to the main storm K, as shown by other aircraft and the Doppler analysis, but the sailplane track is not such as to define updraft diameter very well. Droplet concentrations indicated in Fig. 15.2, recorded from a Particle Measuring Systems (PMS) FSSP probe, were about 600 cm⁻³, and are expected to be

somewhat too low because of coincidence errors, but are not corrected. Measured mean droplet diameters from the FSSP probe range from about 12 to 16 μm and are claimed to be reliable within ± 2 μm. There is a good qualitative agreement between the liquid water content calculated from the FSSP size and concentration data and that measured directly, using a Johnson-Williams hot-wire device, and fair quantitative agreement between the liquid water values when the FSSP concentrations are corrected for the apparent instrument errors. See Appendix B of Breed (1978) for details, and the Appendix to this chapter for the kind of analysis used. The liquid water content values measured from the sailplane are considered to be relatively trustworthy because of this agreement, the consistent instrument performance on this and other days, and the fact that the amounts seem reasonable. The particle camera aboard the sailplane recorded small graupel and rimed dendrites, in total concentrations up to several per liter. These particles were most numerous at the higher altitudes, during the later portion of the penetration.

The N10UW and sailplane observations of updrafts at the western end of this particular cell are consistent with the discussion of the air flow in the early storm stage given in Chapter 14. The low-level inflow is from the south-southeast relative to the storm, and the main updrafts are on the western side of the radar echo, with ventilation by middle and upper level winds from the west. The microphysical data from this first echo study are expected to be characteristic of cells similar to this one on this day, but not necessarily of all of the cells that contributed to the mature storm.

15.2.2 T-28 Penetrations During the Mature Storm Study

The coordinated, mature storm study lasted from about 1600 to 1700, during which time N10UW and the NCAR Queen Airs were mapping storm inflow (Chapter 13). The T-28 was the only aircraft to penetrate the storm during this period, making three penetrations. The large extent of the storm demanded exceptionally long periods within cloud, causing major icing problems that resulted in gaps in the microphysical data, especially in the second penetration.

Almost worse than the gaps, however, are the basic uncertainties in the fundamental data. (The following assessment of data quality applies to the state of the art of obtaining basic measurements within clouds at the time of this study and is not necessarily a consensus view. Others may present the uncertainties in different ways.) Updraft estimates from powered aircraft without inertial navigation systems are quite crude, and the problems are compounded in severe icing environments by possible icing interference with the pilot performance. Temperature measurements are uncertain because of potential problems with wetting or icing of the sensor of reverse flow probes. Both the hot-wire liquid water meter (J-W) and the FSSP performance are uncertain and inconsis-

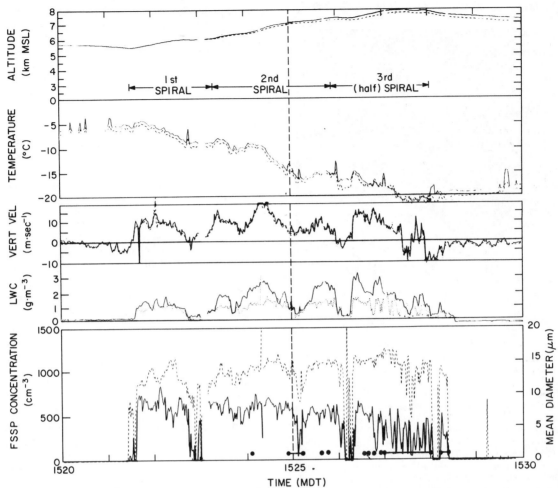

Figure 15.2 Sailplane data from the first penetration, 1520 to 1530, showing altitude, temperature, vertical airspeed, liquid water content, and droplet concentration and mean diameter. For altitude, the solid line is from the Hamilton Standard pressure transducer and the dashed line is from the variometer. For temperature, solid is from the reverse flow probe, and dashed is from an exposed "window" probe. For LWC, solid is from J-W probe, and dashed is from FSSP integrated spectra. For FSSP plots, solid is total concentration with scale on the left and dashed is mean diameter with scale on the right. First, second, and third spirals are indicated below the altitude plot, and times of ice particle image detection are denoted at the bottom by dots (single events) and lines (continuous events).

tent. The precipitation particle data are more trustworthy, though parts of the PMS 2-D probe data are confused by artifacts, both obvious and subtle. The hail spectrometer data on this day appear to be reliable on the basis of cross-checks with 2-D data and (when available) foil impactor data.

The data must be used with considerable circumspection and judgement. The trends of the measurements relate to each other in a reasonable way often enough that one can believe that the sensors are responding to real conditions. For instance, reasonable correlations between the trends of θ_e, vertical wind, and both J-W and FSSP liquid water content are often found. However, detailed study of the data reveals many reasons for uneasiness, and at the time of this study, the basic data within cloud, with the partial exception of data on the precipitation particles, have unevaluated uncertainties. The data are presented and will be used in only a semi-quantitative way, but without error bars since the extent of reliability is not known.

Penetration 1

Figure 15.3 shows the T-28 track superimposed upon a composite CAPPI representation of the radar echo at the approximate penetration level, altered to account for changes as the aircraft progressed along the track (see figure caption). The aircraft entered the storm at 1602, and penetrated the storm along the north side of the highest reflectivity, on the far side from the bounded weak echo region. It left "continuous cloud" at 1609:30, and then, still within the radar echo, turned south out of the storm some 15 km east of the high reflectivity and vault. The altitude, temperature (reverse flow), equivalent potential temperature (θ_e), vertical wind speed, Johnson-Williams liquid water content, droplet concentration, and mean droplet diameter are given in Fig. 15.4. θ_e is calculated assuming saturation with respect to supercooled water, and is only given where liquid water is present, since the T-28 had no humidity sensor. Vertical velocity is

63

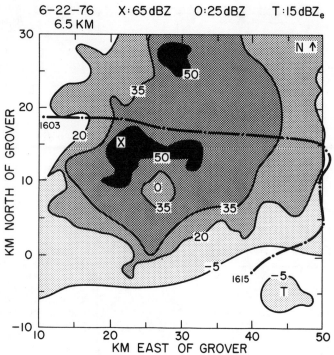

Figure 15.3 The track of the first T-28 penetration is overlaid on a 6.5-km CAPPI of radar reflectivity. The coordinates are kilometers north and east of the CP-2 radar, and the reflectivity shadings represent, from lightest to darkest, greater than -5, 20, 35, and 50 dBZ$_e$. Local extremes of reflectivity are marked by symbols and keyed at the top of the figure. The lack of shading at the northwest corner indicates poor data because of strong horizontal gradients and steep radar scans; that at the south side represents less than -5-dBZ$_e$ reflectivity. From 10 to 35 km east, the CAPPI derives from the 1605:47-1607:46 scan; from 40-50 km east, the 1613:34-1615:37 scan, and the transition (35-40 km east) is hand-drawn. The approximate strengths of two echo maxima and of the echo minimum within the vault are indicated. The closed -5-dBZ$_e$ contour at the southeast corner is reflectivity descending from aloft. The dots along the track are the locations at even minutes; 1603 and 1615 are indicated.

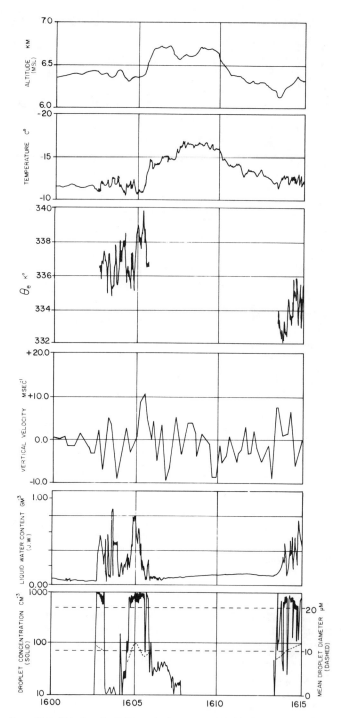

Figure 15.4 Data from the T-28, penetration 1, as explained in the text.

calculated from a crude model using the change of height of the aircraft and true airspeed (Sand, 1976), and should be regarded only as a rough approximation. (How rough is not well known, but a fairly pessimistic view would be ±2 m s^{-1} or 20% of the value, whichever is largest.) The highest liquid water content during this penetration was about one-third of adiabatic, assuming that the cloud base conditions appropriate to the main storm (Chapter 13) also apply here, on the back side of the high-reflectivity core. The droplet concentration and mean diameter are from the PMS FSSP probe after an objective correction for coincidence errors that is explained in the Appendix.

The larger particles were detected by a PMS 2-D imaging probe and the hail spectrometer (Shaw, 1974). The latter is a large, 1-D optical array that furnishes a large sample volume for particles with diameters bigger than about 3.5 mm. The foil impactor was not operating. Data handling is described by Heymsfield and Parrish (1979). Figure 15.5 presents concentration data for several sizes of particles during this penetration, along with vertical velocity once again, and Fig.

15.6 gives the radar reflectivity along a vertical surface, unfolded along the T-28 track shown in Fig. 15.3. The time scale is adjusted so that, viewed as a vertical section with a horizontal distance coordinate, there is little distortion. Radar reflectivity calculated from the large particle data used for Fig. 15.5 agrees well with the actual radar reflectivities. The "total 2-D" concentration plotted in Fig. 15.5 is probably unrealistically

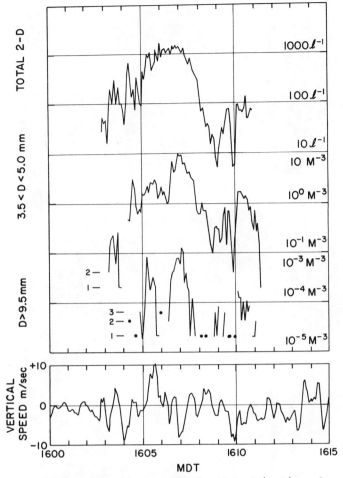

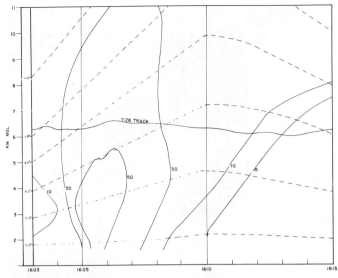

Figure 15.6 Contours of radar reflectivity factor, in dBZ_e, on a vertical surface along the the T-28 track. The T-28 track itself is shown along with the scan elevations that provided the reflectivity data. The sector scan sequences occurred at about 2-min intervals, but always at the same elevation angles. The bends in the scan elevation curves relate to the variations in the aircraft's course.

Figure 15.5 Representation of the concentrations of precipitation-sized particles as a function of time for penetration 1, with vertical velocity repeated from Fig. 15.4, for comparison. Three size ranges given are "total 2-D," from less than 1000 ℓ^{-1} to more than 1000 ℓ^{-1}; 3.5 < D < 5.0 mm, from less than 10^{-1} m^{-3} to about 10 m^{-3}; and D > 9.5 mm, from about 2 × 10^{-5} m^{-3} to about 10^{-3} m^{-3}. D is particle diameter. All data are 5-s (about 0.5-km) averages, and the concentrations corresponding to 1, 2, and 3 counts in 5 s for D > 9.5 mm, and 1 and 2 counts in 5 s for 3.5 < D < 5.0 mm are indicated. The small detached segment at about 1603 belongs with the middle-sized particles. Total 2-D is the total concentration recorded by the PMS 2-D imaging probe, and is always completely dominated by diameters between about 20 and 200 μm. Since there are very large uncertainties in the sample area for the smallest sizes (Heymsfield, private communication), the absolute concentrations should not be taken literally. The trend for the small particles is probably correct, especially in view of the good correlation with the concentration trends of the larger particles. The two large sizes are entirely from the hail spectrometer.

high (Heymsfield, private communication; see figure caption). However, the trends of the data certainly reflect the real trends.

The shapes of the particles recorded on the 2-D probe are irregular until 1608, indicating that they are all ice at diameters above 200-300 μm. It is not possible to distinguish ice from liquid reliably from particle shape at smaller sizes. Occasional hexagonal stellar shapes are recognizable before 1605, but most of the particles appear to be graupel. The smaller particles are more equidimensional, but are distinctly not smooth spheres. Between 1608 and 1611 the 2-D probe

recorded sparse concentrations of remarkably smooth and spherical particles, usually 1 to 2 mm in diameter, interspersed with even smaller numbers of irregular graupel and some artifacts. Figure 15.7 shows the rounded shapes, which are found in concentrations ranging up to about 10 m^{-3} during this 3-min period. The roundness of these images is rare in T-28 penetrations at sub-freezing temperatures in northeastern Colorado, and it indicates liquid water drops or, as seems more likely at a temperature below −15 °C and mostly in downdraft air (Fig. 15.4), frozen water drops. The fact that none of these round images had a small "hole" in the center indicates that they probably are frozen, since liquid drops of this size often display such holes.

Penetration 2

The second storm penetration of the T-28 lasted from 1625:39 to 1639:27. During the penetration the altitude of the aircraft ranged from about 5.5 to 7 km, and the temperature from −5 °C to −17 °C. The penetration was planned to intersect a prominent weak echo vault that had existed for some time when the penetration started but had largely collapsed by the time the T-28 arrived at its location. The radar reflectivity pattern during this penetration changed too rapidly and significantly to allow the whole track to be represented on a single composite CAPPI. Therefore the track is presented in five overlapping segments corresponding to five successive radar volume scans in Fig. 15.8 (see also Figs. 13.7 and 13.8 and Fig. 15.18, below). For the final 4.5 min of the penetration, not shown in Fig. 15.8, the course was maintained SSW. Measurements for this penetration are presented

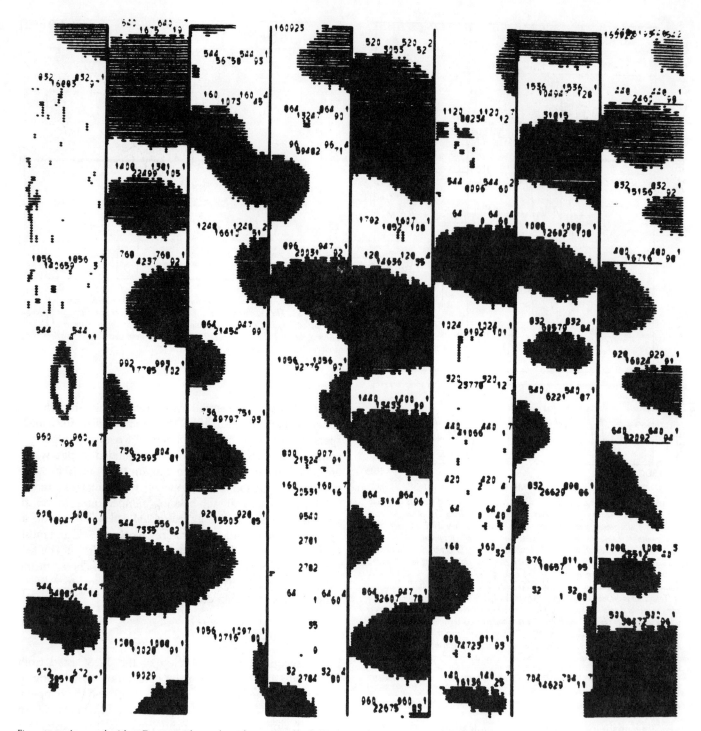

Figure 15.7 An example of the 2-D images of the round particles encountered by the T-28 between 1608 and 1611. Time increases from left to right in each horizontal strip and from top to bottom through the whole figure, and the scale across each strip is 800 μm. The elongation perpendicular to the strips is a distortion arising from the use of an inappropriate lens in the probe. The images should be circular.

in Fig. 15.9. Drop concentrations and mean diameters from the FSSP probe are not presented, but are similar to those in Fig. 15.4. The liquid water content ranges up to about 75% of adiabatic: close enough that, given the uncertainties in LWC, the actual value may have reached adiabatic, especially since θ_e exceeds 342 K in several regions, indicating relatively little dilution (see Chapter 13). Note that while the track is displayed on 6.0-km CAPPI's, the T-28 altitude varied between 5.5 and 7 km. Figure 15.10 shows the vertical reflectivity structure along the T-28 track, and Fig. 15.11 displays some of the larger particle data in the identical format to that used for penetration 1. The 2-D and FSSP probes failed com-

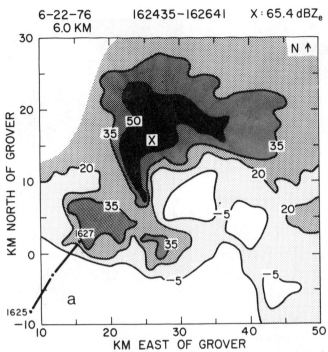

6-22-76   162435-162641   X : 65.4 dBZₑ
6.0 KM

Figure 15.8 The second T-28 penetration, displayed in a format similar to that of Fig. 15.3, except that five successive volume scans are used because the reflectivity patterns change very quickly in important regions. The vault that is especially clear in Fig. 15.8b "collapses" from above as the T-28 approaches it (see also Fig. 15.18). The levels of all the CAPPI's are 6.0 km MSL, to maintain the continuity of the reflectivity changes, though the T-28 rose nearly 7 km toward the end of the penetration. (a) 1624:35-1626:41.

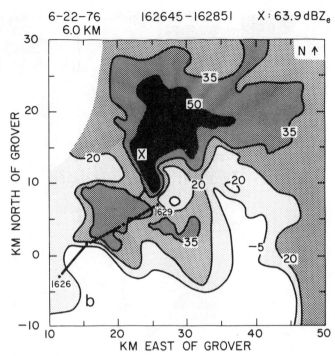

6-22-76   162645-162851   X : 63.9 dBZₑ
6.0 KM

Figure 15.8b 1626:45-1628:51.

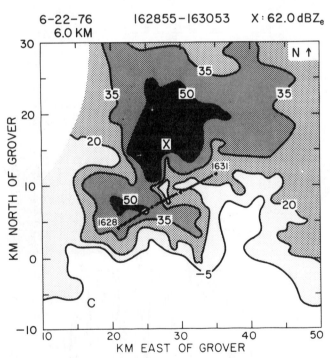

6-22-76   162855-163053   X : 62.0 dBZₑ
6.0 KM

Figure 15.8c 1628:55-1630:53.

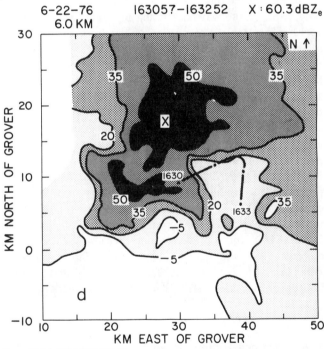

6-22-76   163057-163252   X : 60.3 dBZₑ
6.0 KM

Figure 15.8d 1630:57-1632:52.

some 30 s (3 km) and a peak in large particles sensed by the hail spectrometer just preceding it.

Penetration 3

After the second penetration, the T-28 had to descend below the melting level to shed a heavy load of ice. The third penetration, 1648:21 to 1656:50, was accomplished south

pletely at times during this long penetration.

At about 1628 the T-28 penetrated the edge of the radar echo core of cell C₆. Figure 15.11 shows an updraft lasting

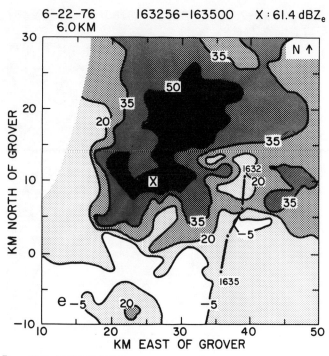

6-22-76   163256-163500   X : 61.4 dBZ$_e$
6.0 KM

Figure 15.8e 1632:56-1635:00.

and west of the high-reflectivity core, between 5.5 and 6 km MSL, on the way back to the airport at Laramie. Figure 15.12 shows the major portion of the track overlaid on a 5.5-km CAPPI that is representative of the whole penetration. The T-28 entered continuous cloud just at the edge of the figure. Figure 15.13 shows the measurements, and Fig. 15.14 the vertical reflectivity structure at the start of the penetration, where updrafts to about 10 m s^{-1} were found within the radar echo. Liquid water contents again ranged to about 3/4 adiabatic, and at one point θ_e barely exceeded 342 K. The hail sensor showed no particles bigger than 3.5-mm diameter, its lowest detection limit, and the 2-D probe gave no useful data until 1654, after which time it recorded some rounded graupel and rimed snow up to several millimeters in diameter. It is possible that the hail sensor was inoperative during this penetration, because the 2-D probe did detect some ice particles several millimeters in diameter. After 1652 the T-28 passed near cells C$_{13}$, C$_{14}$, and C$_{17}$ near the end of their lifetimes, finding only weak updrafts.

Between penetrations 1 and 2

During the 10-min period between the first and second penetrations* the T-28 circled out to the south, staying above

*Note that we have changed the exit time for the second penetration, from 1609:30, the time the pilot judged that he left "continuous cloud," to 1615, a more convenient break for presentation of the data. This is a case of the penetration time being quite arbitrary. As will be seen, the T-28 never exited the storm between penetrations 1 and 2, when one considers "storm" in its broadest, inclusive sense.

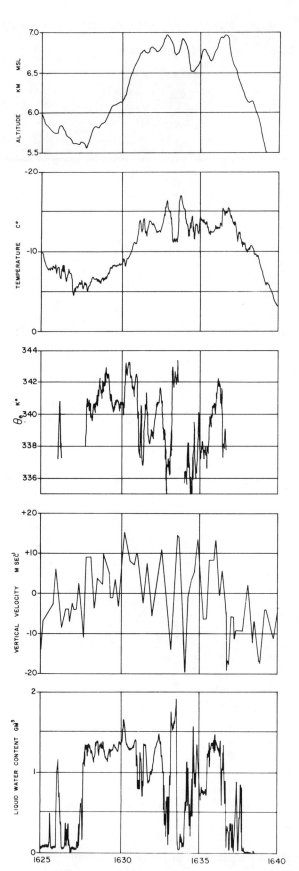

Figure 15.9 Altitude, temperature, θ_e, vertical velocity (5-s averages), and liquid water content (J-W) for the second T-28 penetration.

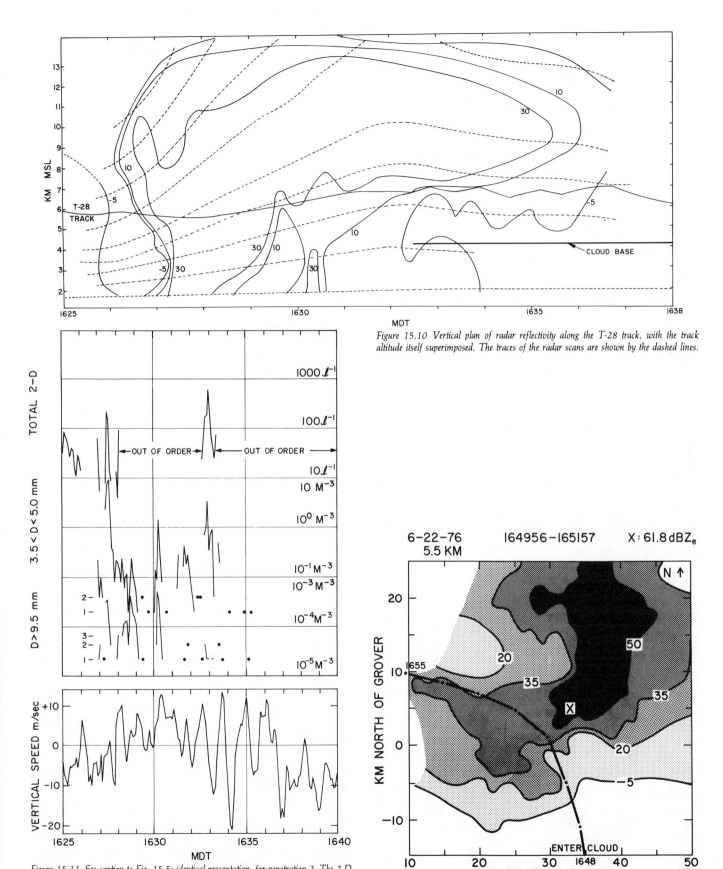

Figure 15.10 Vertical plan of radar reflectivity along the T-28 track, with the track altitude itself superimposed. The traces of the radar scans are shown by the dashed lines.

Figure 15.11 See caption to Fig. 15.5: identical presentation, for penetration 2. The 2-D probe was not functioning in the time periods indicated. Note the very different relative concentrations of the two largest size ranges between penetration 1 (Fig. 15.5) and penetration 2, shown here.

Figure 15.12 The track of most of T-28 penetration number 3, overlaid on a 5.5-km CAPPI that is representative of the echo during the penetration.

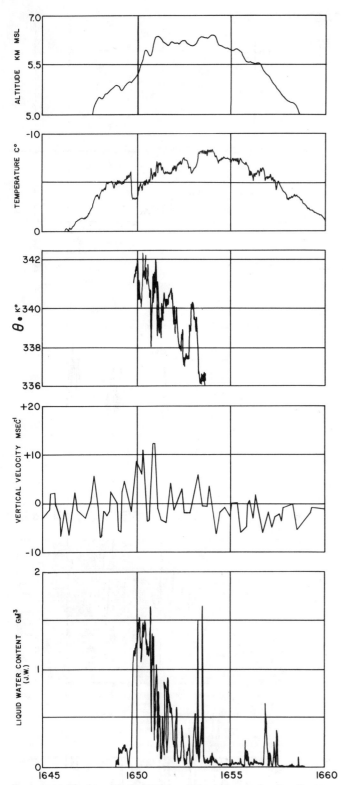

Figure 15.13 *The data of the third T-28 penetration in the same format as Fig. 15.9, again with droplet data deleted because of their similarity to penetration 1.*

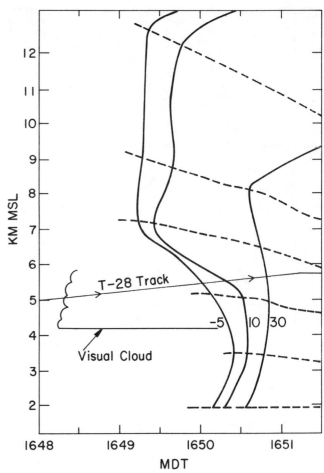

Figure 15.14 *The start of penetration number 3 in vertical section along the T-28 track, with reflectivity contours, aircraft altitude, and scan lines.*

5.5 km. The track is shown in Fig. 15.15, in relation to a 6.1° PPI and rough traces of the southernmost extremity of the continuous echo, at a reflectivity of −5 dBZ$_e$, at 5, 7.5, and 10 km MSL. After the T-28 left the continuous echo, the

pilot reported going "in and out of growing towers . . . around the south to southwest side of the storm." The liquid water content, FSSP and vertical velocity data (Fig. 15.16; FSSP data not shown) show evidence of these turrets. Up until 1621, these turrets were associated with small patches of radar reflectivity at −5 to +5 dBZ$_e$, but after 1621 the track took the aircraft outside of the edge of the sector scan. These clouds contained liquid water contents on the order of 1 g m^{-3} (about ⅓ adiabatic, again assuming the cloud base conditions appropriate for the main storm inflow), and updrafts to about 10 m s^{-1} as shown in Fig. 15.16. The hail sensor recorded no particles during this period, but the 2-D probe did record very sparse snow and graupel particles of millimetric size. Though the 2-D probe was not functioning well during this time, very rough estimates of concentration can be made using the times between particles, and these estimates range from a low of 10 m^{-3} at about 1622, to as much as 100 m^{-3} at other times. Concentration of total 2-D particles increased greatly at 1625, near entry into penetration 2, and many of the larger particles were recognizably hexagonal. Some were unrimed or lightly rimed stellars or dendrites.

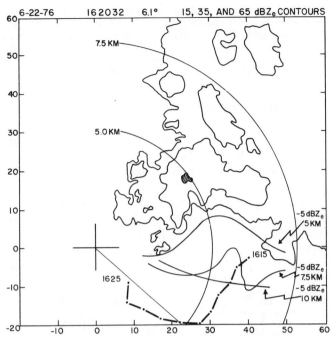

Figure 15.15 The T-28 track is shown during the time between penetrations 1 and 2, overlaid on a 6.1° PPI. The 7.5- and 5.0-km MSL arcs on the PPI surface are shown, and the line to the southeast from the radar at (0,0) indicates the edge of the scan. The edge of continuous echo, at a value of − 5 dBZ$_e$, is shown at 5, 7.5, and 10 km to illustrate the extent of overhang to the south. The two contours are at 15 and 45 dBZ$_e$, and the small shaded core is above 65.

15.3 PRECIPITATION DATA

The 22 June storm was located over a very dense network of hailpads and wedge precipitation gages (Long, 1978) for much of its lifetime. Quantitative, timed hailstone collections were also made at two locations 1 km apart. The data from one of these collections as well as from several of the time-resolving hail separators are presented in this section. The radar context for the observations, to be used later in this section and in the discussion to follow, is given in Figs. 15.17 and 15.18. In all of the maps, the unifying fiducial marks that allow comparison of locations are north-south and east-west lines through the point just between the two timed hail collections, at 26 km east and 13.5 km north of the Grover radar, the origin in most map presentations in this case study.

Figure 15.17 shows a complete series of PPI's from the Grover radar at a 0.5° elevation angle, spanning the time of hail collections. Only the 55- and 65-dBZ$_e$ contours are given. Dye and Martner (1978) have shown that the 55-dBZ$_e$ radar swath for this storm corresponds exceedingly well with the hail swath. Figure 15.18 is a series of 11.8° PPI's, with the coverage displaced toward the south—the direction of the inflow—and including the 15-dBZ$_e$ contour to show the vault development and structure. The 11.8° angle is steep, so that the elevation at the southwest corner of the 20-km square is about 5 km MSL, and that at the northeast corner about 10 km, but most of the features of interest are at or near the 7.5-km-MSL arc, shown in each PPI.

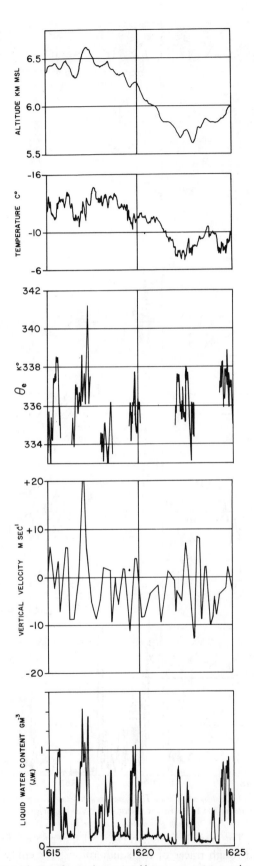

Figure 15.16 The T-28 data for the time period between penetrations 1 and 2, along the track shown in Fig. 15.15, presented in the same format as Fig. 15.13.

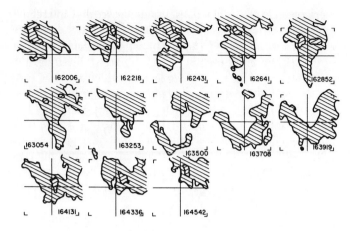

Figure 15.17 The 55- and 65-dBZ$_e$ reflectivity areas are shown for all the 0.5° elevation radar scans between 1620 and 1645, for a 10-by-10-km area centered over the site of the timed hail collections, 26 km east and 13.5 km north of the radar. The north-south and east-west lines intersect at that spot. The small patch of 55 dBZ$_e$ off the southwest corner of the square at 1632:53 grew into the large crescent at 1635:00. The two bursts of hail at the collection site (Fig. 15.21) correspond well with the reflectivity history.

15.3.1 The Dense Network

The dense network of hailpads and wedge gages [Fig. 15.18; see also Long (1978) for a more detailed description and analysis of the dense network data from this day and Fig. 13.1 for the dense network location relative to the other observing systems] was installed for the purpose of evaluating the gage density needed to define hail amounts adequately over an area. The fact that the sensors were not time-resolving is an important disadvantage in interpreting the data for a nearly stationary storm with different distinct

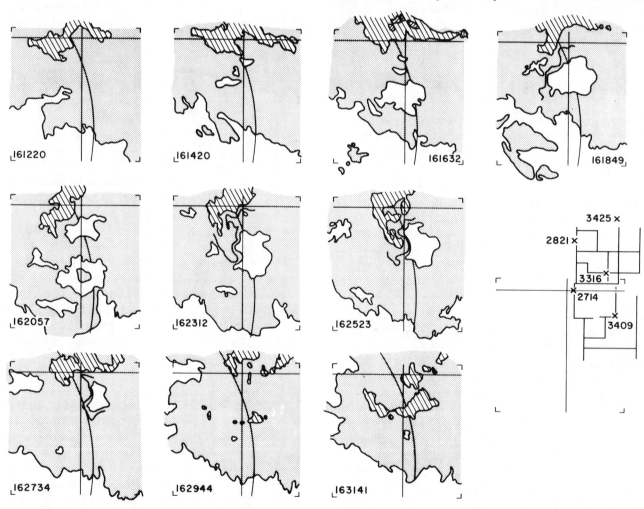

Figure 15.18 The 15-, 55-, and 65-dBZ$_e$ contours are shown for each 11.8° elevation radar scan between 1612 and 1632, for a 20-km-by-20-km area over and to the south of the collection site. The 7.5-km height on the scan is given by the arc in each PPI, and the vault development is clearly evident. See also Fig. 13.8 for a series of north-south and east-west vertical sections through the vault. However, in viewing that figure, generated by computer from the data, one should keep in mind that the next elevation scan above 11.8° was 17.4°, with the result that the top of the vault is usually very poorly resolved by the data. A portion of the 35-dBZ$_e$ contour is given between the vault and the high reflectivity, in *several of the PPI's, to show the reflectivity gradient more accurately. The southwest corner is 15 km east and 5 km south of the Grover radar. Also shown at the same scale and with the same fiducial cross are the locations of the lines of hailpads and wedge precipitation gages, in the southwest quarter of the dense network, and the five hail separator stations that recorded relevant data. The pads and wedge gages were spaced about 1 km apart along east-west roads and three per km apart along north-south roads. The four-digit numbers refer to the hail separator locations marked by x's (see Fig. 15.20).*

periods of hailfall at the same place. Nevertheless, data from this network are presented in Fig. 15.19: three contours of total precipitation (40-, 60-and 80-mm depth; the maximum was 110) and the 3-mm contour of equivalent water depth of hail (the maximum was 10) are given, and the area of maximum detected hail size ≥ 2 cm (the maximum was 3) is shown by shading. A more detailed map of hail amount (Fig. 6 of Long, 1978) shows a lot of very fine-scale variability, and the size data contain similar complexities in contours between 2 and 3 cm maximum diameter. The radar data show that the large, discrete patch of more than 3 mm equivalent hail depth to the southeast fell after 1700, from a later cell or cells than are depicted in Figs. 15.17 and 15.18 or discussed in Chapter 13, and that there was considerable hail to the west of the dense network.

Two prominent conclusions from the dense network data are that the rain greatly exceeded the hail (in fact, hail was less than 3% of total precipitation over the dense network), and the maximum hail size did not exceed 3 cm, but approached or equaled it in a number of places.

15.3.2 Time-Resolved Data

The mesonetwork of conventional meteorological stations has been described in Chapter 13. In addition, a number of hail-rain separators (Nicholas, 1977) were located in the net-

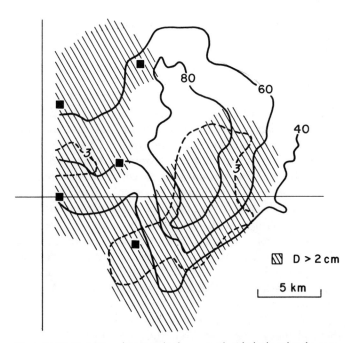

Figure 15.19 Precipitation data from the dense network of hailpads and wedge gages. Contours of total precipitation depth of 40, 60, and 80 mm (solid lines) and of equivalent hail depth of 3 mm (dashed line) are shown, and the shaded area represents maximum hail diameter of greater than 2 cm. The five hail separator sites are shown by the solid black squares, and the north-south and east-west lines correspond to those of Figs. 15.17 and 15.18. The western boundary of the data is the edge of the dense network, and the 40-mm precipitation depth contour ends at the eastern boundary.

work on an experimental basis, using new sensing and rebuilt recording systems. Locations of five relevant stations are given in Figs. 15.18 and 15.19. The separator data from four of the stations are shown in Fig. 15.20, which illustrates the long duration of hailfall and the lack of evidence of size sorting, as would have been indicated by an earlier onset of hail than of rain. The tendency is for maximum hailfall rate to coincide in both space (Fig. 15.19) and time (Fig. 15.20) with the maximum rainfall rate. However, some size sorting on a very large scale is indicated by the displacement of the area of fall of hail larger than 2 cm to the south of the area of maximum rainfall (Fig. 15.19), since, as shown in Chapter 13, cell motion is from south to north.

The weighing raingage data agree in total amounts recorded with the wedge gage data that were the basis of the total precipitation amounts in Fig. 15.20. The maximum rate of precipitation at a point was 10 cm in a little less than an hour; enough to cause appreciable local flooding. (See Chapter 13 for the detailed rain rate analysis that entered into the estimate of precipitation efficiency.) A comparison of data from adjacent weighing raingage and hail separators indicates that the separators markedly underestimate rainfall amount but do give the time variation faithfully (A. Huggins, Electronic Techniques Inc., Auburn, Calif., personal communication).

The two usable, timed hailstone collections were made at (26, 13) and (26, 14) km east and north of Grover, near the intersection of the fiducial lines in Figs. 15.17-15.19. The collections were obtained from a mosquito-net funnel with an opening of 0.40 m² that separated the rain and directed the hail into jars of hexane at dry ice temperature. All the collected hailstones were later photographed, measured, and weighed, and some were sectioned. Data from the more detailed of the two series of collections are presented in Figs. 15.21 and 15.22. Figure 15.21 shows the times of the thirteen collections and the mean diameter and total concentration of hailstones for each. Fig. 15.22 gives size distributions and the total sample size of each. The concentration and mean diameter calculations were made using fall velocities calculated in the following way. The weight and maximum linear dimension D of 451 of the stones were determined, and a plot of mass vs. the volume of a circumscribed sphere (having a diameter equal to D) was found to follow a linear trend, with some scatter. The slope of that line gave a density within the spheres of 0.54 g cm⁻³. The average density of the hailstones was estimated at 0.85 (see, e.g., List et al., 1970) because much of the ice was quite bubbly. Using these two estimates of density, the equivalent spherical diameter D_e can be calculated to be $0.86D$. Terminal velocity was then obtained from the formula of Matson and Huggins (1980), which was derived from direct measurements of hail of the same size range and in the same region, $V_T = 11.45 D_e^{1/2} = 11.45 (0.86D)^{1/2}$.

The collection data of Fig. 15.21 show two bursts of hail. The same two bursts are shown in Fig. 15.20, in the time-

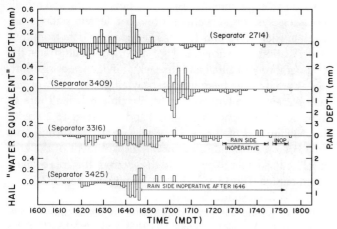

Figure 15.20 Records of hail (down, scale right) and rain (up, scale left) amounts from four of the five hail separator stations identified in Fig. 15.18 and also shown in Fig. 15.19. The bars are 1-min totals. The rain amounts shown probably err in being too small, according to comparison of pairs of adjacent separators and weighing raingages.

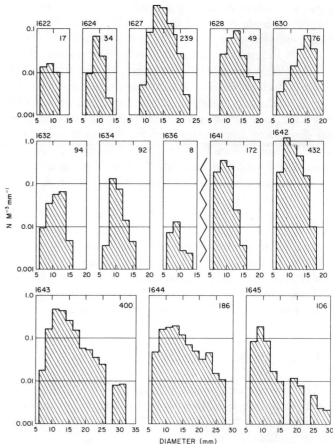

Figure 15.22 Hail size distributions (per cubic meter) for the thirteen collections of Fig. 15.21. The sequence is left to right, top to bottom, and the total number of stones collected is given on each distribution. All of the collected stones were measured. The zig-zag line separates two bursts of hail evident in Fig. 15.21, and times to the nearest minute are given above each histogram.

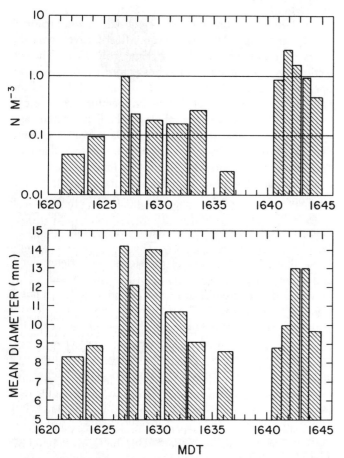

Figure 15.21 Total hailstone concentration per cubic meter of air (using calculated terminal velocities) and mean diameter from the thirteen collections at 26 km east, 14 km north of the Grover radar.

resolved data from nearby separator 2714, and are indicated in the near-surface radar reflectivity data of Fig. 15.17. The first burst of hail relates to cell C_5 of Chapter 13, though it is peripheral. (Recall that the definition of "cell" was made as objective as possible, and that a great deal of area of the radar echo of this storm was not a part of any of the cells as defined in Chapter 13.) The second burst of hail came from cell C_8.

Some of the larger hailstones of this timed series of quenched collections have been used for deuterium analyses, reported by Knight et al. (1981). Structural studies have been made on over 400 stones, 321 from the detailed, timed series of collections and some from several other collections, quenched and otherwise. Typical air bubble and crystalline structures of the hailstones are as shown in Fig. 15.23. They are classified as having graupel embryos (List, 1958; Knight and Knight, 1970). Graupel embryos were found in 97% of the hailstones examined from the one series of thirteen samples. The five stones that had clearly identifiable, frozen drop embryos were all small and were all from the first two of the thirteen collections. Fig. 15.24a shows one small hailstone with a broken, frozen drop embryo, along with three ordinary graupel embryos, all from the first of the thir-

Figure 15.23 Examples of the structures of the hailstones, between crossed polarizers (left), and with ordinary transmitted light (right). The roughly conical growth symmetry and the inconsistent variations of crystal size are representative of all of the stones sectioned. These stones are classified as having graupel embryos, although, as is also common in north-eastern Colorado, no distinct embryo stage is definable, and embryo size is not an applicable concept. Classification of the growth layers, as dry, spongy, and/or with small crystals, is indicated. In these examples and in general (see Table 15.1), crystal size is nearly always either small or large as defined in the text. (a) Large crystals followed by a layer of small crystals in dry growth, followed by a layer of spongy growth.

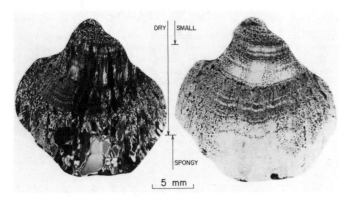

Figure 15.23b A thin layer of small crystals followed by large crystals in dry growth, with a layer of slightly spongy growth at the end.

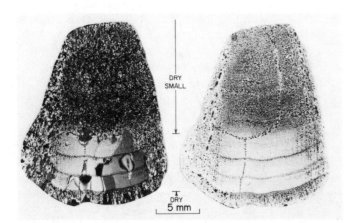

Figure 15.23c A thick layer of small crystals in dry growth followed by large crystals in wet growth, with a final layer of dry growth.

teen collections. Fig. 15.24b includes all three drop embryo examples from the second collection.

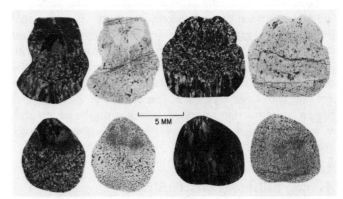

Figure 15.24 Examples of structural features of hail from the first two collections of the series of thirteen samples of Figs. 15.21 and 15.22. The broken frozen drop embryo in (a) and the three drop embryos in (b) are four of the five such embryos found (the other was also in the first collection) in the over 400 stones sectioned and photographed. (a) From the first sample: 1621:20 to 1623:13.

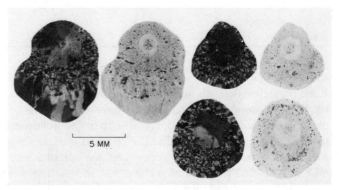

Figure 15.24b From the second sample: 1623:33 to 1625:03.

The selection of the hailstones studied in this section was made to cover the size range relatively evenly, and was therefore weighted unrealistically toward the largest stones.

The similarity of structure of these stones allows one to look at general characteristics, and seek meaningful changes in internal structure as a function of both stone size and time of fall. The classification used here is very simple, and its application to obtain quantitative data is illustrated in Fig. 15.23. Using the air bubble structure, three growth categories are delineated: "dry," meaning the densely bubbly ice characteristic of dry growth; "spongy" (usually only slightly so), meaning nearly clear ice containing air bubbles that formed during the freezing of liquid water included during growth; and "everything else," which includes completely bubble-free ice and ice with alternating layers of clear and bubbly ice at too small a scale to be separated usefully. Crystal size is classed as small, large, or "other," with the terms defined by the long axes of the grains in thin section: "small" is used for equidimensional grains with mean diameter $\leqslant 1$ mm and "large" for grains with long diameter > 1 mm.

Table 15.1 shows the general character of the hailstones (from the thirteen collections only), with the data separated into two size categories, one for hailstones with longest dimension $D < 12$ mm, and the other for those with $D \geq 12$ mm. The data unit is millimeters of growth along the central axis of the cone (see Fig. 15.23). Most of the growth is dry and with large crystals. As expected, the smaller stones have a larger fraction of dry growth than do the larger, and, as well, the larger stones have a larger fraction of spongy growth.

TABLE 15.1
Average hailstone characteristics

| | D < 12 mm | | D < 12 mm | |
	mm	%	mm	%
Dry growth	1540	80	976	69
Small crystals	652	34	449	32
Large crystals	1102	57	818	58
Slightly spongy	31	1.6	142	10.1
Number of stones	227		94	
Total mm of growth	1934		1410	

In seeking general trajectory information from these hailstone characteristics, it is necessary to combine the data from the different stones of a collection. We have not found an entirely satisfactory way of doing this, but the following method does give usable results. The problem is to combine data from different sizes of stones, and we adjusted (normalized) the data to a common stone size and plotted it in terms of tenths of the total hailstone growth.

The results for the whole collection of 321 stones, separated into the two size classes as in Table 15.1, are given in Fig. 15.25. As would be expected, there is a very strong tendency for the early growth to be dry. However, the crystal size is remarkably independent of either position in the stone or of stone size. The usually large crystal size indicates that most of the hailstone growth occurred below approximately the $-20\,°C$ level (Ashworth et al., 1980).

Figure 15.26 shows the data, calculated using ten equal growth intervals in each stone so as not to bias the results toward the larger stones, combined according to collection but without regard to position within a stone. The changes from collection to collection are strikingly large, but do not fit any simple pattern. Furthermore, the two separate bursts of hail show rather different changes with time.

Plots of the type of Fig. 15.25 have been made for both size categories of each of the thirteen collections. The results are like those of Fig. 15.26, in that there are large variations, but not enough consistency to indicate simple interpretations.

15.4 RADAR REFLECTIVITY: THE CELLS

The cellular nature of the 22 June storm has been discussed in and has formed the basis for a major part of the analysis in

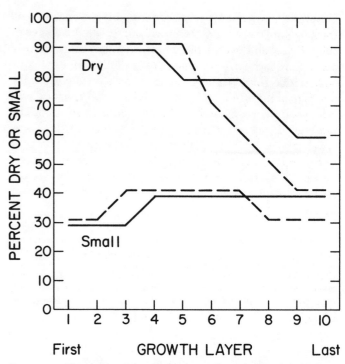

Figure 15.25 The percentage of dry growth and of small crystals (see text) is plotted, for hailstones with D < 12 mm (solid lines) and D ≥ 12 mm (dashed lines), as a function of the stage of growth. All 321 stones are included, and the combined data are obtained by normalizing to the same diameter. Since in almost all cases the stone growth symmetry was conical, "first" applies to the tip of the cone, and "last" to the big end (see Figs. 15.23, 15.24). Growth at small sizes tends to be dry, but the crystal size is remarkably independent of either stone size or location within the stone.

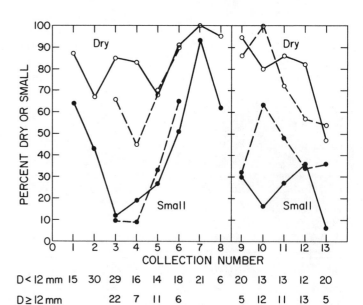

| D < 12 mm | 15 | 30 | 29 | 16 | 14 | 18 | 21 | 6 | 20 | 13 | 13 | 12 | 20 |
| D ≥ 12 mm | | 22 | 7 | 11 | 6 | | | | 5 | 12 | 11 | 13 | 5 |

Figure 15.26 The percentage of dry growth and of small crystals is plotted as a function of collection number for the thirteen sequential collections of Figs. 15.21 and 15.22. The solid lines represent stones with diameter < 12 mm, dashed lines those with D ≥ 12 mm. The first eight and the last five collections represent the two distinct "bursts" of hail from two different cells (see Fig. 15.21). Note both the large changes with time and the complexity. The data used for this figure were normalized to uniform diameter, as in Fig. 15.25. The sample sizes are given below the abscissa.

Chapter 13. Anyone who has performed the exercise of cell tracking within any large storm will appreciate the degree of subjectivity, and even arbitrariness, that inevitably forms a part of any such analysis. One can have little confidence that a radar reflectivity feature called a cell by a research group in one part of the world will correspond closely to that called a cell by a different group in a different place. Even within the same group there can be substantial differences in usage, which depend both upon scientific emphasis and the devoted to communicating how the word "cell" is used, and Chapters 14 and 15 conform to that usage. The cellular nature of the 22 June storm and its implications will be discussed further below.

The radar data that have been discussed in Chapters 13 and 14 refer to areas of the storm with reflectivity factor ≥ 20 dBZ_e. With this threshold, they are comparable with much of the historical storm radar data. However, the nearness of this storm to the CP-2 radar allowed a lower threshold of -5 dBZ_e, and from the standpoint of precipitation formation it is valuable to use these more sensitive data. While parts of the low-reflectivity data are untrustworthy because of side lobes, the southern extent of reflectivity at -5 dBZ_e and weak patches of -5 dBZ_e to $+5$ dBZ_e to the south of the main storm are valid. They are consistent with the T-28 data between penetrations 1 and 2, discussed above, and some of the weak patches of echo can be tracked forward in time as they grow into the much more intense cells.

Figure 15.15, already used in discussing T-28 data to the south of the storm, also shows, very schematically, the southern extent of the continuous radar echo at the -5 dBZ_e level, at 5, 7.5, and 10 km MSL, superimposed on the 6.1° elevation PPI. The great extent of the southward (forward) overhang of this storm is well illustrated, though the locations of the three contours are only roughly defined because of the few and steep elevation scans by the radar. As noted above, there are small, discrete patches of radar echo to the south of the continuous echo, not shown in Fig. 15.15, though the scan coverage did not extend far enough south to intercept them all. The southern edge of the radar sector scan is indicated at the time of the PPI, and it was moved slowly northward during the storm's intense phase, such that at 1640 it intersected a point 35 km east and 20 km south of the radar. In tracking the small, weak patches of echo back in time, it was found that they had often come from south of the radar sector.

We attempted to track the cells identified in Chapter 13 back to their -5-dBZ_e origins, but were successful in only three cases, and partially in a fourth: cells C_9, C_{10}, C_{11}, and (in part) C_{15}. All four time-height diagrams are given in Fig. 15.27, though the data for C_{15} are less reliable than for the others because the determination of cell continuity involved more subjective judgement in that case. Cell C_8 first appeared as a distinct entity within the continuous radar echo and thus could not be traced back to low-reflectivity factors. Other cells either behaved similarly to C_8 or originated beyond the

edge of the scan.

Cells C_9, C_{10}, and C_{11} formed from patches of low-reflectivity factor at about 8 km MSL (about -20°C) that had remained between -5 and 0 dBZ_e for about 5 min. The time from 0 dBZ_e to 20 dBZ_e was again on the order of 5 min.

It seems probable that the very slow echo growth at the lowest reflectivity value represents vapor growth of ice crystals and the slow initial stage of accretion. (Note that the first echo behavior of the case described in Chapter 23 is similar in this respect, though the echo covers a much greater depth.) The fact that it appears in all three cases near the -15°C level could be related to the fast vapor growth of ice crystals at that temperature. However, it is odd that it stays at that level because the cells closer to the storm core are assuredly the location of strong updrafts, according to both aircraft and radar data. Recall that the T-28 did encounter patches of weak radar echo associated with local convective activity and sparse snow and graupel in this general region.

15.5 DISCUSSION

The data on the microphysical aspects of the 22 June 1976 storm have been presented. What remains is to analyze and interpret these data in terms of understanding how the storm produced its precipitation and, insofar as possible, draw conclusions about any simplifying models that might be applicable and about expectations of success for hail suppression techniques. These were the purposes that dictated the case study approach in the first place. The data set for the 22 June 1976 storm is exceptionally complete with respect to the precipitation at the ground—the hailfall and rainfall and the nature of the hailstones themselves. These aspects can be utilized in deducing how precipitation forms in this storm more than in the other two storms discussed in this volume. This storm is also distinguished as exceptionally severe, among the two or three most severe storms observed in four summers, out of roughly 75 thunderstorm days, in the NHRE observational area.

As a first approach to analysis, we take a rather gross, coarse-scale look at the storm data, and test whether there are any major problems in reconciling the observed hail and rain formation with the Doppler and reflectivity data from the storm. Second, we combine a simple, numerical hail growth model with the Doppler wind fields and deduce specific growth trajectories. Finally, an attempt is made to synthesize the data into a conceptual model of the precipitation formation in this storm. Hail suppression is discussed in terms of this model and the storm data.

15.5.1 The Hail Growth Mechanism

Assuming that the hail formation is accomplished through ice nucleation followed by vapor growth and accretion of

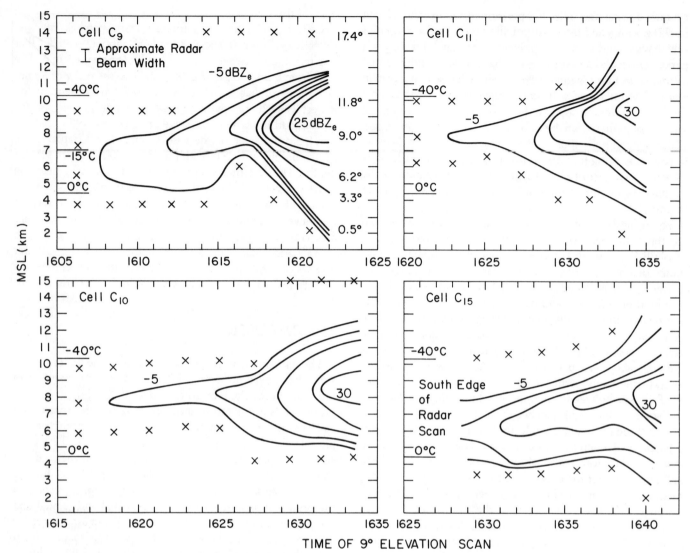

Figure 15.27 Time-height diagram of four cells, from first detectability to attainment of 25 or 30 dBZ$_e$. Except for cell C$_9$, the data were taken from contoured PPI's with a 5-dB contour interval. A special high-resolution display was used for C$_9$, with 2-dBZ$_e$ contour intervals and a 10-km-by-10-km map area. The contours on the figure for C$_9$ have a 5-dB interval, whereas the other three have 10-dB intervals. The X's represent the "no reflectivi- ty" observations that define the boundaries, and their spacing indicates the data spacing. The plot for C$_9$ gives the elevation angles of the scans and a bar that shows the approximate beam width of the radar. The 0°C and −40°C levels are indicated, along with −15°C on C$_9$.

supercooled droplets, three tests can be made:

(1) A comparison of the estimated time for formation of the largest hailstones with an estimate of the maximum residence time expected in the storm.

(2) A comparison of calculated particle growth rates with the rates of growth of the radar reflectivity factor.

(3) A comparison of the growth modes of the hailstones (dry, wet, spongy) with those which would be expected from the microphysical conditions anticipated (observed only in a very small part) within the storm.

The initial assumption of hail formation by accretion of cloud droplets on ice particles formed initially through the ice process is amply justified for the NHRE region as a whole (see Chapter 7) and for this storm from the hail embryo type and the T-28 data. Some frozen drops are present, but are evidently minor factors in the hail formation.

Comparison of the estimated time required for hail formation with that estimated to be available

The time needed to grow a hailstone can be estimated by separating the growth into three stages:

(1) 0 to 1 mm diameter, involving ice crystal growth from the vapor and subsequent riming.

(2) 1 mm to 2 cm diameter, calculated assuming symmetrical accretion on spheres of appropriate density and fall speed.

(3) $D > 2$ cm, ordinary hail growth at essentially solid ice density.

The time required for the first stage, forming a 1-mm diameter graupel at temperatures in the range −5 to −25 °C or so, has been estimated by others, using numerical models, to be in the range of 10-15 min, based upon measured diffusional ice crystal growth rates, riming size thresholds, and estimated accretion rates. For example, Scott and Hobbs (1977) give a good discussion of the times involved for this first stage.

For the second and third stages, accretional growth can be estimated in terms of estimates of the conditions within this storm. The rate of growth of a particle by accretion is

$$\frac{dD}{dt} = \frac{\rho_d V_T W_e}{2\rho_p}, \qquad (15.1)$$

where D is diameter, ϱ_d the density of the water droplets, V_t the terminal velocity of the collector particle (that of the droplets is approximately zero), $W_e = WE$ is the effective liquid water content, i.e. ,the actual liquid water content W times an average collection efficiency E, and ϱ_p is the density of the collector particle. There is a rather wide range of choices of V_t and ϱ_p, but estimates for the northeastern Colorado area are available. Figure 15.28 presents a range of accretional growth rates calculated from a variety of different assumptions, with W_e always taken to be $1\,g\,m^{-3}$. The various points are calculated for either pairs of values of assumed ϱ_p and drag coefficient or pairs of ϱ_p and V_t as taken from an empirical study of graupel from clouds of this area (Heymsfield, 1978). In this latter case, the paired values chosen are extremes, with the lowest (or highest) ϱ_p for a given diameter D paired with the lowest (or highest) V_t. These curves may start to be applicable to particle growth in the 22 June storm at a diameter of about 1 mm, though they are extended down to $D = 0.4$ mm in Fig. 15.28. The curves drawn can be integrated directly to give growth times by accretion from $D = 1$ mm to $D = 2$ cm for (1) $(\varrho_p, C_D) = (0.9\,g\,cm^{-3}, 0.55)$, 33 min; (2) $(\varrho_p, C_D) = (0.9, 1.0)$, 45 min; and (3) the empirical straight line $dD/dt = .0043D + .0057$, in units of centimeters and minutes, 52 min. Clearly a drag coefficient of 0.55 is unreasonably low, especially at the small end of the range calculated, and we may take 50 min as a rough growth time, from $D = 1$ mm to 2 cm, for $W_e = 1\,g\,m^{-3}$.

Adiabatic liquid water content of the 22 June storm, using the sounding in Chapter 13, is about $1.7\,g\,m^{-3}$ at the $-5\,°C$ level within cloud and rises to a broad maximum of roughly $3\,g\,m^{-3}$ that extends from about $-15\,°C$ to $-35\,°C$. The average adiabatic liquid water content in the hail growth layer is about $2.5\,g\,m^{-3}$. On this basis the 50 min estimate for growth from 1 mm to 2 cm can be reduced to 20 min, assuming the very best growth environments short of super-adiabatic concentrations of liquid water, for which there is no observational evidence, and assuming a collection efficiency of about one. But this time is probably too short, because W_e is in reality reduced by both depletion and the fact that the average collection efficiency is below 1. For instance, using a droplet concentration of $600\,cm^{-3}$ (Fig. 15.2) and a maximum adiabatic liquid water content of $3\,g\,m^{-3}$, and assuming (as appears to be the case) that a significant large droplet tail is absent, one gets a median volume droplet diameter of $22\,\mu m$ and from that an estimated mean collection efficiency of 0.8

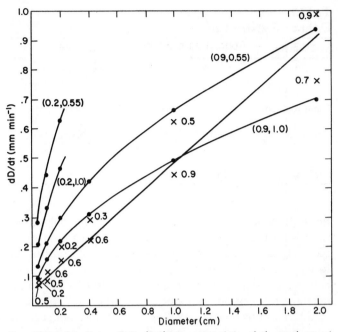

Figure 15.28 Assuming an effective liquid water content of 1 g m⁻³, the growth rates of graupel and small hail by accretion are calculated with several values of density and drag coefficient and/or terminal velocity. Four curved lines are given for (ϱ_p, C_D) of (0.9, 0.55), (0.2, 0.55), (0.9, 1.0), and 0.2, 1.0). The X's are empirical values taken from a compilation of densities and deduced fall velocities given by Heymsfield (1978) and are labeled with the density alone. A straight line is drawn by eye that represents a reasonable fit to these points in terms of the calculations given in the text. It has been assumed that the slowest terminal velocities given by Heymsfield correspond to his lowest densities. All calculations are adjusted to the air density at 400 mb and $-20\,°C$, the middle of the hail growth layer in the 22 June storm.

for a hailstone between 1 and 2 cm in diameter, from Macklin and Bailey's (1966) calculations and measurements. Thus we estimate a reasonable range of minimum required growth time at 20-30 min, for growth from 1 mm to 2 cm diameter. The word minimum is emphasized, since growth at such a fast rate would require a large fraction of the adiabatic liquid water content over nearly all of the time period. As discussed below, this is an improbable occurrence in this storm.

The largest hail from this storm is 3 cm in diameter. An estimate of the growth time for the final centimeter of growth is 5 to 10 min, using the curve of Fig. 15.28 with $\varrho_p = 0.91\,g\,cm^{-3}$ and $C_D = 0.55$. Less than adiabatic effective liquid water content is assumed here, since there must be considerable depletion in this storm on the average, and as noted above, E is approximately 0.8.

The results of the growth estimates for the three stages and the total are summarized in Table 15.2. The estimated total time to reach 3 cm diameter is thus 35-55 minutes, with the shorter time based upon very optimistic assumptions about the growth environment.

The multiple-Doppler analysis presented in Chapter 14 showed all winds within the storm to have southerly components, with the exception of the surface outflow and a very minor fringe in the divergence at the storm top, well above

TABLE 15.2
Estimated growth times

Diameter	Time required	Mechanism
0 to 1 mm	10 to 15 minutes	Vapor growth of snow crystal followed by initial riming
1 mm to 2 cm	20 to 30 minutes	Accretional growth at varying density, $W_e = 2$ to 3 g m^{-3}
2 cm to 3 cm	5 to 10 minutes	Accretional growth, $\rho_p = 0.9$ g m^{-3}, $W_e = 1$ to 2 g m^{-3}
0 to 3 cm	35 to 55 minutes	

the $-40\,°C$ level. It was concluded that there could be no southward movement of particles, and therefore no major recycling, that contributed to hail formation. It is instructive to try to apply Browning's idea about the role of the angle between the inflow direction and the mid-level winds on the potential for recycling of particles and echo vault formation to this storm (Browning, 1977, p. 37). It does not seem to apply easily, using data of Chapter 13 and 14, mostly because of the great extent of this storm complex to the south and west of the main updraft. Thus an estimate of maximum available growth time can be made using horizontal motions alone. This estimate is a maximum because of the possibility of vertical transport out of the layer where supercooled water can exist, reducing the possible growth time during transit through the storm. A similar circumstance has been exploited by Browning et al. (1976) to estimate residence times in a multicell storm. There is, of course, a possibility of local, temporary southward transport at scales too small to be resolved by the Doppler radar analysis.

While maximum residence time can be estimated from the horizontal winds in principle, in practice (in this case) the estimate must be quite rough. For one thing, the region of coordinated coverage by the three Doppler radars is restricted to a portion of the storm near the echo maximum, and to only three 5-min time periods. For practical reasons it could not be extended upwind to the south and southwest nearly as far as the edge of the storm cloud. The conclusion that the winds are from the south to southwest within the hail growth layer in the part of the storm south of Doppler coverage appears certain, however, in view of the cell tracks (Fig. 13.9) and the environmental sounding (Fig. 13.4) given in Chapter 13. The cell velocities, the environmental winds and, where present, the local horizontal winds derived from the Doppler analysis are all consistent in indicating a horizontal velocity averaging about 15 m s^{-1} from the south or southwest within the layer between about $-10\,°C$ and $-30\,°C$, in which the major particle growth is expected. The extent of cloud in that direction combined with this translation velocity gives an estimate of the maximum total residence time available for hail growth from the first possible ice nucleation events.

The T-28 data between penetrations 1 and 2 (Figs. 15.15

and 15.16) show vigorous convective clouds at least 10 km south of the edge of the continuous -5-dBZ$_e$ contour at about 1620. Thus, a horizontal distance of roughly 35 km to the reflectivity core is available for hail growth at this time. Evidence from the three Queen Airs consistently places the southern boundary of the shelf cloud some 25 km south of the reflectivity maximum (see Figs. 13.18b and 13.19c of Chapter 13), and the GOES-E satellite photograph at 1600 is in agreement with the aircraft data.

Using the most generous estimate of 35 km, the average cell velocity of 15 m s^{-1} gives about 40 min as the residence time, and the slowest measured cell velocity gives nearly 60 min.

The required time to grow a hailstone 3 cm in diameter is at least 35 to 55 min, estimated by use of the best expected growth conditions, while the estimated residence time is about 40 min, at most up to 60. According to those estimates, indeed, 3 cm appears to be a reasonable maximum hail diameter for the storm: a stone would need a very lucky trajectory, in terms of encountering exceptionally good growth environment for an exceptionally long time, to grow that large. But the 3-cm stones were not frequent, and no extraordinary circumstances such as accumulation zones of supercooled water or giant solid particles to bypass the vapor growth stage [Rosinski et al., (1979)] appear to be required to explain the largest hail, according to these estimates.

Particle growth rates and echo increase rates

For a collection of particles of uniform diameter D and density ϱ_p, growing at the rate dD/dt, the rate of change of the equivalent reflectivity factor dBZ$_e$ is, assuming Rayleigh scattering,

$$\frac{d(dBZ_e)}{dt} = \frac{d(dBZ)}{dt}$$
$$= \frac{60 \log e}{D}\frac{dD}{dt} + \frac{20 \log e}{\rho_p}\frac{d\rho_p}{dt}, \qquad (15.2)$$

where the second term will be neglected as normally being small, compared with the first, because ϱ_p refers to the density of the entire particle, which changes very slowly. It is interesting to note that it is not a function of the reflectivity itself, in that the concentration does not enter in. In reality, of course, radar echo comes from a distribution of particle sizes. However, given the knowledge that the size spectra are highly variable (compare, for instance, Figs. 15.5 and 15.11, and see Chapter 7, Vol. 1), calculations from monodisperse spectra can be used to compare with the measured rates of change as a rough check. As a rough approximation, the reflectivity growth rate for a particle size distribution is about that calculated according to equation (15.2) for the size that dominates Z in the real distribution, which is toward the large end of the spectrum because of the D_M^6 term in Z_e (D_M is the melted diameter). Figure 15.29 shows the reflectivity factor growth rates corresponding to the particle characteristics and growth rates of Fig. 15.28, calculated by use of the first term in the equation only.

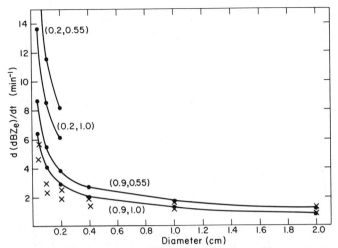

Figure 15.29 *Reflectivity growth rates for monodisperse particle assemblages as a function of the size, according to equation (15.2), and the set of dD/dt values of Fig. 15.28. Since the calculated* d(dBZ_e)/dt *is independent of concentration and therefore also of dBZ_e itself, these rates are applicable to a particle distribution when applied to a certain size within that distribution. The appropriate size will not be far from the size that dominates Z itself.*

Figure 15.30 shows the reflectivity growth rates of a number of discrete reflectivity maxima that can be tracked long enough to cover a growth of at least 20 dBZ_e. These maxima are not restricted to the cells defined in Chapter 13. $d(dBZ_e)/dt$ is shown as a function of mean dBZ_e over the interval. While the maximum size that corresponds to a particular reflectivity is not known very well [Dye and Martner (1978) give 55 dBZ_e as the threshold for hail, probably meaning a maximum diameter of at least 6 to 8 mm in this storm], it is clear that the data on reflectivity growth rates are not inconsistent with growth rates expected according to Fig. 15.29. (In fact, the data of Fig. 15.30 match well with the calculations if N is taken to be 1 m^{-3}.) Since Fig. 15.29 is calculated for $W_e = 1$ g m^{-3}, all the rates could be increased by a factor of about 3, as an extreme upper limit, using maximum adiabatic liquid water contents and a collection efficiency of unity.

The reflectivity growth rate values are consistent with ordinary hail growth by accretion in the expected environments within cloud.

Spongy growth of the hail

Estimates of critical liquid water content for spongy growth as a function of hail size and cloud temperature are given by Macklin and Bailey (1966, 1968). They show that for hail of 2 to 3 cm diameter and $W_e \approx 3$ g m^{-3} the growth only becomes spongy at a temperature warmer than about $-20\,°C$. As illustrated in Fig. 15.23 and true in general for the hail from this storm, detectable spongy growth in the collected particles occurs usually at sizes nearly 2 cm in

diameter, and nearly always at the end of growth. This also is consistent with the expectation that superadiabatic liquid water contents are not present, based upon aircraft observations in this storm and in storms of the High Plains in general, (e.g., Chapter 7).

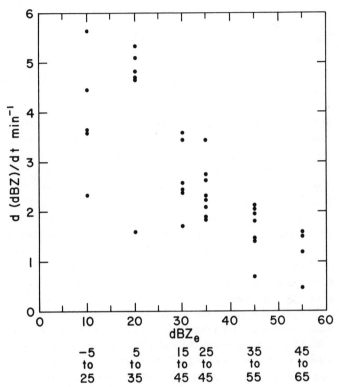

Figure 15.30 *Values of* d(dBZ_e)dt *plotted against mean dBZ_e, for local reflectivity maxima that are traceable in time over an increase of at least 20 dBZ_e. The dBZ_e range actually used is indicated.*

15.5.2 Hail Growth Trajectories

Ice particle trajectories were calculated with a hail growth model adapted from Paluch (1978) and described in some detail in Chapter 20. Six sizes of graupel were placed at several starting locations within the Doppler-derived wind fields, and allowed to grow by accretion of cloud droplets of uniform size* at a concentration of 600 cm^{-3}. Particle growth was computed assuming spherical particles, and the fallspeed was determined from the Reynolds number and an empirical drag law (see Chapter 20). The growth equations (Mason, 1971) were solved for particle mass and temperature subject to local ambient conditions of liquid water content, pressure, and air temperature. The movement of particles was deduced from the fallspeed and the measured airflow, with the 1620 Doppler radar scan assumed to apply for 35 min. Although the 1625 and 1635 Doppler scans were used for the next 8 and 9 min, respectively, most of the precipitation fell out in

---

*Cloud droplet size was computed from the liquid water content and concentration data.

the first 30 min so that trajectories represent growth during and shortly after the approximately steady phase of the storm, the period from about 1550 to 1630.

Adiabatic liquid water content was calculated from the environmental sounding at Potter, Nebraska, released at 1450 (Chapter 13). Since the T-28 apparently never encountered regions with over about 75 % of the adiabatic liquid water content, the values used in the model were not allowed to exceed 0.8 times adiabatic and reached this only if the particles were within the main updraft core. Outside this core the reduction in water content by lateral mixing was approximated by decreasing these values as the square root of the measured vertical air motion, thereby making the horizontal profile of liquid water somewhat flatter than that for vertical motion. For example, at 7 km the maximum updraft speed was about 40 m s^{-1} (Chapter 14). Where the vertical motion exceeded 25 m s^{-1}, the water content was set to 0.8 times the adiabatic value of 2.8 g m^{-3}. Otherwise, the fraction was $0.03 + 0.94 \, w/w_c$, where $w_c(z) = 12.31 \, (z - 1.5) - 0.95 \, (z - 1.5)^2$ closely approximated the measured core profile at 1620. Liquid water content was further linearly reduced to zero at $-40\,°C$ from the amount at $-25\,°C$ in order to represent depletion by ice particles growing below the $-40\,°C$ level. The graupel and hail were assumed to grow entirely from liquid water.

Graupel embryos whose diameters ranged from 1 to 10 mm were started at heights of 5, 7, and 9 km in the main updraft region, in an area from 12 to 48 km east and 0 to 14 km north of Grover (Fig. 15.31). Starting embryo sizes and locations were chosen to be consistent with the observed reflectivity values in cells as they approached the southern boundary of the main updraft. Densities of these starting embryos were calculated from $\varrho_i(D) = 0.08D + 0.1$ g cm^{-3} (D is in millimeters), which is broadly consistent with observed graupel densities (e.g., Heymsfield, 1978), and gives a density of 0.9 g cm^{-3} when the diameter is 1 cm. The accretion density formulation is given in Chapter 20. Drag was changed from high for the smallest sizes to low for the largest sizes by use of the formulation given in Chapter 20, Section 20.2.2. Changing the drag from the high to the low values increased the fallspeed at a single size by about 50 % to 60 %. Melting during fall was included, the water being shed immediately. Considering this kind of sensitivity and the uncertainties in the measured flow field, trajectories should be considered as only qualitatively correct. Despite this, calculated trajectories do appear to represent hail growth in this storm and allow us to conclude something about the main functions of the intense updraft. Figure 15.31 shows the final horizontal positions and diameters (in millimeters) of hailstones that grew from 6-mm graupel particles started at 7 km within the box on the south side of the storm. About 25 % of the 100 embryos that were placed in the wind field grew and reached the ground as hail, ranging in size from 2 to 14 mm. The remainder melted before reaching the surface, where their positions are indicated by zeros. The fallout pattern that resulted

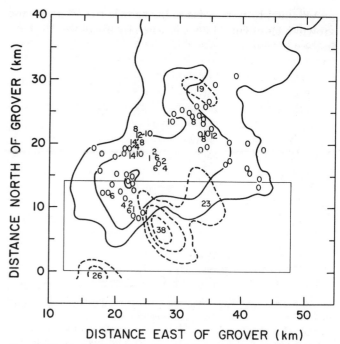

Figure 15.31 The final horizontal locations and diameters (mm) for hailstones that grew from 6-mm graupel embryos injected at 7-km altitude within the box (light solid lines). Starting positions of embryos were every 1.5 km east-west and every 4.5 km north-south. Radar data are from the 1620 scan. Heavy solid lines are the 20-, 40-, and 60-dBZ$_e$ isopleths of measured reflectivity at 2 km, while the heavy dashed lines are the 15, 22.5, and 30 m s^{-1} isopleths of vertical air motion at 7 km, with local maxima of updraft velocity labeled in m s^{-1}. Diameters marked as zeros represent hail that melted before reaching the ground.

from the other graupel sizes (1, 2, 4, 8 and 10 mm in diameter) were similar to the 6-mm one except that nearly all hail grown from the smallest sizes melted. Those small particles that did get inside the updraft core where speeds were greater than about 15 m s^{-1} were carried upward and exhausted into the anvil to the northeast. None of these graupel reached sizes bigger than about 5 mm.

The combined fallout pattern replicates some of the features in the reflectivity structure and surface observations. The southward extension of radar echo west of the main updraft core was apparently a result of hail falling from starting locations west of about 25 km east of Grover. Particles that had not attained sizes larger than about 10 mm as they rose west of the 38 m s^{-1} updraft core were carried northward and fell in an area elongated north-south. These trajectories are also consistent with tracks of the westernmost cells described in Chapter 13, such as C_4-C_7, in Fig. 13.9. The cyclonic trajectories east of the storm core are also consistent with the tracks of cells C_9-C_{11}, though they occurred near the end of the calculated trajectory times. A second area of fallout was north-northwest of and between the two main updraft cores, in the area of the southeastward bulge of reflectivity into the vault. Finally, a third area broadly consistent with the echo structure was located north-northeast of the main updraft. The spatial distribution of hail falling with rain is similar to the observations within the surface rain gauge and hailpad network.

In all three areas of hailfall in the model the largest hail was located in the middle of the fallout zone with smaller particles and rain toward the edges of the zone. There was no marked preference for the largest hail to be located closest to the vault as suggested by Browning and Foote (1976) in their discussion of the "hail cascade." Indeed, there was no evident size sorting of any kind in the sense of Young (1977). The trajectory calculations produce results consistent with the "hail core" observations of Admirat and Goyer, summarized by Goyer (1977). An increasing amount of surface data supports the contention that hailfall and rainfall maxima are usually nearly coincident in time and space, (e.g. Fig. 15.20), though some of the more intense but rare storms do exhibit size sorting. We attempted to determine the conditions that might lead to the largest hail falling adjacent to the vault and found that the starting embryos at the locations used had to be larger than 1 cm for this to happen. However, such large sizes followed trajectories northward across the vault below the height of the 20-dBZ$_e$ contour, and were therefore inconsistent with the observed echo structure using the approximation of near steadiness over the period of the trajectories.

Some representative growth trajectories are shown in plan view in Fig. 15.32. Nearly all are for 6-mm graupel started at 7-km altitude (about −12 °C), but there are also two 8-mm particles, one started at a height of 7 km to the southeast of the 38 m s^{-1} core and the other at a height of 9 km to the south of the core. This first 8-mm graupel embryo grew to a diameter of nearly 2 cm and fell close to coordinates (28, 21) km. Two 1-mm embryos were started within strong updraft at a height of 5 km and grew to 3 mm but were exhausted into the anvil at an altitude of 12 km. Obviously any particle that had to grow from an ice crystal nucleated at −5 °C within the main updraft was carried into the anvil before attaining hail size. Embryos that began east of the updraft maximum followed cyclonic paths around the southeast side of the vault and apparently resulted in some of the largest hail. Browning and Foote (1976) speculated that similar paths were followed in what they called the "embryo curtain," the embryo source region. Heymsfield et al. (1980) have also described similar trajectories in the 22 July 1976 case study, which is discussed in Chapter 20. Browning and Foote deduced a recirculation through the updraft of ice particles grown initially in a first ascent in the main updraft. However, no such recirculation was possible in this case or in that of Heymsfield et al. (1980): embryos must have grown south of the main updraft in our case and subsequently moved into the updraft.

The sensitivity of trajectories to the beginning size of embryos is depicted in Fig. 15.33, where the paths of five particles are shown, with diameters of 2, 4, 6, 8, and 10 mm and started at (27, 0) km from Grover and 7 km altitude. As expected, the 2-mm embryo grew to only 4 mm before being carried into the anvil at an altitude of 11 km. The remaining embryos followed cyclonic paths with the path progressing westward as the starting size was increased until the 10-mm

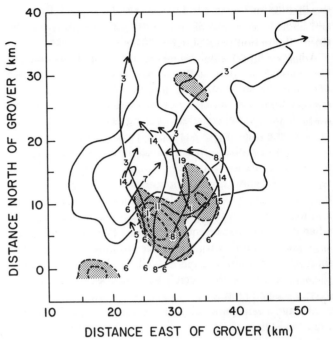

Figure 15.32 Plan view of selected trajectories overlaid upon reflectivity and updraft isopleths as in Fig. 15.31. Hailstone diameters are shown every 10 min along the paths. All growth trajectories were begun at 7 km and from a diameter of 6 mm, except the ones near (30,5) and (27,0) km, which were started as 8-mm diameter particles, and the two 1-mm embryos that were started at 5 km. All hailstones reached the surface except the two that began at a size of 1 mm; they were carried into the anvil near an altitude of 12 km.

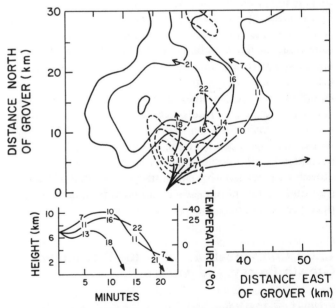

Figure 15.33 Same as Fig. 15.32, except graupel particles started at (27, 0) km and 7 km altitude with diameters of 2, 4, 6, 8, and 10 mm. The insert shows time-height plots for the 4-, 8-, and 10-mm embryos.

embryo followed a nearly straight-line path northward across the vault. Time-height plots shown in the insert indicate that periods of 12 to 20 min were required for hail to reach the ground.

The representative growth trajectories suggest that one function of the vault was to produce the final growth increment on hailstones. If this region of high liquid water content and large vertical velocities had not been present, the several-millimeter-sized embryos produced in the cells may not have been able to grow to the 2- and 3-cm hailstones that were observed. Furthermore, so few particles grew to large sizes and fell out unmelted after traveling across the vault that much of the water vapor transported into the storm by the intense updraft probably was exhausted into the anvil and lost to the precipitation process. The inefficiency of the vault region in the conversion of cloud water to precipitation contrasts interestingly with the relative efficiency of the storm as a whole, discussed in Chapter 13. In spite of the fact that when the vault was present the moisture flux through it was a large part of the flux for the whole storm (Chapter 14), the short lifetime of the vault in its well-developed state did not allow it to reduce the overall efficiency significantly. The storm's precipitation efficiency after the vault ceased to exist must have been quite substantially higher than during the vault's existence.

15.6 CONCLUDING DISCUSSION

The overall conclusion of the analysis and discussion in Section 15.5 is, in a sense, that the entire data set on the 22 June storm is coherent in terms of precipitation production by the storm. Conventional estimates of the expected cloud conditions and processes in northeastern Colorado (no superadiabatic liquid water contents, no coalescence growth) are adequate to explain the hail and rain formation, and growth trajectories within the Doppler-derived wind fields, according to a very simple growth model, are "consistent enough" with the surface precipitation data and the near-surface radar echo data. (Indeed, a good deal less consistency would not have been alarming in view of the number and seriousness of the approximations involved in deriving the growth trajectories.) This circumstance engenders a generalized trust in the entire data set, and provides a basis for the final, concluding discussion in this section.

15.6.1 Conceptual Model of the Precipitation Growth Process and Trajectories

Figure 15.34 is a summary of both the range of the data on this storm and the concept of precipitation formation within it that has emerged from the data. The radar data are given in a south to north vertical section that shows a prominent new cell south of the echo core. The temperature levels 0°C, −20°C, and −40°C are given in the top panel along with the radar scan lines, indicating the relatively sparse radar coverage above the −20°C level. In the middle and lower panels of Fig. 15.34, three reflectivity contours are retained

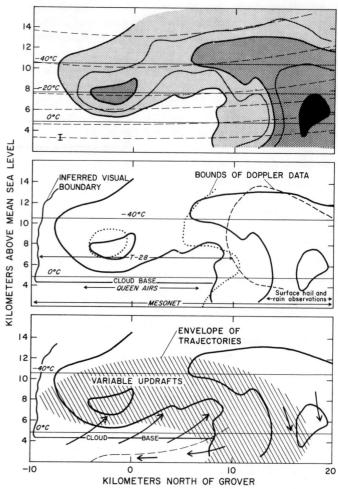

Figure 15.34 The three parts of this figure are the basis for the discussion of the concept of precipitation formation within the 22 June 1976 multicell storm. At the top, contours of equivalent radar reflectivity factor of -5, 20, 35, 50, and 65 dBZ_e are shown, as well as the 0°C, −20°C, and −40°C levels (temperatures appropriate to pseudoadiabatic ascent), which delineate the extreme bounds and the center of the possible hail growth layer. The radar scan traces are shown as dashed lines. The bar at the lower left corner represents the approximate beam diameter at this range from the radar. In the middle, the areas of Doppler radar coverage adequate for horizontal and for vertical air velocity deduction are shown by dotted and dashed lines, respectively, with the locations of the other measurements as well as the cloud base. The inferred visual cloud boundary is also shown. According to T-28 data, a few isolated cumulus congestus farther to the south are sometimes present. The bottom drawing shows the airflow information in broad outline, and the envelope of trajectories. See text for complete discussion.

along with the 0°C and −40°C levels, which represent the absolute limits of accretional growth of ice particles.

The middle panel of Figure 15.34 shows the data types and locations. The areas of both horizontal and vertical Doppler data (the dotted and dashed lines) are fairly representative: the data are at some times a little more complete, and at other times and places less so. The positions of T-28 and Queen Air data are also given, and surface data sources are indicated.

The aircraft data show perturbations in vertical velocity below cloud base superimposed upon a very broad, storm-wide updraft (see Chapter 13, Fig. 13.19). The smaller-scale variations tended to occur at a scale, about 8 to 10 km, that is comparable to that of the radar echo maxima higher in the cloud. Much greater variability was observed at 6.5 km, the T-28 penetration altitude. In penetration 2 (Fig. 15.9), a common scale of updraft is about 5 km (a little less than 1 min of flight). These features are convincingly real because of the correlated θ_e variations, but they cannot be placed in direct correspondence with radar cells. The velocity fields deduced from the Doppler radar data are smoothed and do not show these smaller-scale variations, but are consistent with the aircraft observations once the smoothing is taken into account.

Given the northward horizontal winds within the hail growth layer and the northward cell motion, a trajectory envelope for the growth of rain and hail can be drawn, and is shown by the shaded area on the bottom panel of Fig. 15.34. The two-dimensionality of this representation is obviously a simplification, but it is not fundamentally misleading in this case because of the relatively uniform wind direction with respect to the storm as a whole (see also the trajectories in Figs. 15.32 and 15.33). The envelope is not always within the reflectivity contours because the reflectivity itself is an instantaneous map. Most of the cells did not travel over the reflectivity vault, but moved into or around it.

This trajectory envelope is very wide. The natural next step is to try to be more specific, as is Browning et al.'s (1976) deduced hail growth trajectory, for instance. We start with the evidence provided by the hailstones themselves.

While there is no consensus at present on detailed interpretations of hailstone structures, all investigators appear to agree that in dry growth, crystal size reflects the growth temperature, with large crystals (> 1 mm in diameter) implying growth at an air temperature warmer than $-20\,°C$, and small crystals ($\ll 1$ mm in diameter) implying temperatures below $-15\,°C$ (Levi and Aufdermaur, 1970; Rye and Macklin, 1975; Knight et al., 1978; Levi et al., 1978; Levi and Prodi, 1978; summarized in Ashworth et al., 1980). According to this criterion, and referring to Table 15.1 and Figs. 15.23-15.26, the hailstones show (a) most growth below (warmer than) the $-20\,°C$ level, about 8 km, (b) a remarkable variety of trajectories, at least as indicated by crossings of the level from $-15\,°C$ to $-20\,°C$, and (c) fairly common slightly spongy growth at the largest sizes. The "variety of trajectories" need not involve large excursions from the $-20\,°C$ level, but the hailstone structures do not preclude the possibility of large excursions.

The results from measurements of the deuterium-to-hydrogen ratios on a very small number of hailstones from this storm (Knight et al., 1981) show growth within a rather restricted temperature range, estimated to be $-15\,°C$ to $-28\,°C$. There are strong qualifications on how seriously the absolute values of temperature should be taken, but, as is argued by Knight et al. (1981), the rather narrow range seems to be a more solid conclusion. This deduction is compatible with both the unsystematic nature of the crystal size transitions in the hailstones and with the cell histories given in Chapter 13, Fig. 13.10: namely, the cells' maximum reflectivity tends to straddle the 8-km ($\approx -20\,°C$) level throughout their evolution until descent to the ground.

The deduction is also consistent with the trajectory calculations. Note in Fig. 15.33 that the trajectory that produced the largest hailstone at the ground did accomplish most of the growth near 8 km, while smaller hailstones grew along both the higher and the lower trajectories (from the smaller and larger starting embryo sizes). Of course there is an element of circular reasoning here, insofar as the model contained a linear depletion of liquid water from no depletion at $-25\,°C$ to complete depletion at $-40\,°C$. This was justified by the relatively high precipitation efficiency of the storm, about 50%. In the vault itself the depletion was probably much smaller, and results for trajectories in that region could be questioned on that basis. However, similar trajectory calculations were made with no depletion, and the actual differences in final hail size that resulted from motion across the vault region were only a few millimeters at most. This is because the reflectivity vault was rather small and the horizontal airflow through it was rapid, limiting the time a hailstone could spend growing there. Thus the effect of the probable lack of depletion of liquid water in the upper portion of the echo vault on the amount of hailstone growth was relatively modest.

The concept that emerges from these data has a statistical component. The residence time of an ice particle within the storm is limited by the horizontal motion through the storm, and its ultimate size is determined by the environment it encounters along its path. Within the complex, turbulent environments within the storm, the probability of encountering a highly favorable growth environment rises from zero at $0\,°C$ to a maximum at about $-15°$ to $-20\,°C$, and decreases again to zero at $-40\,°C$. In general, both depletion and entrainment become more important with height, and the adiabatic liquid water content increases rather rapidly with height up to about $-15\,°C$ to $-20\,°C$.

As has been concluded in Section 15.5.1, the biggest hailstones must have been very lucky in encountering high liquid water content for nearly all of their residence time in the storm, and in encountering the vertical and horizontal winds that allowed them to do this. This implies, in terms of overall probability, that a large hailstone on the average encountered increasing updrafts along its path as it grew, thereby keeping it in the middle levels where the probability of encountering the best growth environments would be highest. Less lucky particles grow to smaller hail and the least lucky end up as rain or evaporate.

Within the trajectory envelope of the 22 June storm (Fig. 15.34, bottom panel), the most favorable trajectories for producing large hail would be just those kinds suggested by the deuterium analyses, though not all such trajectories would necessarily be favorable. Most of the unfavorable ones for this storm would be too high, within depleted regions and even above the $-40\,°C$ level. A much smaller proportion of the unfavorable ones would be too low in this general echo pattern. (Recall that a very small proportion of the hail collected, and only small stones, had frozen drop embryos. These presumably resulted from descent of graupel below the 0°C level and re-ascent in updraft, not possible in the schematic of Fig. 15.34. The echo overhang did descend below the 0°C level at other times.) It would appear that much of the rain from this storm must have come from particles with very high trajectories, since there is so much echo above the −40°C level. This would account for the general precipitation pattern of Fig. 15.19, with heavy rain to the north of the hail.

Recall that only about 3% of this storm's precipitation was hail, the rest rain, and the large hail was a very small fraction of that 3%. It is intuitively reasonable that the large hail maintained its favorable altitude for growth by residing much of the time at moderate updraft speeds, which often occur in strong horizontal gradients of updraft velocity. The trajectories of Figs. 15.32 and 15.33 show this rather well, and the T-28 data from other storms often show hail in such regions (Musil et al., 1973; Sand, 1976). The smoothing involved in the trajectory analysis of Section 15.5.2 has not altered the fundamental nature of the hail growth trajectories, in spite of the expectation that much of the actual hail growth took place in sharper updraft gradients than are preserved in the velocity fields deduced from the Doppler data.

There are indications that this concept may be rather generally applicable to the great majority of hailstorms of moderate or low severity. The "hail cores" have been observed a number of times and the low ratio of hail to rain is common. Perhaps most importantly, deuterium studies indicate that the altitude range of hailstone growth is usually quite limited especially for hailstones only a few centimeters in diameter (Knight et al., 1981), and the great diversity of hailstone layering transitions even in single collections is very common. The one NHRE storm that was a striking exception, on 8 June 1976, also produced the heaviest hailfall that was sampled within NHRE. It possessed a very steady echo vault that was well-developed for about 35 min, and its echo signature was much steadier than that of the 22 June storm.

Elements of this kind of semi-statistical concept of hail formation have been evolving in discussion among and in the thoughts of NHRE investigators for some years. Such ideas have appeared in research by Young (1978) and Sand (1981).

15.6.2 The Role of the Cells

Cells have played a large role in the thinking about thunderstorms since the Thunderstorm Project (Byers and Braham, 1948). Insofar as the cells do tend to encompass the high-reflectivity cores, much of the hail growth certainly takes place within them, but there is no need to argue either that the entire growth history of a hailstone had to be within one cell, however that cell's boundary may be defined, or, for that matter, the contrary. The existence of unsteady radar echo perturbations at some scale is of course a signal that a storm's precipitation production may be well represented by a probabilistic model, as suggested above. In that sense one should be more comfortable applying such a concept to what are called multicell storms than to supercell storms, though it might also be useful in some of the latter.

The present storm provides clear evidence that not all of its cells were closed entities, as does the 22 July case. Precipitation streaming from one cell into another was shown in Chapter 14, and a number of the cells first appeared within pre-existing echo, so that inheritance of pre-existing particles (as suggested by Ludlam, 1952 and 1958) seems certain. In analyzing the important factors in precipitation formation, the exact definition of a cell is not especially relevant. While cell tracks were used as a rough measure of the residence time for hailstone growth within the storm, this depended upon the existence of a correlation between cell motion and mid-level winds, not on any specific definition of a cell. It is also not especially relevant to the inquiry into precipitation formation, whether a distinct cell scale exists or whether the cells are just a conveniently identifiable scale within a continuum.

It may be worth emphasizing that these remarks are not intended to imply that the cells themselves, however defined, are unimportant. The cells undoubtedly contain a major part of the upward flux of air and water substance, and are the sites of a major part of the supercooled cloud water, and hence of the precipitation particle growth, especially in the early stages. Obviously the existence of strong, localized updrafts is crucial to the phenomenon of thunderstorm precipitation. The point being made is that there seems to be no point in worrying about whether the largest hailstones reside within single cells for their entire growth history, or about what the best objective definition of a cell may be.

Given the coverage and resolution of the data on this storm, it would be impossible in any case to examine trajectories at scales much smaller than that of the cells. One might argue that the probabilistic view being advanced is needed only because of insufficient data to provide better detail. A germane question of considerable practical importance is the scale of detail that is really important for understanding the precipitation formation. That more detail exists is beyond question, as the hailstones, the T-28 data, and the radar echo history all attest. While this question cannot be answered with confidence yet, the general agreement of the trajectory calculation with the observations of the hailfall in the present case is gratifying. On this basis, we feel that more important shortcomings of the present data set are, first, the limited coverage in the low-reflectivity part of the inflow sector to the south, and, second, the relatively poor coverage and resolution time.

15.6.3 The Role of the Vault

The evidence is clear that the vault structure did not

dominate hail formation in this storm in the same way that it has been argued to do in some supercell storms by Browning and Foote (1976). There is no hail cascade in their sense, according to either the hailstone data, the radar reflectivity, or the trajectory analysis. The echo gradient at the back of the vault is rather gradual. As discussed above, the vault is evidence of a localized, rather transient inefficiency in the conversion of the inflow vapor to precipitation, but there is no evidence that it has any further, strong significance for precipitation. In writing this, we do not mean that the major updraft region that the echo vault accompanies is not significant. Obviously it is: it is the core of the storm circulation, dominates the upward flux of matter, and is important in the final growth of hail to the largest sizes observed (see Fig. 15.33). It is the fact that it happens to be the location of a bounded weak echo region for about 20 min that does not seem to be very important. If the echo vault were not there either because the updraft maximum was a little weaker or because of a different distribution of precipitation elements, hail formation would be about the same. The largest hail does not grow in the strongest updraft anyway, but forms in trajectories around the edge of the vault. A significant reduction of the maximum updraft velocity need not markedly influence maximum hail size. It is interesting to recall that a later period of this storm also produced 3-cm hail, as was noted in connection with Fig. 15.18. In this period, the reflectivity pattern of the storm contained no vault, but was otherwise similar to that described herein.

A "steady," vaulted period of the storm has been noted in Chapters 13 and 14 as an approximation. While the vault may be an identifiable feature for about an hour (Fig. 14.6), it was "well-developed" for only about 20 min (Fig. 13.8), and the radar configuration is not, in fact, very steady. Figure 15.18 was constructed to emphasize the vault's unsteadiness in one segment of not-smoothed data, but echo patterns are an uncertain indication of airflow, and the data do not show just how steady the vault circulation really was. Better time and space resolution in the Doppler coverage would be very valuable for questions like this one.

Nearly all of the particle growth in the trajectory model was in a steady flow field appropriate at a time close to the end of the vault's well-developed phase. Since the trajectory model also did not show the salient hail-growth features of Browning and Foote's (1976) concept, the inapplicability of that concept to this storm may be caused by the storm's organization as much as by a lack of steadiness.

15.6.4 Hail Suppression

The exceptional detail of the data on precipitation at the ground in this case is useful for discussing hail suppression possibilities. It has been shown that the overall precipitation efficiency of this storm was fairly high (approximately 50%), that only about 3% of the precipitation was hail, with a small proportion of this having substantial size, and that the hail accompanied heavy rain at the ground. The conclusion is inescapable that, on the storm scale, most depletion of the cloud water was accomplished by graupel smaller than 1 cm in diameter, and that the hail itself had an insignificant role in depletion. This is consistent with the general conclusion reached in Chapter 7 of Volume 1 that, per equal linear size interval, the middle-sized particles of most of the measured T-28 particle size spectra do the most depleting of cloud water. For this storm, further evidence is furnished by the correlation of the hailfall with heavy rainfall at the ground. This indicates that the cloud water depletion was primarily through the smaller particles, not only on the storm scale but also locally, where hail was present. Of course this need not always be the case, and in fact Chapter 19 documents an exception in the 22 July storm.

The idea for hail suppression in NHRE (Foote and Knight, 1979), adapted from Soviet and other sources, was to induce "beneficial competition" by seeding with ice nuclei. According to this idea the introduction of more ice particles (more hailstones) would deplete the liquid water to the point where there would not be enough to grow hail as large as would have been the case without seeding. The hailstones, especially the largest hailstones on 22 June, are most likely to have grown from ice crystals nucleated near the farthest southward extremity of the overhang, so as to allow maximum residence time within the storm. The idea then would be to seed there (not, incidentally, as was actually done in NHRE) so that the artificially induced ice would follow along about the same paths and provide the beneficial competition with the embryonic, natural hail. However, since depletion by the natural hail is insignificant, adding more ice crystals in this way carries the danger of increasing the hail by increasing the population of the early ice particles that have a chance to follow favorable trajectories and grow to large hail.

The approximately 1- to 5-mm graupel that accomplish most of the depletion may originate over a wide region. They may form from primary nuclei or from (hypothetical) secondary processes that act within a broad extent of the overhang. They may be particles that nucleated far to the south but had relatively unlucky trajectories, or ones that nucleated closer to the storm core and had a shorter residence time but were luckier. While all of these arguments entail a great deal of conjecture, a less risky seeding procedure for a storm of this type would be to seed a substantial distance in from the farthest extent of the overhang, to lessen the risk of increasing the large hail while maintaining the chance of adding to the population of effective depletors. However, the relatively efficient natural depletion does not bode very well for the success of this procedure either.

In conclusion, it appears that the time for confident evaluation of hail suppression possibilities has not yet arrived, though this case study analysis would tend to make one pessimistic rather than optimistic. At present, none of the techniques that have been advanced inspire much confidence when mentally applied to the 22 June storm. A true evaluation of the possibility of hail suppression by seeding probably must await a future generation of numerical model that includes a combination of microphysical and dynamical complexities beyond the present state of the art.

Appendix

Treatment of Cloud Droplet Concentration Data

The raw data on droplet concentration in the range $2 \mu m < d < 30 \mu m$ from the PMS FSSP probe on the T-28 showed a maximum of about 200 cm⁻³, and the liquid water content calculated therefrom was much lower than that from the J-W hot-wire probe. These low concentrations were deemed unreasonable, and a suggestion from W. A. Cooper of the University of Wyoming, that a major problem was probably coincidence errors, was followed to provide an objective correction. The coincidence error problem is described by the equation

$$N_M = N_T e^{K N_T}, \qquad (A15.1)$$

where N_M is the raw, measured concentration, N_T is the actual concentration, and K is the product of sample area, true airspeed, and electronic dead time of the probe after sensing one droplet. This relationship gives N_M a maximum possible value for all values of N_T, and provides two possible values of N_T for each N_M (both values exceeding N_M).

The sailplane data on this day and general experience in northeastern Colorado cumulus lead one to expect droplet concentrations of the order of 1000 cm⁻³ within active updrafts, so that there is a strong presumption that the FSSP values are too low. Since K is unknown, the 1-s accumulated data were treated as follows: the measured values of N_M were accepted without correction except when N_M exceeded 100 cm⁻³ for three successive seconds or more, corresponding to at least 300 m of path. For each such period, it was assumed that the highest value of N_M (N_M') was the maximum permitted by the exponential relationship given above; this fixed the value of K as being equal to $(e \, N_M')^{-1}$. For those 1-s intervals when N_M was less than N_M', the value of N_T was taken as being the larger of the two roots of

$$K = \frac{-1}{N_M \max^e}. \qquad (A15.2)$$

For this flight's data, the larger root is expected to be the correct one nearly always, since this does give concentrations in the expected range (see Chapter 7, Volume 1), about 1000 cm⁻³. A new N_M' was used for each separate period of $N_M > 100$ cm⁻³ because N_M' appeared to increase steadily during the flight.

The corrected droplet spectrum was derived by increasing the count of all size categories in the same proportion. A

typical result of this treatment is given in Fig. 15.35. It results in much better agreement of the liquid water content values, and in deduced maximum droplet concentrations (N_T) of somewhat over 1000 cm⁻³, which does appear reasonable. While this procedure appeared to work well on this day, as Fig. 15.35 attests, on some other days it did not seem nearly as satisfactory. Droplet concentrations presented in the text result from this correction procedure.

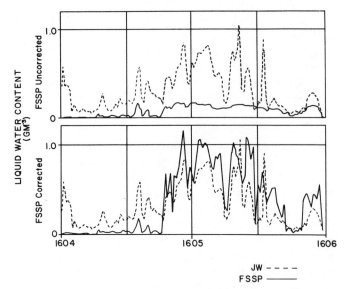

Figure 15.35 Two minutes of typical liquid water content data within cloud, from the Johnson-Williams hot wire device (dashed line) and the FSSP probe (solid line). The top frame shows the FSSP data without corrections, and the bottom frame shows the same data after the objective correction procedure described in the text.

References

Ashworth, E., T. Ashworth, and C. A. Knight, 1980: Cylindrical accretions as simulations of hail growth: III. Analysis techniques and application to trajectory determination. *J. Atmos. Sci.* 37, 846-854.

Breed, D. W., 1978: 22 June 1976: First echo case. NCAR/TN-130+STR, NTIS# PB295754/AS. National Center for Atmospheric Research, Boulder, Colo., 49 pp.

Browning, K. A., 1977: The structure and mechanisms of hailstorms. *Meteorol. Monogr.* 38, 1-43

-----, and G. B. Foote, 1976: Airflow and hail growth in supercell storms and some implications for hail suppression. *Q. J. R. Meteorol. Soc.* 102, 499-533.

-----, J. C. Fankhauser, J.-P. Chalon, P. J. Eccles, R. G. Strauch, F. H. Merrem, D. J. Musil, E. L. May, and W. R. Sand, 1976: Structure of an evolving hailstorm: Part V. Synthesis and implications for hail growth and hail suppression. *Mon. Weather Rev.* 104, 603-610.

Byers, H. R., and R. R. Braham, 1949: *The Thunderstorm*. U. S. Govt. Printing Office, Washington, D. C., 287 pp.

Dennis, A. S., C. A. Schock, and A. Koscielski, 1970: Characteristics of hailstorm of Western South Dakota. *J. Appl. Meteor.* 9, 127-135.

Dye, J. E., and B. E. Martner, 1978: The relationship between radar reflectivity factor and hail at the ground for northeast Colorado thunderstorms. *J. Appl. Meteor.* 17, 1335-1341.

Foote, G. B., and C. A. Knight, 1979: Results of a randomized hail suppression experiment in northeast Colorado: Part I. Design and conduct of the experiment. *J. Appl. Meteorol.* 18, 1526-1537.

Goyer, G. C., 1977: Response to "The climatology of hail in North America." *Meteorol. Monogr.* 16, 129-133.

Heymsfield, A. J., 1978: The characteristics of graupel particles in northeastern Colorado cumulus congestus clouds. *J. Atmos. Sci.* 35, 284-295.

-----, and J. L. Parrish, 1979: Techniques employed in the processing of particle size spectra and state parameter data obtained with the T-28 aircraft platform. National Center for Atmospheric Research, Boulder, Colo.

-----, P. N. Johnson, and J. E. Dye, 1978: Observations of moist adiabatic ascent in northeast Colorado cumulus congestus clouds. *J. Atmos. Sci.* 35, 1689-1703.

-----, A. R. Jameson, and H. W. Frank, 1980: Hail growth mechanisms in a Colorado storm: Part II. Hail formation processes. *J. Atmos. Sci.* 37, 1779-1807.

Knight, C. A., and N. C. Knight, 1970: Hailstone embryos. *J. Atmos. Sci.* 27, 659-666.

-----, T. Ashworth, and N. C. Knight, 1978: Cylindrical ice accretions as simulations of hail growth: II. The structure of fresh and annealed accretions. *J. Atmos. Sci.* 35, 1997-2009.

-----, N. C. Knight, and K A. Kine, 1981: Deuteruim contents of storm inflow and hailstone growth layers. *J. Atmos. Sci.* 38, 2485-2499.

Levi, L., and A. N. Aufdermaur, 1970: Crystallographic orientation and crystal size in cylindrical accretions of ice. *J. Atmos. Sci.* 22, 443-452.

-----, and F. Prodi, 1978: Crystal size in ice grown by droplet accretions. *J. Atmos. Sci.* 35, 2181-2189.

-----, L. Lubart, and E. M. de Achaval, 1978: Crystal structure of ice accretions. *Nuovo Cimento* 1C, 86-92.

List, R., 1958: Kennzeichen atmospharischer Eispartikeln, I. *Z. Angew. Math. Phys.* 9a, 180-192.

-----, R. B. Charlton, and P. I. Buttels, 1968: A numerical experiment on the growth and feedback mechanisms of hailstones in a one-dimensional steady-state cloud model. *J. Atmos. Sci.* 25, 1061-1074.

-----, J.-G. Cantin, and M. G. Ferland, 1970: Structural properties of two hailstone samples. *J. Atmos. Sci.* 27, 1080-1090.

Long, A. B., 1978: Design of hail measurement networks. *Atmos.-Ocean* 16, 35-48.

Ludlam, F. H., 1952: The production of showers by the growth of ice particles. *Q. J. R. Meteorol. Soc.* 78, 543-553.

-----, 1958: The hail problem. *Nubila* 1, 12-96.

-----, 1976: Aspects of cumulonimbus study. *Bull. Am. Meteorol. Soc.* 57, 774-779.

Macklin, W. C., and I. H. Bailey, 1966: On the critical liquid water concentrations of large hailstones. *Q. J. R. Meteorol. Soc.* 92, 297-300.

-----, and -----, 1968: The collection efficiencies of hailstones. *Q. J. R. Meteorol. Soc.* 94, 393-396.

Mason, B. J., 1971: *The Physics of Clouds.* Oxford Press, London, England, 671 pp.

Matson, R. J., and A. W. Huggins, 1980: The direct measurement of the sizes, shapes and kinematics of falling hailstones. *J. Atmos. Sci.* 37, 1107-1125.

Musil, D. J., W. R. Sand, and R. A. Schleusener, 1973: Analysis of data from T-28 aircraft penetrations of a Colorado hailstorm. *J. Appl. Meteorol.* 12, 1364-1370.

-----, E. L. May, P. L. Smith, Jr., and W. R. Sand, 1976: Structure of an evolving hailstorm: Part IV. Internal structure from penetrating aircraft. *Mon. Weather Rev.* 104, 596-602.

Nelson, S. P., and R. R. Braham, Jr., 1975: Detailed observational study of a weak echo region. *Pageoph* 113, 735-746.

Nicholas, T. R., 1977: A review of surface hail measurement. *Meteorol. Monogr.* 38, 257-267.

Paluch, I. R., 1978: Size sorting of hail in a three-dimensional updraft and implications for hail suppression. *J. Appl. Meteorol.* 17, 763-777.

-----, C. A. Knight, C. T. Nagamoto, G. M. Morgan, N. C. Knight, and F. Prodi, 1979: Further studies of large, water-insoluble particles within hailstones. *J. Atmos. Sci.* 36, 882-891.

Rye, P. J., and W. C. Macklin, 1975: Crystal size in accreted ice. *Q. J. R. Meteorol. Soc.* 101, 207-215.

Sand, W. R., 1976: Observations in hailstorms using the T-28 aircraft system. *J. Appl. Meteorol.* 15, 642-650.

-----, 1981: 22 July 1975: mature storm study. A conceptual model synthesized from microphysical and dynamic observations of a multicell hailstorm. NCAR/TN-184+STR, National Center for Atmospheric Research, Boulder, Colo., 260 pp.

Scorer, R. S., and F. H. Ludlam, 1953: Bubble theory of penetrative convection. *Q. J. R. Meteorol. Soc.* 79, 94-103.

Scott, B. C., and P. V. Hobbs, 1977: A theoretical study of the evolution of mixed-phase cumulus clouds. *J. Atmos. Sci.* 34, 812-826.

Shaw, W. S. 1974: Development of an airborne optical hailstone disdrometer. Rep. 74-16, Inst. of Atmospheric Sciences, South Dakota School of Mines and Technology, Rapid City, S. Dak.

Summers, P. W., G. K. Mather, and D. S. Treddenick, 1972: The development and testing of an airborne droppable pyrotechnic flare system for seeding Alberta hailstorms. *J. Appl. Meteorol.* 11, 695-703.

Young, K. C., 1977: A numerical examination of some hail suppression concepts. *Meteorol. Monogr.* 16, 195-214.

-----, 1978: On the role of mixing in promoting competition between growing hailstones. *J. Atmos. Sci.* 35, 2190-2193.

The 22 July 1976 Case Study

CHAPTER 16

The 22 July 1976 Case Study: Radar Echo Structure and Evolution

G. Brant Foote and Charles G. Wade

16.1 INTRODUCTION

On 22 July 1976 thunderstorms developed and moved southward across the NHRE observing network, as shown in Fig. 16.1. Research was focused on Storm III, which was observed by conventional and Doppler radars throughout most of its lifetime (1500-1730 MDT) and by two aircraft, the NCAR Queen Air (N306D) in the sub-cloud region and the T-28 penetrating aircraft during one particular phase (1615-1730 MDT). The NCAR Sabreliner was also flying in a related research mission in the near vicinity of the storm and took a number of useful pictures.

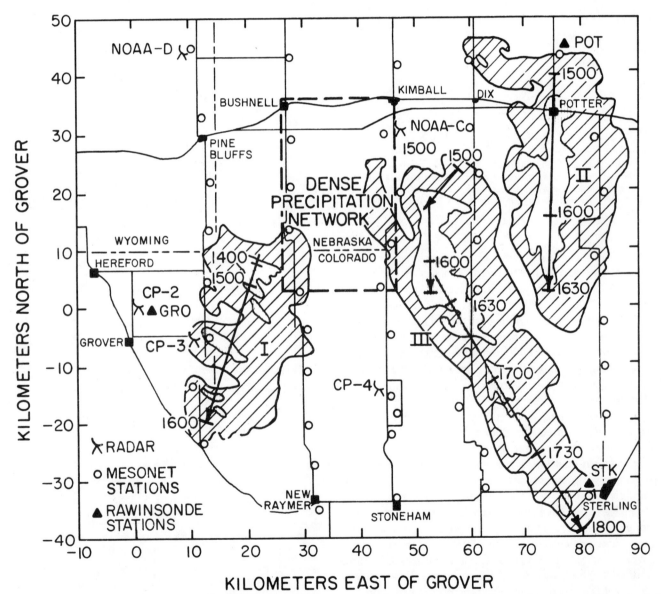

Figure 16.1 Map showing the location of the 1976 NHRE research network over portions of northeastern Colorado, southeastern Wyoming, and southwestern Nebraska. Special rawinsonde sites are located near Grover (GRO) and Sterling (STK), Colorado, and north of Potter (POT), Nebraska. Reflectivity swaths are also shown for the three major storm systems observed over the NHRE research network on 22 July 1976, with the envelopes of 35- and 55-dBZ$_e$ contours near the 3-km level indicated. The southern end of Storm I is incomplete due to restricted sector scans by the Grover radar after 1530.

16.2 ENVIRONMENTAL CONDITIONS

Large-scale features over the western U.S. on 22 July 1976 showed a fairly typical summertime pattern. At 500 mb (Fig. 16.2) the main zone of westerlies was located along the U.S.-Canadian border while a large anticyclone was centered over southern Utah. Cool air advecting southeastward

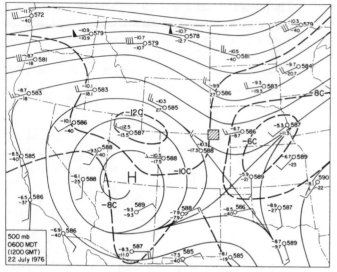

Figure 16.2 *500-mb map for 0600 MDT, 22 July 1976, showing the mid-tropospheric temperature and streamline pattern over the western U. S. Data plotted at each station include temperature and dewpoint (°C), height (decameters), and wind (full barb = 5 m s⁻¹). The location of the NHRE research network in northeastern Colorado is also shown.*

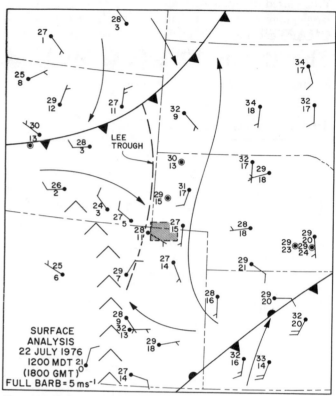

Figure 16.3 *Surface map for 1200 MDT showing synoptic-scale features that influenced convection over the NHRE research network on 22 July 1976. Temperature and dewpoints are shown at each station in °C with a full wind barb equal to 5 m s⁻¹.*

around this anticyclone enhanced the convective potential of the air mass lying over northeastern Colorado. At the surface (Fig. 16.3) a stationary front was located across Kansas, beneath the 500-mb diffluent region, while a weak Pacific cool front approached from the northwest. Light southeasterly winds behind the stationary front provided an influx of low-level moisture to the NHRE region. Convection was initiated during the early afternoon, apparently as a result of localized heating and convergence associated with the lee trough located near Grover.

Upper-air soundings were obtained at 90-min intervals from each of the three sites during the afternoon. The Sterling 1605 sounding, released approximately 40 km southeast of the storm in question, is reasonably representative of the pre-storm environment in the boundary layer, and is shown in modified form in Fig. 16.4. The temperatures above cloud base in Fig. 16.4 have been taken from the Grover 1627 sounding, which was released to the west of the storm in a relative upwind location. In the 350- to 600-mb layer the Grover sounding was 1°C to 2°C warmer than the Sterling sounding. In the lower levels the air was relatively well-mixed, with a nearly constant potential temperature of 315.5 K and an average mixing ratio in the lowest 150 mb of 10 g kg⁻¹. These values agree closely with those observed by the Queen Air aircraft flying near cloud base on this day.

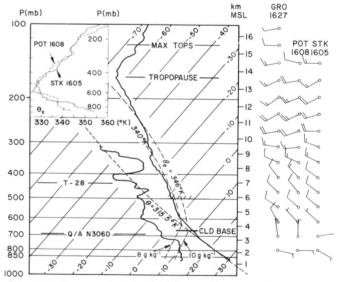

Figure 16.4 *Thermodynamic diagram showing a representative sounding based on the STK 1605 MDT and GRO 1627 releases. The temperature and dewpoint traces are from the STK sounding, except that above 600 mb the somewhat warmer temperatures measured by the Grover sounding are shown. Solid curves are the temperature and dewpoint traces while the labeled dashed curves designate moist and dry adiabats and lines of constant mixing ratio. The values shown represent the range of cloud base conditions observed within updraft air. To the left of these lines the typical altitudes flown by the two research aircraft are shown. Winds at 1-km intervals from three soundings are shown at the right (full barb = 5 m s⁻¹). The inset in the upper left corner shows the profile of equivalent potential temperature from soundings released ahead of and behind the storm.*

When lifted to saturation these values produce a cloud base temperature of $+8\,°C$, a cloud base height of 3.6 km MSL (670 mb), and a parcel lifted index at 500 mb cooler at cloud base than the environmental temperature, resulting in only a small "negative energy" at cloud base and a relatively small amount of lifting required to reach the level of free convection.

The inset in the upper left-hand corner of Fig. 16.4 shows the vertical distribution of equivalent potential temperature (θ_e) for the Sterling 1605 and the Potter 1608 soundings. At the time of release these two sounding sites were located 40 km ahead of and 35 km behind the main thunderstorm activity, respectively. The relatively higher potential instability below cloud base is apparent on the Sterling sounding, while the two traces assume a nearly identical profile above cloud base. The level of potentially coldest air ($\theta_e \approx 328$ K) lies between 500 and 600 mb (≈ 5 km) on both soundings.

The environmental wind profile is shown on the right side of Fig. 16.4 for the three soundings released between 1600 and 1630. Each sounding shows a wind field that backs with height from easterly near the surface to southwesterly near the tropopause. This is illustrated somewhat better by the hodograph for the Sterling sounding in Fig. 16.5. Maximum observed winds on this sounding are 18 m s^{-1}, with a wind shear for the cloud-bearing layer (3.5 to 13 km) of 1.8×10^{-3} s^{-1}. The storm motion vectors, V_1, V_2, and V_3, discussed later, show that the storm's motion was to the right of all winds in the troposphere. If the winds are considered with respect to the storm, rather than with respect to the ground, then it is clear from the hodograph that the winds below 9 km actually veered rather than backed with height in this more appropriate coordinate system.

In summary, the storms on 22 July 1976 developed under relatively weak synoptic conditions. The low values of wind shear, boundary layer winds, and negative energy at cloud base, coupled with relatively high potential instability are characteristic of an environment that tends to support multicellular thunderstorms (Marwitz, 1972c).

16.3 GENERAL STORM FEATURES

16.3.1 Echo Evolution at the Large Scale

Three major storm systems developed and moved across the research network on 22 July 1976. The envelope of the low-level reflectivity contours is shown in Fig. 16.1. The first system, Storm I, developed in the early afternoon over relatively high ground just east of the Grover radar site. It was studied by research aircraft and Doppler radars between 1420 and 1530 (Breed, 1978). After 1530 Storm I began to weaken and the research emphasis shifted to the more intense Storms II and III, which had developed to the northeast. The history of these two storms between 1444 and 1730 is illustrated in Fig. 16.6 by a sequence of low-level PPI's at 15-min intervals. The storms moved along parallel tracks southward until about 1630, when Storm II began to weaken and Storm III shifted to a more southeasterly heading. The storms continued southeastward out of the NHRE network after 1800 and finally dissipated in east-central Colorado around 2100.

Storm III was selected for research because of its better position with respect to the Doppler radars, and the remaining discussion in this paper is concerned with Storm III. The

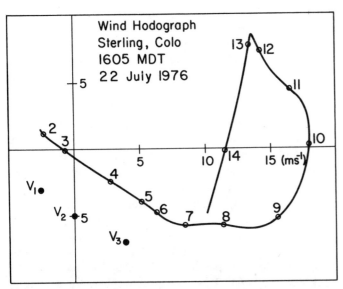

Figure 16.5 Wind hodograph for the STK 1605 MDT sounding. Storm motion vectors for the three phases of Storm III's development are shown as V_1, V_2, and V_3. During phase 3 relative low-level flow is from the southeast at 10 m s^{-1}, while mid-tropospheric flow is less than 5 m s^{-1} from the west-southwest.

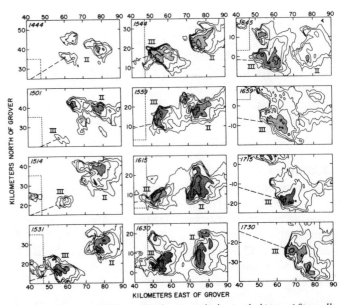

Figure 16.6 Sequence of PPI's at 0.5° elevation angle showing the history of Storms II and III between 1444 and 1730 MDT. Reflectivity contours are at 10-dB intervals with values greater than 45 dBZ$_e$ shaded. The short dashed line is the eastern boundary of the dense precipitation network while the long dashed line is the radial to the Grover radar.

possible effects of the first two storms on the evolution of Storm III, however, will be considered in Chapter 17. Special attention is given here to the 1615-1730 time period, when Storm III was being studied by research aircraft and Doppler radars.

16.3.2 Surface Hail Measurements

Although Storms II and III did not pass over the high-density precipitation network they did pass over portions of a less dense network of "hailcubes" (devices with hailpads mounted on their four vertical sides and tops) operated by NCAR to study the relationship between hail and crop damage. The locations of these hailcubes with respect to the low-level 45- and 55-dBZ$_e$ reflectivity envelopes from Storms II and III are shown in Fig. 16.7. The numbers that appear next to the darkened circles represent the maximum observed hail size for each site, obtained with the hailpad calibration technique described by Long et al. (1979). Stations marked with a G indicate cubes hit by soft hail or graupel. The termination of the hailcube network along the Colorado-Nebraska state line reflects the shift in agricultural emphasis from farming on top of the bluffs to ranching south of the bluffs.

Also shown in Fig. 16.7 are the locations of hail samples obtained by University of Wyoming and NCAR hail chase crews. The maximum hail diameter in these samples was 1.7 cm, similar to that of the hailpad network.

Crop damage in general was minimal because the primary crop, wheat, had been harvested about a week earlier. One report of an 80% wheat loss is shown in the figure, while damage to corn in the Potter, Nebraska, area ranged up to 33%.

Work presented by Dye and Martner (1978) suggests that there is generally a low correlation between low-level reflectivity and hail at the surface in northeastern Colorado. A similar comparison with the hail data on 22 July 1976 shows a considerably better correlation. Table 16.1 shows the frequency of hail occurrence as a function of the maximum observed low-level (0.5° elevation) radar reflectivity factor

TABLE 16.1
Relation between hail occurrence and reflectivity

Maximum reflectivity	Number of pads		Fraction of pads with hail	Average maximum diameter	Average number of hailstones
	With hail	No hail			
≥60 dBZ	13	0	1.00	1.8 cm	542
55-60	20	0	1.00	1.4	173
50-55	11	7	0.61		
45-50	2	11	0.15	1.2	116
40-45	2	15	0.12		
	48	33			

over 81 hailcubes. The fraction of pads that were struck increases monotonically with the maximum reflectivity. All pads with at least 55 dBZ$_e$ overhead experienced hail. No pads with less than 40 dBZ$_e$ overhead experienced hail. Table 16.1 also shows that the maximum stone size and the number of impacts on the pad (of area 0.1 m²) tend to increase with increasing reflectivity.

16.3.3 General Airflow Pattern

Before considering in more detail the structure of this storm, it is useful to consider first the general pattern of air motion. In persistent storms on the plains the overall pattern of inflow and outflow of air can often be deduced quite reliably by knowing only the position of the persistent updraft at cloud base and the horizontal wind profile, as described for example by Chisholm (1973). In the present case the low-level wind relative to the storm was southeasterly (Fig. 16.5) so that one would expect the updraft at cloud base to be on the southeast side of the storm.

Figure 16.8 (top) shows the measured updraft position relative to the radar echo. The updraft shown is a composite based on several passes by the Queen Air near this time. (A detailed discussion of the Queen Air measurements is given in Chapter 17.) The updraft maximum of 5-7 m s^{-1} was observed on each of 12 passes made just under the flat cloud base over a 70-min period. The updraft area shown, about 65 km², is typical for hailstorms in northeastern Colorado (Auer and Marwitz, 1968).

At the bottom of Fig. 16.8 is a vertical section along the inflow direction showing the cloud base position and the relative wind there measured by the Queen Air. The upward flux of moist air first passes through a region with no echo detectable by normal radars, encounters a forward overhang of echo thought to be indicative of growing hydrometeors, and finally, by making a turn roughly 90° to the right (in plan view), as indicated by the hodograph, exhausts into the anvil toward the northeast, into the plane of the figure. The figure also shows schematically a downdraft in the region of heavy precipitation and an inflow/outflow boundary. The position of the latter is determined by the surface network and the Queen Air, as discussed in Chapter 17.

In the following chapters this general circulation pattern will be refined considerably, but for the present discussion, this overview will suffice. Actually, this simple pattern is quite typical of the overall airflow in persistent, right-moving storms (see for example, Marwitz, 1972a; Fankhauser, 1976; and many others). This comes about because soundings associated with most traveling storms display roughly 90° of veering between the low-level and upper-level winds when viewed in a coordinate system moving with the storm.

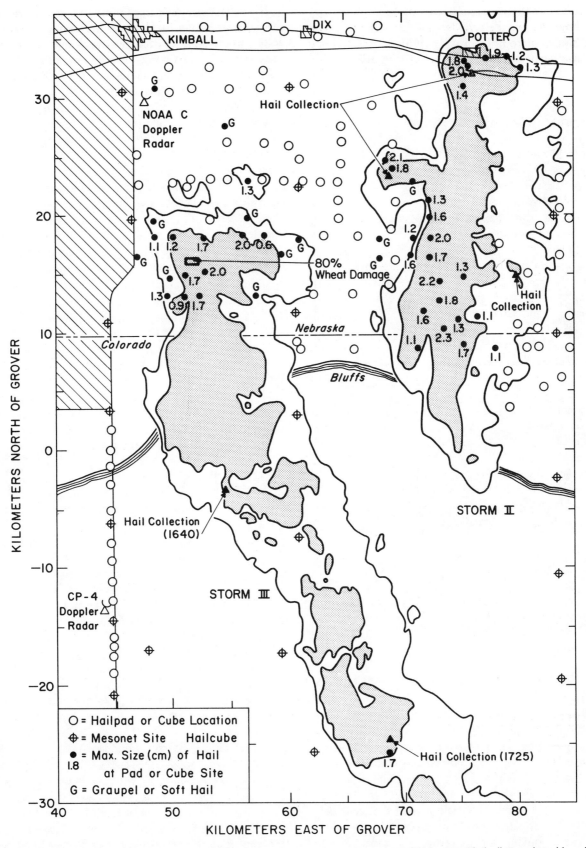

Figure 16.7 *Map showing the location of surface hailpads and hailcubes in relation to the low-level 45- and 55-dBZ$_e$ reflectivity envelopes swept out by Storms II and III. Sites experiencing hail are shown as darkened circles with the maximum observed diameter in cen-* *timeters written next to it. The locations of hail collections obtained by mobile hail chase crews are also shown. The hatched region southwest of Kimball, Nebraska, is the dense precipitation network that Storm III narrowly missed.*

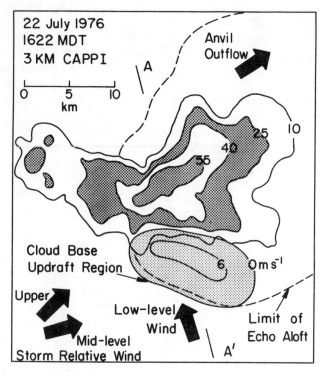

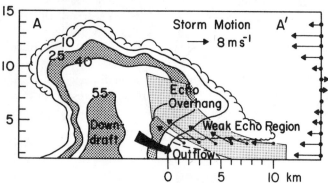

Figure 16.8 Horizontal and vertical sections through Storm III at 1622 MDT illustrating the storm's general airflow pattern. Storm motion was toward the south-southeast (along AA') at 8 m s⁻¹. The components of the relative winds in the plane of AA' are shown along the right side of the bottom panel. The horizontal extent of the cloud base updraft region is based on a composite from the subcloud research aircraft, obtained between 1618 and 1650 MDT, and adjusted for storm motion. The visual cloud outline is only roughly estimated.

16.4 CELLULAR CHARACTER AND MOTION

Reflectivity data from the Grover S-band radar provide the basis for the discussion of storm structure and evolution in this and later sections. This radar was operated in a sector scan mode with 1.5 to 2 min needed for a complete scan of the storm volume. The antenna scanned in azimuth at constant elevation angle, with increments of 0.7° in elevation between successive scans. The analysis used the radar data displayed in PPI format as well as in horizontal and vertical sections. The technique for converting the radar data to Cartesian coordinates has been described by Mohr and Vaughan (1979).

The reflectivity envelopes presented in Fig. 16.1 showed the track swept out by Storm III as it moved southeastward across the research network. A more detailed summary of the radar return is given in Fig. 16.9. The upper panel shows the height of the storm's top (defined by the 5-dBZ$_e$ contour) as a function of time. The three middle panels show time histories of the maximum reflectivity in the storm at the 3-, 7-, and 11-km MSL levels, while the bottom panel shows the area enclosed by various reflectivity contours at an altitude of 7 km. The figure illustrates the storm's evolving intensity and implies that it was comprised of a series of convective impulses, which developed at intervals averaging about 15 min. As will be seen later, these impulses can generally be identified in radar PPI's with individual cells in the storm complex.

Because usage of the word "cell" is sometimes ambiguous, perhaps it is worthwhile to explain our usage here. It is pos-

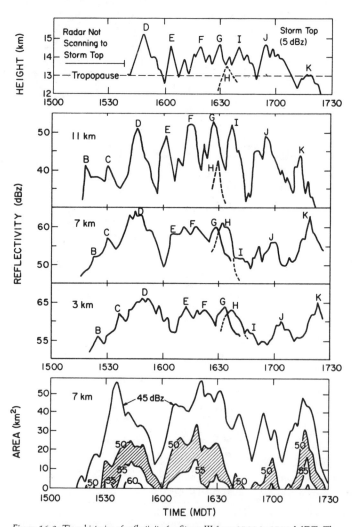

Figure 16.9 Time histories of reflectivity for Storm III from 1500 to 1730 MDT. The upper panel shows the evolution of the storm's top (5-dBZ$_e$ echo), while the three middle panels show the variation of maximum reflectivity in the storm at 3, 7, and 11 km. The bottom panel shows the areas enclosed by various contours at 7 km MSL. Major impulses are labeled for later reference (B, C, D, etc.).

sible for some purposes to produce an objective definition for the term "cell" from radar measurements, though a common definition is not always useful. In using the term one generally refers to a compact region of updraft that gives rise to a perturbation in the radar reflectivity pattern such as those responsible for the pulsations shown in Fig. 16.9. Cells considered in this way generally show up on PPI's as closed contours, and can be uniquely defined solely in terms of a reflectivity contour of some intensity, as in the study of Foote et al. (1979b). However, as one turns to radar observations of finer scale, more and more detail is seen in the reflectivity pattern. What one might call a cell with coarse observations is then seen to have a good deal of substructure and evolution, prompting one to reconsider usage of the term "cell."

We use the term here without precise definition to refer to reflectivity perturbations that seem to have a dominant influence on the development of the storm, with the underlying thought that we are trying to identify major updraft regions. There is no strong distinction between these cells and what we refer to as "minor impulses"—small updrafts, generally short-lived, which, as judged by their associated reflectivity pattern, do not have a major influence on the development of the storm. These small impulses may have an influence, though, on such things as the motion of the echo centroid, by virtue of their relatively high frequency of occurence. They may also be important in other ways, for example, as sources of embryos for hail destined to continue growing elsewhere in larger updrafts, a topic discussed in later chapters. In the current context it is not productive to ask whether some small feature in a storm is or is not a cell. Since the local intensification of echo generally depends on the availability of water that is condensed and carried in updrafts, then presumably all such perturbations reflect the presence of updrafts (though the reflectivity perturbation may last much longer than the updraft that produced it). Thus it is only the scale and longevity of the updraft that separates "cells" and "minor impulses," as we use these and other terms, and there is no generic difference between them, just as there is no sharp boundary. It might also be noted that Foote and Mohr (1979) show that the sizes and lifetimes of convective cells (as defined by the 45-dBZ$_e$ contour) tend to be lognormally distributed. There are many more of the smaller, shorter-lived cells, but the distribution to larger sizes is continuous and has only a single mode.

By reference to PPI's the perturbations shown in Fig. 16.9 have been related to dominant cells, with a letter attached to each cell. The labeling starts with the initial echo around 1500, Cell A, which was very weak and is off scale in this figure. Reflectivity maxima in the various cells may be traced from aloft, where they are generally distinct, to the lower levels, where they become less obvious as a result of the merger and mixing of precipitation between various cells. Maximum storm tops averaged about 1.5 km above the tropopause, and low-level reflectivity ranged between 55 and 65 dBZ$_e$.

As judged by the reflectivity at 7 km, the storm intensity was at a maximum near 1545 (Cell D), and at a minimum about an hour later (Cell I). It is interesting, however, that the height of the storm top does not confirm that the updraft in Cell I was weaker. It is also interesting to contrast the behavior of Cell K, which had a low storm top but high reflectivity in the middle level. Since echo intensity relates well with precipitation intensity, and since there is some basis for relating echo top with maximum updraft speed (see Foote and Mohr, 1979), it appears that factors other than updraft intensity are important in rain and hail production. One such factor is the availability of growth embryos in cold-based clouds such as these, as discussed by Knight et al. (1979). An explanation of the observed behavior in these terms is considered by Heymsfield et al. (1980).

The bottom panel of Fig. 16.9 shows the area enclosed by various reflectivity contours as a function of time. For the 45-dBZ$_e$ contour the area is generally a maximum, on the order of 40-50 km^2, when two or more cells are in existence at the same time. The sizes of individual cells on this day (as judged by their 45-dBZ$_e$ echo area at 7 km) are more typically in the range of 15-30 km^2. Such sizes are comparable to those for cells occurring on the ten most severe hail days during the seeding program of the National Hail Research experiment, as tabulated by Foote et al. (1976). (See also Chapter 5.) Typical sizes for the larger cells (those in the upper 25th percentile) from that sample are between 20 and 50 km^2, though a few are over 100 km^2. The latter tend to be the vaulted cells (supercells), and have been discussed further by Foote and Mohr (1979). The cell sizes in the present storm are thus typical of moderate to severe hailstorms in northeastern Colorado.

The area enclosed by the 55-dBZ$_e$ contour has also been integrated over time to give an estimate of the overall storm intensity, termed P$_{55}$ by Foote et al. (1979a). The value of P$_{55}$ for this day would rank eighth in the three-year hailstorm sample of Foote et al., reinforcing the view that this storm is typical of moderate to severe hailstorms in Colorado, though it is not one of the most severe.

The tracks of the various cells are shown in Fig. 16.10 superimposed upon the low-level reflectivity envelope. The solid lines are the tracks over the ground of the most intense echo associated with each cell. The long dashed line from about 1600 onward is the track of the 15-dBZ$_e$ echo centroid as discussed presently.

It is convenient to describe the storm in three phases, as indicated in the figure. During the initial phase of development the cells formed preferentially on the west side of the existing storm. When combined with the general southeastward drift of the cells and the dissipation of older cells this resulted in an overall storm motion toward the southwest (220°) at 4 m s^{-1} (shown as V$_1$ in Fig. 16.5).

After the storm reached full maturity at about 1545 the new cells began forming on the south side of the system and two branches developed: a western branch composed of

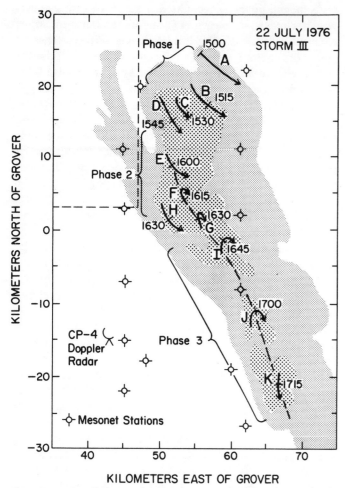

Figure 16.10 Ground-relative tracks of the major convective impulses superimposed on the low-level (3-km) reflectivity envelopes. Contour levels are 35 and 55 dBZ. The long dashed line from about 1600 MDT onward is the track of the storm's echo centroid.

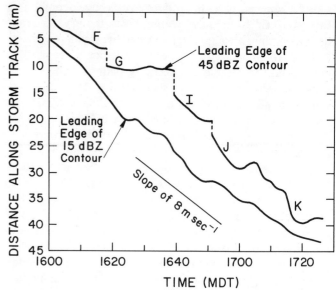

Figure 16.11 Diagram showing the comparative motions of the 15-and 45-dBZ$_e$ echo contours along the storm track, at 7 km MSL, as a function of time.

motion. Abrupt changes in its position occur with the formation of cells G, I, and J, as a new 45-dBZ$_e$ contour appears separately ahead of the older one, within the 15-dBZ$_e$ envelope. As shown in Fig. 16.10, though the storm as a whole moves to the southeast, the echo core associated with the updraft pulses can remain virtually stationary, or even move back toward the northwest.

When comparing the overall motion of the three storm systems on this day it is apparent that the early motion of Storm III toward the southwest and south (1500-1630) was similar to the motion of Storms I and II, and that the development of the eastern branch toward the south-southeast (1615-1730) was somewhat anomalous. Possible reasons for this behavior are considered in Chapter 17.

16.5 DETAILED ECHO STRUCTURE AND EVOLUTION

We now consider the echo structure in more detail for each of the three phases, documenting the manner in which the cellular evolution takes place.

16.5.1 Phase 1: Development of the Mature Storm (1500-1545)

The initial phase represents the storm's early development from the "cumulus stage" (Byers and Braham, 1949) at 1500 to the formation of the intense Cell D at about 1545. Evolution during this early period is illustrated in Fig. 16.12 with a sequence of horizontal sections through the storm at 7 km.

major cells E and H, and an eastern branch composed initially of cells F and G, later including cells I, J, and K. The two branches were somewhat divergent in their motion, with the western branch developing toward the south-southwest and the eastern branch developing toward the south-southeast. This produced a net storm motion for the second phase of 180° at 5 m s^{-1} (V$_2$ in Fig. 16.5).

Actually, during Phase 3 and slightly before, the storm was much steadier in overall configuration than Figs. 16.9 and 16.10 would indicate. The long dashed-line track during this period is meant to show the movement of the storm during a period in which something more like pulsations of a persistent updraft occurred, rather than the evolution of individual cells with separate updrafts. The pulsations in updraft strength aloft led to cyclical changes in the high-reflectivity core, but produced essentially no change in the outer reflectivity contour of the storm. Figure 16.11 illustrates this in a time-distance plot. The lower line shows the steady advance of the 15-dBZ$_e$ contour toward the southeast. In contrast, the leading edge of the 45-dBZ$_e$ contour shows a more erratic

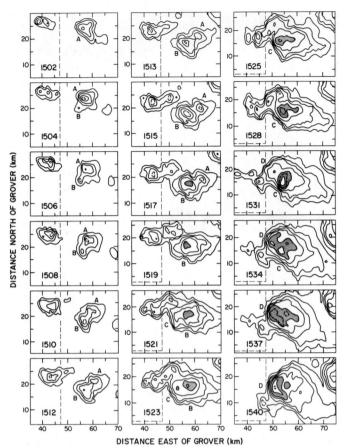

Figure 16.12 Sequence of CAPPI's at 7 km MSL showing the development of Storm III between 1502 and 1540 MDT. Contour intervals are 10 dB, starting at 5 dBZ_e, with values in the range of 45-55 dBZ_e shaded. A few intermediate contours are dashed (5-dB interval). Letters refer to the major convective impulses discussed in the text.

precipitation core on the storm's west side (with strong reflectivity gradients located around this core on the south, west, and north sides), a middle-level forward overhang extending southward over the main inflow region for 4 to 5 km, and a storm top penetrating above the tropopause by 2 to 2.5 km and located almost directly over the main precipitation core. Though the cloud-base aircraft had not arrived at this time, we can be confident that the general airflow model presented earlier applies here, with low-level inflow on the south side, exhausting in the anvil to the northeast. The echo from the anvil is shown in Fig. 16.13 extending some 20 to 25 km toward the northeast from the echo core. A photograph of the anvil taken slightly later at 1630 is shown in Fig. 16.14. The photograph was taken from the NCAR Sabreliner on the northwest side of the storm, and illustrates the great horizontal expanse of the anvil outflow, similar to that reported by Fujita and Byers (1962) for a storm in the same region.

16.5.2 Phase 2: the Period 1545-1630

Following the development of Cell D the storm began to evolve in a more southerly direction. New cells formed initially on the storm's southwestern or right flank, and then, about 1605, began forming concurrently on both its right and forward flanks. This dual mode of cell development resulted

Cells B and C may be seen as they developed on the storm's southwestern flank with each contributing successively higher reflectivities to the storm. A reflectivity of 55 dBZ_e was first produced in Cell B at about 1521, suggesting that sporadic hailfall may have begun in the storm by this time. Flanking radar echoes began merging with the storm at about 1515, and by 1535 Cell D had begun to form out of these mergers on the storm's northwest side. This cell dominated the storm during the next 25 min and is believed to be responsible for most of the observed hailfall from Storm III north of the Nebraska-Colorado state line (Fig. 16.7).

The three-dimensional structure of the storm at 1545 when Cell D was near its maximum intensity is shown in Fig. 16.13. The lower panels show horizontal slices through the storm at 3, 6, 9, and 12 km, while the upper panels illustrate the storm's vertical structure along the two intersecting, perpendicular lines. Reflectivity contours and shading are the same as in Fig. 16.12. The storm was near its maximum intensity of the day at this time, with cell tops exceeding 15 km in height and low-level reflectivities approaching 70 dBZ_e. Notable features in the storm reflectivity structure include an intense

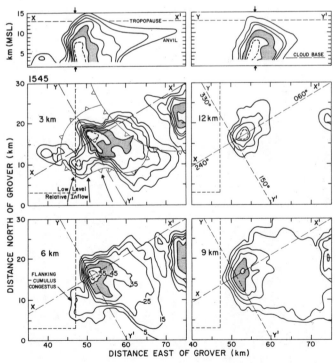

Figure 16.13 Horizontal and vertical sections at 1545 MDT showing the structure of Storm III near the time of its maximum intensity. The locations of the surface outflow boundaries from Storms II and III are shown in the 3-km panel based on mesonet data discussed in Part II.

Figure 16.14 Composite photograph of Storm III, taken at 1630 MDT from a forward-looking camera mounted on the NCAR sabreliner. The aircraft was executing a turn on the northwest side of the storm at an altitude of 9 km, and was looking generally toward the southeast. The storm's anvil is estimated to extend some 100 to 150 km to the northeast.

in a gradual splitting of the storm into two coexisting cells (G and H) by 1630. Figure 16.15 illustrates this evolution with a sequence of CAPPI's at 9 km. The figure shows that Cell D evolved into two cells, D_1 and D_2, about 1550. Cell D_1 developed on the north side of the storm while Cell D_2

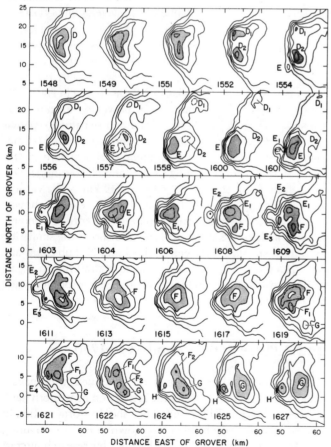

Figure 16.15 Sequence of CAPPI's at 9 km showing the development of convective impulses on the storm's southeast through southwest flanks during phase 2. Reflectivity contour intervals are the same as in Fig. 16.12.

formed closer to the main inflow region on the south side of the storm. This was followed at 1553 by the formation of a strong impulse, Cell E, on the southwest side of the storm. The development of E is illustrated further in Fig. 16.16, which shows the formation of the cell's high-reflectivity core and its subsequent descent and merger with remnants of Cell D_2. As seen in Fig. 16.16, the early development of E is identifiable only in the upper levels of the storm. That is the reason for showing 9-km observations in Fig. 16.15, rather than those at the 7-km level as in Fig. 16.12. One would probably not discover this evolving aspect of the storm structure by considering 7-km data alone.

Cell E was the first of a series of convective impulses to develop on the storm's right flank and merge with the existing echo core. Others following in sequence between 1600 and 1620 are labeled as Cells E_1 through E_4 in Fig. 16.15. These later impulses were somewhat less intense than cell E but were sufficiently vigorous to produce reflectivities in the range of 55 to 60 dBZ_e.

The visual appearance of Cell E as it developed on the west flank of the storm about 1600 is shown in Fig. 16.17a and b, taken from the NCAR Sabreliner flying at an altitude of about 8.8 km. The photo in Fig. 16.17a was taken at 1558:30 from the northwest side of the storm looking to the southeast, and shows Cell E developing to the right (southwest) of Cell D_2. Other congestus towers on the west and northwest sides of the storm may be seen in the foreground of the photo. The photo in Fig. 16.17b was taken about 4 min later just west of the storm, looking up at Cell E (see the upper right panel of Fig. 16.16). The aircraft was flying over the top of Cell E_2 before it had an echo and the picture shows Cell E_1 in the foreground developing alongside of E. The radar echo from E was near its maximum height of 14.5 km at this time, while the top of E_1 was at approximately 9.5 km. The photographs suggest that there is a correspondence between the radar echoes from E, E_1, and E_2 and identifiable cumulus towers, as noted previously by Warner

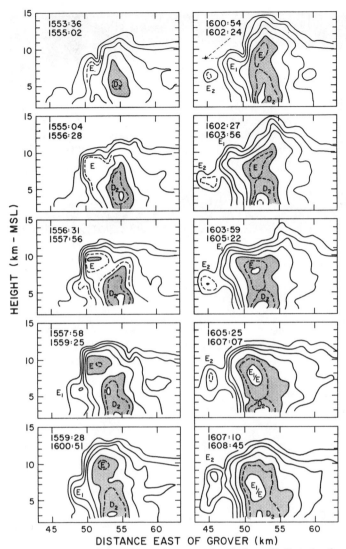

Figure 16.16 Vertical sections oriented east-west through Storm III illustrating the development of Cells E, E₁, and E₂ from 1553 to 1608 MDT. Contour intervals are the same as in Fig. 16.12. The position of the NCAR Sabreliner at 1602:15 (see photograph in Fig. 16.17b) is shown in the upper right panel.

identifiable entities within the forward overhang at altitudes of 6.5 to 7.5 km, and in a region where reflectivities of 35 dBZ$_e$ or more already existed. The cell's core then developed vertically toward the left in the figure, reaching a peak height in the 9- to 10-km range before descending to the ground. The path of Cell G through the storm is plotted at 3-min intervals on the 1625 section. The elapsed time from initial detection in the forward overhang to fallout at the ground was about 20 min. Note that the core moves backward (i.e., to the northwest) through the storm while the storm as a whole advances to the southeast, as previously illustrated in Figs. 16.10 and 16.11.

There was little change in the gross features of the storm during the development of F and G. In fact, while we refer to F and G as distinct cells, one is led to wonder whether the transition here really represents a succession of distinct, separate updrafts, or whether the updraft could have been fairly steady, with a somewhat discontinuous patch of embryos being introduced into it to produce the reflectivity perturbations, or whether the perturbations were produced by only minor modulations of the updraft strength. There is an indication of a separate echo top associated with G (more pronounced out of the plane of Fig. 16.18), supporting the idea of separate updrafts, at least in the upper region of the storm. As will be seen, the evolution of Cells I, J, and K is similar to that of F and G, and we shall discuss this topic more later.

During the period when Cells F and G were developing on the storm's forward flank, the storm also continued to develop on its western or right flank. For the most part the convective impulses there tended to merge with the main storm complex and lose their identity, as shown in Fig. 16.15 for E₁ through E₄. About 1623, however, a somewhat stronger impulse, Cell H, developed and remained separate from Cell G's high-reflectivity core throughout its lifetime. Figure 16.19 shows the three-dimensional structure of the storm at 1630 with coexisting Cells G and H. Four horizontal slices through the storm, at 3, 6, 9, and 12 km, and a vertical section oriented east-west through the two cells are shown. The figure shows the separate reflectivity cores extending through all levels of the storm, and also shows the similarity in vertical structure between Cells G and H. Sub-cloud aircraft data show the presence of a broad updraft region ahead of the storm with separate updraft maxima associated with the two cells.

16.5.3 Phase 3: the Period 1630-1730

During the third phase the storm continued to propagate on its southeastern flank, and the activity on its western flank gradually died out. This resulted in a shift in storm motion to a more southeasterly heading, as shown in Fig. 16.1. During this phase the storm was also observed by research aircraft and Doppler radars, and for this reason we will present the echo evolution in somewhat more detail.

(1971). It is unknown how long the towers existed before having detectable echoes.

About 1605 new cells also began to develop on the storm's southeast flank. The first of these, Cell F, grew to dominate the storm's reflectivity pattern between 1610 and 1620 and was followed by Cell G between 1620 and 1630. As with the E series, the development of F and G is somewhat ambiguous in plan view and is more easily visualized with the aid of the vertical sections in Fig. 16.18. These sections are oriented from northwest to southeast (330° to 150°) through the position of the newly developing cells and the storm's high-reflectivity core, and generally along the inflow direction. Again, the storm's forward overhang extends to the right or southeast over the sector of low-level inflow. Cloud base height is shown with a dashed line. The cells first appeared as

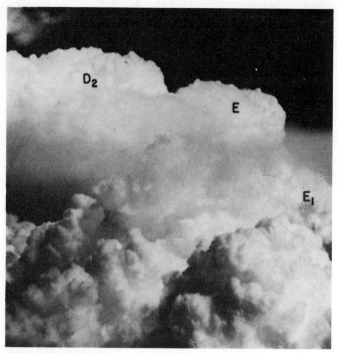

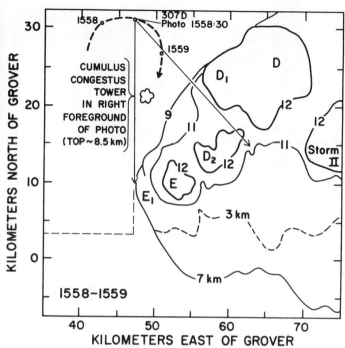

Figure 16.17a Photograph of Storm III taken from the NCAR Sabreliner at about 1558:30, showing the discrete cellular evolution on the storm's southwest flank. The topography of the echo top is indicated in the associated line drawing.

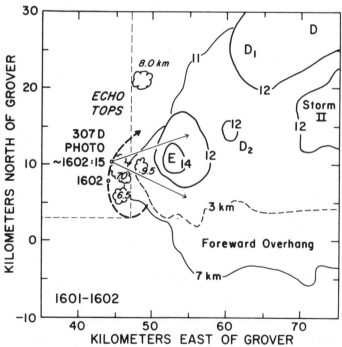

Figure 16.17b As in Fig. 16.17a, but at 1602:15.

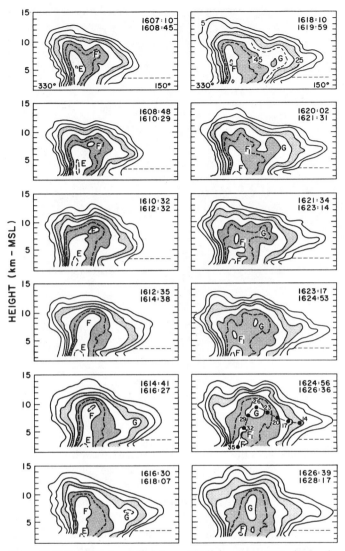

Figure 16.18 Sequence of vertical sections oriented along 330°-150° showing the development of Cells F and G. The relative motion of Cell G with respect to the storm is shown at 3-min intervals on the 1625 section. Horizontal and vertical scales are equal. Cloud base at 3.5 km MSL is indicated by the dashed horizontal line. The contours are the same as in Fig. 16.12.

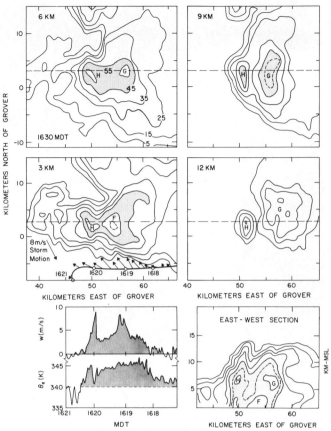

Figure 16.19 Horizontal and vertical sections through Storm III at 1630 MDT showing coexisting Cells G and H. Contour intervals are shown in the upper left panel. The track of the sub-cloud aircraft (Queen Air 306D), flying in the updraft region between 1618 and 1621 MDT, has been adjusted for storm motion. The wind vectors are relative to the ground and are scaled so that 2 m s^{-1} is equal to 1 km horizontal distance. Vertical air velocity and equivalent potential temperature determined by the aircraft near cloud base are shown in the lower left panel. Note the peaks in the cloud base updraft corresponding to Cells G and H.

The history of the storm between 1630 and 1730 is illustrated in Fig. 16.20 with a sequence of CAPPI's at 7 km. The storm was sustained by three major impulses during this period. The first of these, Cell I, developed shortly after 1630 and was followed by Cell J about 1650 and Cell K about 1710. As with the two previous cells, F and G, which formed on the storm's forward flank, Cells I, J, and K developed in such a way that they were first identifiable in the forward overhang of the storm and then merged with the existing high-reflectivity core in the middle and lower levels. The evolution of these cells is somewhat ambiguous if only horizontal sections are examined. Indeed, from Fig. 16.20 there seems to be little evolution at all, and little reason to call the whole sequence of G-I-J-K more than one cell, as a single

echo core dominates during this period. It is believed that in this phase (and actually including cells F and G) the storm is best described in terms of a generally steady airflow pattern on which a modulation of the updraft intensity is superimposed. This view is supported by the persistent character of the echo in Fig. 16.20, and in the vertical sections of Figs. 16.21, 16.22, and 16.27. A degree of unsteadiness, on the other hand, is implied by the surges in the storm top and reflectivity features previously shown in Fig. 16.9 and also by the development of weak, short-lived radar echo vaults preceding each of the cells I, J, and K.

The development of these vaults (or bounded weak echo regions, after Chisholm, 1973) is best illustrated by vertical sections. The left column of Fig. 16.21 shows the development of a vault in the forward overhang of the storm around 1632 and the formation of the echo core of Cell I above it. At this time the core of Cell G is falling to the ground. In the right column the echo of Cell I can be traced falling to the ground also.

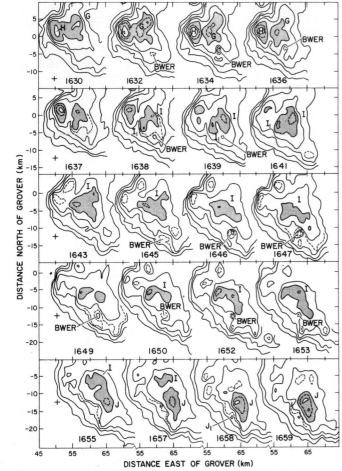

Figure 16.20a Sequence of horizontal sections at 7 km showing the evolution during phase 3 from 1630 to 1700. Reflectivity contours are the same as in Fig. 16.12. Several transient bounded weak echo regions are indicated with "BWER." Times are shown to the nearest minute, although the CAPPI's are spaced approximately equally in time at 90-s intervals.

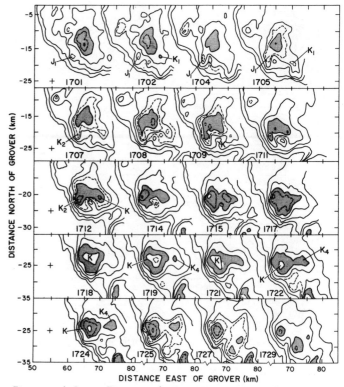

Figure 16.20b Same as Fig. 16.20a, but from 1700 to 1730.

A similar series of events is shown in Fig. 16.22 for Cell J. A vault is seen starting to develop by 1647. The vault lasts somewhat longer than that in Fig. 16.21, but the sequence of events is the same: an implied surge in the updraft, causing particles (and associated reflectivity contours) to be carried to a somewhat higher altitude and leading to an increase in the height of the echo top, and causing a local enhancement of the reflectivity aloft because of the additional particle growth.

Perhaps the more striking feature of Figs. 16.21 and 16.22, though, is not the perturbations in the echo but rather its overall steadiness. For example, there is a persistent high-reflectivity core extending from the upper levels of the storm to the ground, implying a more or less continuous particle growth process. Doppler wind data presented in Chapter 18 also support the idea of a fairly steady, large-scale organization.

The vaults in this storm are quite small in comparison with those of some storms described in the literature (e.g., Chisholm, 1973; Marwitz, 1972a; Browning and Foote, 1976), though Foote and Mohr (1979) report that weak, transient vaults of this general kind are not uncommon. Fig. 16.23 illustrates their temporal history somewhat better. As the updraft surge with each cell penetrates into the storm's upper levels the echo top rises and a high-reflectivity core develops above (and around) the vault and subsequently descends to the surface. Cells I and J develop echo cores near 11 km and tops near 14 km, while for Cell K both features are approximately 1 km lower. In each case the reflectivity in the vault begins to increase in its lower and middle regions before the updraft surge has even penetrated into the upper levels. It is believed that this increase in reflectivity is a result of the growth of particles that have sedimented across streamlines and mixed into the updraft in a predominantly horizontal fashion. Measurements from the T-28 aircraft (presented in Chapter 19) indicate that all the particles in the region of the vault were rising, so the lowering of the vault ceiling in this case is not associated with particles falling with respect to the ground. It is rather a result of the growth of particles in this region.

An estimate of the vertical velocity within the storm can be made using the rate of change of height of the echo minimum in Fig. 16.23. In Cell I, for example, the slope of this feature is 19 m s^{-1}. If one assumed that the terminal

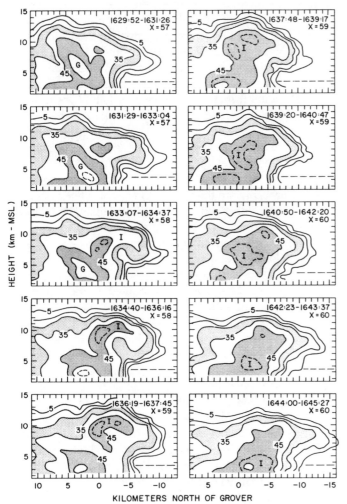

Figure 16.21 Sequence of north-south vertical sections showing the development of Cell I in the forward overhang. First evidence of the new impulse is the appearance of a weak vault (1630-1634 MDT) followed by the development and descent of a reflectivity core aloft (1634-1645 MDT) and a rise in the echo top. The horizontal dashed lines represent the height of cloud base.

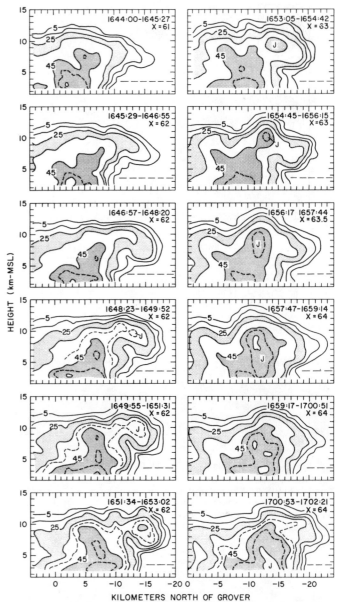

Figure 16.22 Sequence as in Fig. 21, showing the development of Cell J.

velocities of the particles in the updraft were on the order of 5 to 10 m s^{-1}, an updraft magnitude of 25 to 30 m s^{-1} would result. This is in agreement with the maximum values determined from the Doppler radars and the T-28 penetrating aircraft as presented in Chapters 18 and 19.

There is some documentation from ground and airborne photography of the changes in the visual cloud top associated with Cells I and J. Figure 16.24 shows a sequence of 16-mm time-lapse photographs taken from Grover and shown at 1-min intervals. The camera was pointed eastward and the development of Cell I is seen on the right-hand side of the cloud. Though the quality of the photographs is relatively poor, one gains some visual sense of the evolution of the cloud, in particular the protrusion of a large cumulus tower from a common cloud mass in the middle levels. Unfortunately, the interesting early history of the visible tower is

obscured by intervening cloud. The cell to the left of Cell I is Cell H, and it appears to be higher only because it is closer to the camera. Figure 16.9 indicates that Cell H reached its maximum height of about 13.5 km at about 1635, while Cell I reached its peak of 14.5 km near 1640. These times are in agreement with the photographs.

Two composite photographs of Cell J during its growth stage are shown in Fig. 16.25 along with the topography of the echo top. The pictures were taken from a forward-looking camera on the Sabreliner, which was flying at an altitude of about 9 km. Both composites show a view of the storm from the southeast, looking toward the northwest. Remnants of the anvil from Cell I may be seen in the extreme

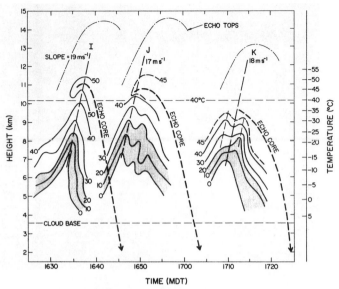

Figure 16.23 *Time-height diagram showing the development of the weak echo regions for each of the major convective impulses during phase 3. The figure was produced by plotting the minimum reflectivity observed within the weak echo regions as a function of height and time. The dashed curves at the top of the figure show the echo top associated with each cell.*

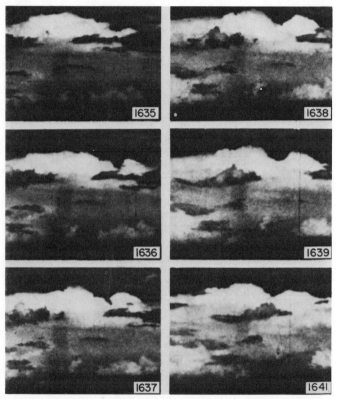

Figure 16.24 *Time-lapse 16-mm photographs of Storm III taken from the Grover radar site looking eastward. The top of Cell I may be seen rising above the cloud mass to the right (south) of Cell H.*

right of both photos, and are identified on the echo map. Figure 16.25a shows J in the center background as a collection of cumulus turrets protruding above a wispy cirrus region. The composite photograph five minutes later in Figure 16.25b, taken from almost the identical position over the ground, shows that a dramatic growth of the cloud has occurred. This intensification associated with J is seen also in the echo topography, where altitude increases of 2-3 km are indicated, and in the vertical sections of Fig. 16.22. There is thus little doubt from these photographs that surges in the intensity of the updraft accompanied the development of I and J.

There are several smaller-scale features in the echo patterns that we have not previously mentioned. At 1653, for example, the echo in Fig. 16.20 shows two small reflectivity maxima that probably are the result of small short-lived updrafts (small turrets) imbedded in or slightly flanking the larger-scale flow. Another example is J_1, shown at 1658 as a slight minimum in the reflectivity pattern, though by 1701 it displays a local maximum.

After 1700 such features become somewhat more prominent, and Fig. 16.26 shows the subsequent echo evolution in more detail. In the scan at 1705:40 two separate weak-echo regions can be seen in the forward overhang. These have both filled with echo, designated K_2 and K_3, by the next scan. A large weak-echo region between them persists for another 10 min or so, presumably reflecting the existence of a single major updraft there. By 1710 K_3 and the earlier impulse K_1 appear to have merged together, losing their identity shortly after this, but K_2 is more persistent. By examining the progressively lower scans in Fig. 16.26, K_2 can be tracked to the surface along with Cell K, the echo associated with the large vault. The dual-cell appearance on the right-hand side of Fig. 16.26 makes it fairly clear that two large updrafts must have coexisted during this phase of the storm.

By 1718 the echo in Fig. 16.20 is dominated by Cell K, though yet another small updraft just to the east of K produces K_4 shortly after this. As will be seen in Chapters 18 and 19, the inference here of several updrafts during the 1700-1720 period is supported by the Doppler radar and aircraft measurements.

The evolution of the dominant Cell K is illustrated in Fig. 16.27, again by use of vertical sections oriented in a north-south direction. Passing through the position of the vault, the sections show that K starts intensifying in the forward overhang at about 1710, in the manner of I and J and the earlier Cells F and G, and subsequently descends to the surface. Observations made at the surface beneath K indicate that it produced a burst of hail that covered an area of several square kilometers and had a maximum diameter of about 1.7 cm.

16.6 DISCUSSION

16.6.1 The Early Phase

In the preceding sections we discussed the storm behavior during different phases of its lifetime. The structure and

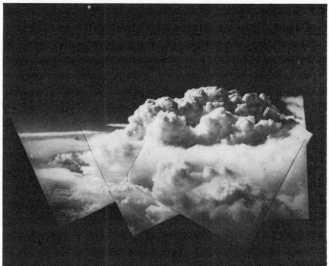

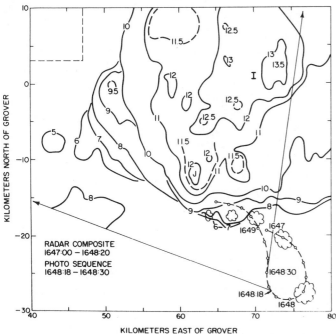

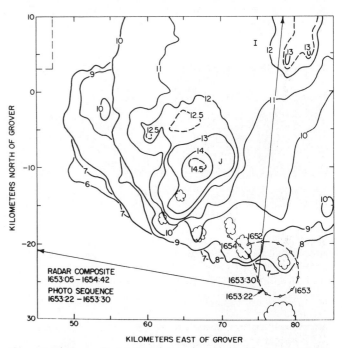

Figure 16.25a Composite photograph of Storm III taken from the forward-looking camera mounted on the NCAR Sabreliner. the pictures were taken at about 1648 as the aircraft flew on the southeast side of the storm looking toward the northwest. Cell J is beginning to develop out of the cloud mass near the center of the photo, while the remnants of Cell I may be seen to the right. Some of the cumulus towers on the storm's southeast flank may be seen in the foreground of the photograph and are indicated schematically in the map below, showing the topography of the echo top (5 dBZ_e). A wispy cloud mass is seen advecting around the storm's southern flank in the middle levels.

Figure 16.25b Same as Fig. 16.25a, only 5 min later. The rapid development of Cell J is evident.

evolution during the first phase of the storm were characteristic of what may be called "organized multicell" storms; that is, storms that evolve in a basically regular, periodic fashion with the formation of apparently discrete updrafts and their associated radar cells on a preferred flank of the parent storm, as opposed to a seemingly random

development of new cells. (For a discussion of multicell storms see, e. g., Marwitz, 1972b and Chapter 5.) In the present case the right flank was the preferred region of new cell formation and the storm as a whole deviated in its motion to the right of the mean winds.

Cell D was the most intense cell of the day, with both the highest echo top and greatest reflectivity. It formed during a period of general intensification. Its formation appears to have been influenced by the merger of Cell C with some weaker cumulus towers on its right rear flank. Though it was

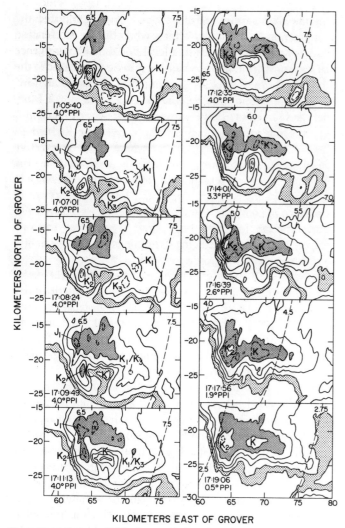

Figure 16.26 Sequence of PPI's showing details of the evolution from 1705 to 1719 MDT. Constant altitude areas (dashed curves) are graduated in kilometers MSL. The larger impulses K and K_2 can be followed as they develop in the storm's middle levels and descend to the ground.

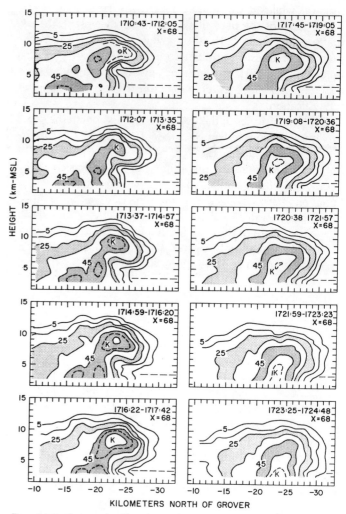

Figure 16.27 Sequence of north-south vertical sections showing the development of Cell K from 1710 to 1725 MDT.

large and intense, it lasted only 15-20 min in its mature state and probably should be considered an ordinary cell according to the scheme proposed by Browning (1977).

The evolution from D to E was also part of the organized multicell phase, with new growth preferentially on the right flank. While the echoes of new cells such as E first appeared very close to, and occasionally within, existing echo in the middle levels, the photographs show that in their upper levels they were composed of cloud turrets growing into clear air, not imbedded in a parent cloud. Probably they started out as discrete turrets separated from the main cloud, as in the description of multicell storms given by Dennis et al. (1970), and merged with the storm only as they were finally developing echoes.

Cell H was the last of the cells to form on the right flank. If it had produced daughter cells the storm probably would

have split. Reasons why this branch of the storm did not continue after Cell H are considered in the following chapter.

16.6.2 The Steadier Phase

Starting with Cell F the storm changed into a steadier regime, with F, G, I, J, and to a lesser extent the K's forming a series of reflectivity and updraft perturbations that we interpret as being superimposed on a steady mean flow. It appears from the data that the pulsations extended through the whole depth of the cloud. The evidence includes the intensification of the echo and the increase in the echo top and the height of the visual cloud top. However, by themselves these could be ambiguous, because the echo intensity could be modulated by microphysical processes, such as changes in the supply of hail embryos, without any dynamical change. The height of the cloud top could fluctuate because of the turbulent breakdown of the updraft into detached eddies at its upper reaches, while in the middle levels, say from 5 to 9 km, where most of the precipitation growth takes place, the updraft could remain fairly steady. But during I, J, and K the presence of transient vaults also indicates updraft perturbations, and the various pieces of evidence fit together in a coherent manner, all pointing toward periodic updraft intensification.

The time histories of the vaults are represented in the series of vertical sections in Figs. 16.21, 16.22, and 16.27 and in the time-height profiles in Fig. 16.23. The earliest indication of the updraft invigoration is the lifting of the underside of the overhang to form a small vault. This implies that the intensification starts from the lower levels. The vault then develops upward, a local maximum of reflectivity forms over it, and the height of the echo top and visual cloud top both increase. The pulsation is superimposed on a persistent reflectivity structure so we conjecture that there was a persistent component to the updraft as well.

The above picture is supported by the Doppler data presented in Chapter 18, where it is also noted that the airflow pattern during phase 3 is in substantial agreement with the supercell model of Browning and Foote (1976).

As regards the updraft, the existence of a steady mean flow constitutes the supercellular component of the storm organization. The perturbations noted are reminiscent of the organized multicell (e.g., Browning et al., 1976), with formation of new echo tops at intervals of 10-18 min. However, according to the present interpretation, the perturbations do not represent new updrafts forming in discrete new locations, as in the classical multicell model (and in contrast to the earlier stage of this storm), but rather pulsations of a continuing updraft, with perhaps only minor changes in form or position.

Few previous investigations have noted the pattern of storm evolution documented here. Chisholm (1973) observed some transitions in the size of the vault as seen in plan view but did not interpret them. Foote et al. (1975) also noted changes in the vault size, which they attributed to a transient influx of particles. Foote and Mohr (1979) discussed several storms displaying transient vaults similar to those of the present case, though none was examined in such detail. In a study of a supercell storm with a large, persistent vault, Nelson and Braham (1975) observed periodic changes in the geometry of the upper part of the vault. They interpreted the observations in terms of precipitation loading with reference to the analysis of Srivastava (1967). The latter is not a possibility here, however, because the perturbations are first seen low in the updraft in a region of rather weak echo where the weight of hydrometeors could not have been substantial. It cannot be determined whether the perturbation in Nelson and Braham's case also started in the lower regions of the cloud.

Orville and Kopp (1977) found behavior apparently similar to that documented here in the solutions from a two-dimensional cloud model. They noted that the computed

motion consisted of "a large circulation cell upon which are superimposed smaller perturbations." They attributed the phenomenon to continuity effects whereby the acceleration in a buoyant updraft leads to a local horizontal convergence that can constrict it, thereby causing it to be modulated in the form of a series of bubbles superimposed on the mean flow.

This explanation may be considered a variation of Moncrieff and Green's (1972) argument that steady convection is only possible over a range of the convective Richardson number, $Ri = CAPE[1/2(\Delta u)^2]^{-1}$, where the convective available potential energy (CAPE) is evaluated as the "positive energy" on a tephigram, and Δu is the difference in the wind speeds between upper and lower troposphere in conditions of constant shear. For Ri too large the inflow is not strong enough to keep the updraft supplied at the rate "natural" to it, the latter being determined by the instability, so that no steady solution is possible.

Browning (1978) has criticized Orville and Kopp for applying their simulation to the Fleming storm described by Browning and Foote (1976), particularly because in their calculations they reduced the environmental wind shear to only 20% of that observed for the Fleming storm. This increased Ri by a factor of 25, and might have tended to make the model updraft more unsteady.

It now appears that as far as the storm dynamics are concerned the simulation of Orville and Kopp could be applied more realistically to the present storm, which has weaker environmental shear than the Fleming storm (about one-fourth as much), but similar instability. It then seems possible that the pulsating nature of the updraft observed in nature and in the model may be a general result, representing a class of behavior characteristic of storms with Ri somewhat larger than that appropriate to a steady regime. Such a "pulsating-steady" behavior was in fact postulated by Newton (1963). Further clarification of the nature of the updraft pulsations is given in Chapter 18, based on multiple-Doppler observations.

16.6.3 Small-Scale Structure

A well-known advantage to numerical simulations is the relative ease with which the results can be analyzed. If one is interested in updraft evolution, for example, the associated buoyancy field can be examined to gain further insight, as in Orville and Kopp (1977). Examination of the updraft evolution in the latter paper, in fact, serves to highlight a limitation inherent in considering only reflectivity measurements: reflectivity does not mirror very faithfully the underlying updraft. Even apart from the effects of pulse-volume averaging, a storm's reflectivity structure tends to show only the integrated effect of particle growth in regions of liquid water, and much of the updraft behavior itself remains hidden. There must be more structure in the updraft than analyses like the present one can detect. In spite of this fact there is often a good deal of structure evident in the data. In the present case, for example, the evolution documented for the E and K series of cells shows that updrafts of smaller size are detected along with the larger ones, and there is a basis for thinking of the storm as composed of a spectrum of scales. At certain times different scales appear to be more dominant in Storm III. For example Cells F, G, I, and J are dominated by the larger (5-to 8-km) scale, while scales of 2-5 km are more important in the K series. Even with I and J, though, small undulations only a kilometer or two across appear in the underside of the forward overhang. They have continuity for several minutes and indicate that small updraft pockets are imbedded within the larger updraft. The possible importance of such small scales of motion has been emphasized by Battan (1975) and Barge and Bergwall (1976), but their general significance will probably prove difficult to establish. Marwitz (1972c) and Nelson and Braham (1975) have also mentioned such features in radar measurements.

It may be possible to view the major updrafts of this storm as being composed of a number of unsteady but fine-scale convective elements, not well resolved by the radar, whose joint action serves to provide a statistically steady large-scale flow (an argument along the lines of Barge and Bergwall, 1976). But this is no more than saying that here is a quasi-steady large-scale motion that is merely turbulent; that is, it has smaller scales of motion imbedded in it. The distinction is perhaps important only if individual turbulent elements themselves have some inherent role to play, as for example would be the case if particle growth were confined to single elements. Calculations reported in Chapter 20 indicate that particle trajectories are not so confined.

Our point here is that motions do occur over a range of scales, including the 5- to 10-km scale. In pursuing a better understanding of convection and precipitation development none of these scales should be dismissed out of hand. With present measurement techniques, the role of the smaller scales will be the most difficult to document. Concerted attempts should be made to obtain more high-resolution radar data such as those of Krehbiel and Brook (1979) and Battan (1975).

16.6.4 Comparison with the Raymer Storm

Particularly during phase 3, the present storm resembled in many ways the Raymer storm discussed by previous investigators (of specific relevance here are Chalon et al., 1976; Fankhauser, 1976; Musil et al., 1976; and Browning et al., 1976). It had a roughly equivalent environmental structure, with comparable wind hodograph and instability. The echo evolution was also similar, with new radar cells appearing on the front flank of the storm (that is, in the direction of storm motion), rather than the right flank as in a number of other studies (e.g., Marwitz, 1972b). The point of interest here is that while we have chosen to emphasize the overall

steadiness of the radar structure and underlying airflow pattern, Chalon, et al., Fankhauser, and Browning et al. chose in describing the Raymer storm to emphasize the discrete nature of the cells and their associated updrafts. This situation warrants some discussion.

It is likely that the two storms were more alike than the two interpretations recognize, and that these previous authors overstressed the independence of individual cells, partly because the radar data they were using had a rather high threshold (approximately 30 dBZ_e) which prevented them from recognizing the structure and, we conjecture, the greater steadiness of the outer boundaries of the storm. While the subcloud data of Fankhauser (1976) seemed to indicate that updrafts of consecutive cells were indeed separate, the T-28 measurements near an altitude of 7 km showed a single large updraft that had two peaks, corresponding to the two cells, that were only some 30% larger than the updraft mean (Musil et al., 1976). The evolution and structure of the Raymer storm may then have been consistent with that deduced for the present study. This will be discussed further after consideration of the storm airflow in Chapter 18.

16.7 CONCLUDING REMARKS

In the preceding sections we have described the behavior of Storm III on the basis primarily of detailed S-band radar observations of its structure and evolution. Few previous studies have had access to data with such good time resolution (90 s), and its availability must be considered a key to studies of this kind.

In terms of severity the hailstorm discussed here is representative of moderate to intense hailstorms in northeastern Colorado. It produced hail up to about 2 cm in diameter, displayed reflectivities in the range of 60 to 70 dBZ_e, and penetrated 1-2 km above the tropopause. Comparison of cell areas and time-area integrals of reflectivity with the NHRE sample of 52 storm days (Foote et al., 1979a; Foote and Mohr, 1979) supports the notion that this was a significant hailstorm, though not one of the most intense. The storm started out as a series of ordinary cells (Browning, 1977), an "organized multicell" storm in the terminology adopted here, and evolved into a steadier phase.

The dichotomy of ordinary cell vs. supercell is often developed in terms of qualitative structural characteristics, for example as detected by radar (e.g., Marwitz, 1972a; Barnes, 1978). On the other hand, Browning (1977) has proposed that classification be according to cell lifetime: either normal lifetime (i.e., short-lived ordinary cells), or long-lived (supercells). This can be somewhat more objective. The shortcoming of a simple dichotomy is also more obvious. Since the cells in many storms will be neither short- nor long-lived, but somewhere in between, the scheme is not likely to encompass all cases. While a continuum in storm morphology probably exists, the nature of transitions along this continuum requires further elucidation. A start along these lines is given in Chapter 18, where further details of the updraft evolution are examined. The present storm has some conventional attributes of both the organized multicell and supercell extremes. When a more complete classification scheme is devised, the mode of behavior of the present storm, termed "pulsating-steady" after Newton (1963), will probably fit best somewhere between these two somewhat idealized models.

References

Auer, A. H., Jr., and J. D. Marwitz, 1968: Estimates of air and moisture flux into hailstorms on the High Plains. *J. Appl. Meteorol.* 7, 196-198.

Barge, B. L., and F. Bergwall, 1976: Fine scale structure of convective storms associated with hail production. Proc. 2nd WMO Scientific Conf. on Weather Modification, Boulder, Colo., 1976, World Meteorological Org., Geneva, Switzerland, 341-348.

Barnes, S. L., 1978: Oklahoma thunderstorm on 29-30 April 1970: Part I. Morphology of a tornadic storm. *Mon. Weather Rev.* 106, 673-684.

Battan, L. J., 1975: Doppler radar observations of a hailstorm. *J. Appl. Meteorol.* 14, 98-108.

Biter, C. J., and M. E. Solak, 1979: On top seeding for hail suppression: an NHRE operational feasibility study. Prepr. 7th Conf. Inadvertent and Planned Weather Modification, Banff, Alberta, 1979, Am. Meteorol. Soc., Boston, Mass., 215-216.

Breed, D. W., 1978: Case studies on convective storms: Case study 2. 22 July 1976, first echo case. NCAR Tech. Note TN-132+STR, National Center for Atmospheric Research, Boulder, Colo., 59 pp. [NTIS TB295753/AS].

Browning, K. A., 1977: The structure and mechanisms of hailstorms. *Meteorol. Monogr.* 16 (38), 1-43.

-----, 1978: Comments on "Numerical simulation of the life history of a hailstorm." *J. Atmos. Sci.* 35, 1553-1554.

-----, and G. B. Foote, 1976: Airflow and hail growth in supercell storms and some implications for hail suppression. *Q. J. R. Meteorol. Soc.* 102, 499-533.

-----, J. C. Fankhauser, J-P. Chalon, P. J. Eccles, R. C. Strauch, F. H. Merrem, D. J. Musil, E. L. May, and W. R. Sand, 1976: Structure of an evolving hailstorm: Part V. Synthesis and implications for hail growth and hail suppression. *Mon. Weather Rev.* 104, 603-610.

Byers, H. R., and R. R. Braham, Jr., 1949: *The Thunderstorm.* U.S. Government Printing Office, Washington, D. C., 287 pp.

Chalon, J.-P., J. C. Fankhauser, and P. J. Eccles, 1976: Structure of an evolving hailstorm: Part I. General characteristics and cellular structure. *Mon. Weather Rev.* 104, 564-575.

Chisholm, A. J., 1973: Alberta hailstorms, Part I: Radar case studies and airflow models. *Meteorol. Monogr.* 14 (36), 1-36.

Dennis, A. S., C. A. Schock, and A. Koscielski, 1970: Characteristics of hailstorms of western South Dakota. *J. Appl. Meteorol.* 9, 127-135.

Dye, J. E., and B. E. Martner, 1978: The relationship between radar reflectivity factor and hail at the ground for northeast Colorado thunderstorms. *J. Appl. Meteorol.* 17, 1335-1341.

Fankhauser, J. C., 1976: Structure of an evolving hailstorm: Part II. Thermodynamic structure and airflow in the near environment. *Mon. Weather Rev.* 104, 576-587.

Foote, G. B., and C. G. Mohr, 1979: Results of a randomized hail suppression experiment in northeast Colorado: Part VI. Post hoc stratification by storm intensity and type. *J. Appl. Meteorol.* 18, 1589-1600.

-----, C. G. Wade, and K. A. Browning, 1975: Air motion and hail growth in supercell storms. Prepr. 9th Conf. Severe Local Storms, Norman, Okla., 1975, Am. Meteorol. Soc., Boston, Mass., 444-451.

-----, R. C. Srivastava, J. C. Fankhauser, F. I. Harris, T. J. Kelly, R. E. Rinehart, C. G. Wade, P. J. Eccles, E. T. Garvey, M. E. Solak, R. L. Vaughan, B. E. Weiss, and R. J. Wolski, 1976: Final Report—National Hail Research Experiment Randomized Seeding Experiment 1972-1974: Vol. IV. Radar summary. NCAR, Boulder, Colo., 326 pp. [NTISPB266199/AS].

-----, R. E. Rinehart, and E. L. Crow, 1979a: Results of a randomized hail suppression experiment in northeast Colorado: Part IV. Analysis of radar data for seeding effect and correlation with hailfall. *J. Appl. Meteorol.* 18, 1569-1582.

-----, C. G. Wade, J. C. Fankhauser, P. W. Summers, E. L. Crow, and M. E. Solak, 1979b: Results of a randomized hail suppression experiment in northeast Colorado: Part VII. Seeding logistics and post hoc stratification by seeding coverage. *J. Appl. Meteorol.* 18, 1601-1617.

Fujita, T., and H. R. Byers, 1962: Model of a hail cloud as revealed by photogrammetric analysis. *Nubila* 5, 85-105.

Heymsfield, A. J., A. R. Jameson, and H. W. Frank, 1980: Hail growth mechanisms in a Colorado storm: Part II. Hail formation processes. *J. Atmos. Sci.* 39, 1779-1807.

Knight, C. A., G. B. Foote, and P. W. Summers, 1979: Results of a randomized hail suppression experiment in northeast Colorado: Part IX. Overall discussion and summary in the context of physical research. *J. Appl. Meteorol.* 18, 1629-1639.

Krehbiel, P. R., and M. Brook, 1979: A broadband noise technique forfast-scanning radar observations of clouds and clutter targets. *IEEE Trans. on Geosci. and Electron.* GE-17, 196-204.

Long, A. B., R. J. Matson, and E. L. Crow, 1979: The hailpad: construction and materials, data reduction, and calibration. NCAR Tech. Note TN-144+STR, NCAR, Boulder, Colo.

Marwitz, J. D., 1972a: The structure and motion of severe hailstorms: Part I. Supercell storms. *J. Appl. Meteorol.* 11, 166-179.

-----, 1972b: The structure and motion of severe hailstorms: Part II. Multicell storms. *J. Appl. Meteorol.* 11, 180-188.

-----, 1972c: The structure and motion of severe hailstorms: Part III. Severely sheared storms. *J. Appl. Meteorol.* 11, 189-201.

Mohr, C. G., and R. L. Vaughan, 1979: An economical procedure for Cartesian interpolation and display of reflectivity factor data in three-dimensional space. *J. Appl. Meteorol.* 18, 661-670.

Moncrieff, M. W., and J. S. A. Green, 1972: The propagation and transfer properties of steady convective overturning in shear. *Q. J. R. Meteorol. Soc.* 98, 336-352.

Musil, D. J., E. L. May, P. L. Smith, and W. R. Sand, 1976: Structure of an evolving hailstorm: Part IV. Internal structure from penetrating aircraft. *Mon. Weather Rev.* 104, 596-602.

Nelson, S. P., and R. R. Braham, Jr., 1975: Detailed observational study of a weak echo region. *Pure Appl. Geophys.* 113, 735-746.

Newton, C. W., 1963: Dynamics of severe convective storms. In *Severe Local Storms, Meteorol. Monogr.* 5(27), 33-58.

Orville, H. D., and F. J. Kopp, 1977: Numerical simulations of the life history of a hailstorm. *J. Atmos. Sci.* 34, 1596-1618.

Srivastava, R. C., 1967: A study of the effect of precipitation on cumulus dynamics. *J. Atmos. Sci.* 24, 36-45.

Warner, C., 1971: Visual and radar aspects of large convective clouds. Ph.D. thesis, McGill University, Montreal, Quebec, 116 pp.

CHAPTER 17

The 22 July 1976 Case Study: Low-Level Airflow and Mesoscale Influences

Charles G. Wade and G. Brant Foote

17.1 INTRODUCTION

The structure and evolution of Storm III was discussed in Chapter 16 in terms of radar observations and a simple airflow model. Here, we treat the structure of the subcloud region of the storm and consider the extent to which the storm evolution can be understood in terms of features that can be identified there. Attention is focused on the low-level air lying ahead of Storms I, II, and III, and its role in influencing storm behavior, including its interaction with the cold outflow produced by the spreading downdrafts.

Storm characteristics near the surface, and in particular the structure and importance of outflow regions, have received considerable attention following the early investigations of Byers and Braham (1949), Tepper (1950), Fujita (1959a and b, 1963) and others. Recent field studies include, for example, the work of Charba (1974), Goff (1976), Barnes (1978), and Brandes (1978) in Oklahoma, Foote and Fankhauser (1973), Ellrod and Marwitz (1976), and Kropfli and Miller (1976) in Colorado, and Miller and Betts (1977), Ulanski and Garstang (1978), and Zipser (1977) in low latitudes. Important laboratory and theoretical investigations of density currents (e.g., Keulegan, 1958; Simpson, 1969, 1972; Benjamin, 1968) have also given valuable insights into the dynamical structure of gust fronts. The role of outflows in maintaining as well as modifying storm structure has also been emphasized in a number of numerical cloud simulations (e.g., Hane, 1973; Orville and Kopp, 1977; Miller, 1978; Thorpe and Miller, 1978), and Mitchell and Hovermale (1977) have attempted a detailed simulation of flow in the vicinity of the gust front. On the basis of satellite observations Purdom (1973) stated that "the majority of new convective activity" forms along outflow boundaries. In Purdom (1976) he expanded the discussion to emphasize the importance of intersecting outflows and convective lines in intensifying convection, and Gurka (1976) considered further the problem of identifying such boundaries and estimating their intensity. These authors did not attempt to explain the physical processes involved, nor has the generality of such principles been determined.

The observations here are unusually complete in that the storm in question was observed by high-resolution radar for its entire life (≈ 2.5 h), during which it was also over the instrumented surface network, thus providing the opportunity to relate detailed storm behavior to meteorological conditions observable at the surface. For a period of 70 min measurements were also collected by an aircraft flying just below cloud base. These are also reported here.

17.2 NATURE OF THE DATA

Measurements of wind, temperature, humidity, and pressure at the surface were obtained from the 45-station mesometeorological network shown in Fig. 16.1. The terrain in this region (see Fig. 13.1) is marked by a broad plateau extending through central portions of the network with an escarpment along its southern and western sides. The elevation ranges from about 1220 m in the southeast, near Sterling, to more than 1600 m over western portions of the area. The mesonetwork is described in Chapter 13. The measurement accuracies of the PAM and conventional stations were roughly equivalent: 0.5 °C for temperature, 1 g kg^{-1} for mixing ratio under typical conditions, 1 mb for pressure, 1 m s^{-1} for wind velocity.

An estimate of the relative accuracies of the measurements over the whole network can be made from Fig. 17.1. The figure shows the surface winds along with mixing ratio and potential temperature on 22 July 1976 at a time (1200 MDT) when the data were undisturbed by deep convection. The average mixing ratio for the fourteen conventional stations in the three western rows was 10.0 g kg^{-1}, which compares well with the average of 9.8 g kg^{-1} for the fourteen PAM stations along the same three rows. A similar comparison of potential temperature shows an average of 315.0 K for PAM and 314.7 K for the conventional stations. Comparisons between these two variables over longer periods of time show similarly small mean differences. Variations between neighboring stations ranged up to 2 g kg^{-1} and 1 °C and were within the expected range of small-scale atmospheric variability and instrument performance. A quantitative comparison of the winds in Fig. 17.1 was not made, but a visual inspection shows a good "fit" between the two sets of data.

Information on the wind and thermodynamic structure was also obtained by an NCAR aircraft, Queen Air N306D, flying just below cloud base. The accuracy of the wind and thermodynamic measurements was checked throughout the summer with a program of tower fly-bys and side-by-side intercomparison flights with other research aircraft. The vertical component of air motion was determined with a method described by Lenschow (1976), utilizing the aircraft as sensor in estimating the angle of attack, in lieu of measuring it directly with a vane system. The vertical air velocity is then computed using the angle of attack along with measurements of pitch angle and the aircraft vertical velocity (form an inertial platform) and true airspeed. The resultant vertical air motion is believed to be accurate to 1 m s^{-1} or so with a horizontal resolution of about 250 m.

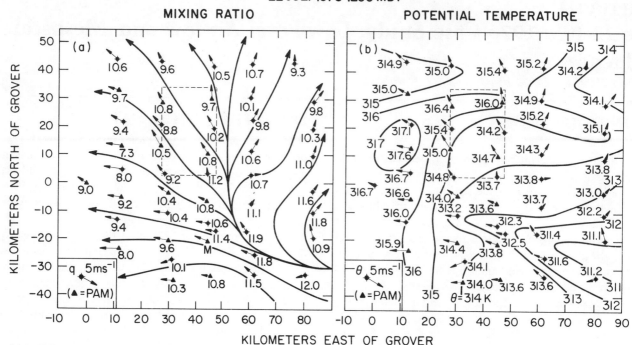

Figure 17.1 Surface meteorological fields at 1200 MDT, 22 July 1976. The figure shows the structure of the mixing ratio, potential temperature, and wind fields approximately 2 h before the development of deep convection over the network. Warmest potential temperatures are found over the higher elevations on the western side of the network, while cool, moist air is being advected into the southeastern portions of the network beneath the remnants of the low-level nocturnal inversion.

17.3 PRE-STORM SURFACE CONDITIONS

The data in Fig. 17.1 show the mesoscale conditions that preceded the development of thunderstorms across the network during the afternoon. The low-level air was characterized by a diffluent southeasterly flow pattern with wind speeds under 5 m s^{-1}. Moist air was being transported at low levels beneath the remnants of a nocturnal inversion situated over the region. The convective potential temperature, the potential temperature at the top of the inversion, was 315.5 K. Stations located at higher elevations over the western portions of the area were the first to exceed this value, and the vertical mixing of the moist air with the drier air aloft as the daytime boundary layer developed is evident in the lower humidities noted at a number of these stations.

The cool air located over the low-lying southeastern portions of the network is a residue of the pool of cold air formed by nocturnal drainage winds. This stable valley air was the slowest to heat up during the day (see, e.g., Lenschow et al., 1979, for a recent discussion of the morning boundary-layer transition under such circumstances). The axis of minimum potential temperature was oriented along the low-level streamline coming out of the valley, and by 1200 the cool air was being advected along with the moisture into central portions of the network. This configuration persisted throughout the early afternoon, and the development of Storm I around 1400 (see Fig. 16.1) occurred along the resulting band of moisture.

17.4 EVOLUTION AT THE SURFACE

As thunderstorms developed during the afternoon the meteorological conditions at the surface became disturbed by the intrusion of rain-cooled air from the storm downdrafts. The outflow gradually spread across the network and had a significant influence on the evolution of the storms. Ahead of the storms, moisture inhomogeneities developed in the southeasterly flow and further influenced their development. We now examine these changes in more detail using a series of surface analyses based on the mesonet observations.

Figure 17.2 shows the surface conditions at 1515 MDT. The boundaries between ambient air and downdraft air are indicated by the cold front symbol, and the surface flow is emphasized by use of streamlines. Separate outflow regions associated with Storms I and II may be seen over the western and northeastern portions of the network, respectively. Over north-central portions of the network the first radar echoes from Storm III had begun to appear along a convergence line separating drier air to the north from more humid air to the south. Low-level winds ahead (i.e., south) of the storms continued to be light and from the southeast, with a diffluent region ahead of Storm I and a developing line of confluence extending southeastward from Storm III. Mixing ratios across the network had decreased typically 2 to 3 g kg^{-1} during the early afternoon, though the values of 9 g kg^{-1} or larger that were present were still sufficient to support deep convection. Near the center of the network a region of locally higher

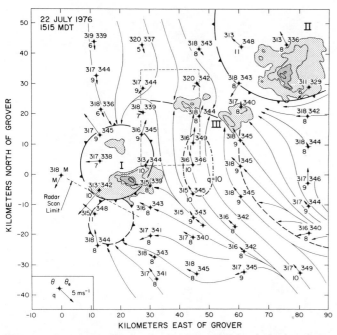

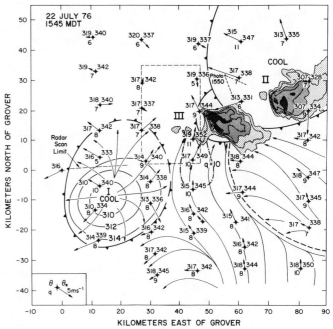

Figure 17.2 Surface meteorological conditions at 1515 MDT. Low-level radar reflectivity data are shown with contour intervals of 10 dB above the 15 dBZ$_e$ threshold. Reflectivity data from Storm I is incomplete due to the sector scanning mode of the CP-2 radar. Potential temperature, mixing ratio, and equivalent potential temperature values have been rounded to the nearest whole integer.

Figure 17.3 Same as Fig. 17.2, except at 1545 MDT.

humidity (greater than 10 g kg^{-1}) is indicated in Fig. 17.2; it would tend to favor the intensification of Storm III.

During the next half-hour the outflow regions from Storms I and II continued to expand horizontally, and a new cold dome began to appear beneath Storm III. The positions of the surface outflow boundaries at 1545 are shown in Fig. 17.3. The moist air located to the south and southwest of Storms II and III, coupled with the local convergence that developed along the outflow boundaries, apparently contributed to an active period of cell development on the southwest flanks of both storms. Figure 17.4 shows the appearance of the subcloud region at 1550. The photograph was taken on the northwest side of Storm II (see position in Fig. 17.3) looking to the southeast. The broad, flat cloud base, extending approximately 10 km to the south and southwest of the precipitation curtain, indicates the updraft region of such storms (Marwitz et al., 1972; see also Browning and Foote, 1976). Dust raised by the outflow winds may be seen near the surface extending out to near the edge of the cloud. The dust appears to be lifted near the position of the outflow boundary by the outflowing air rising up into the "head" of the gust front (e.g., Charba, 1974). At this time Storm III is dominated by Cell D (as discussed in Chapter 16), the largest and most intense cell of this storm. This period of maximum intensity correlates nicely with the ingestion by the storm of air with a high mixing ratio (10 g kg^{-1}).

After 1545 Storms II and III moved southward and roughly

parallel until about 1615 (Fig. 17.5) when changes in the low-level air lying ahead of the storms may have altered the evolution of the storms: surface humidities to the southeast of Storm II had continued to decrease during the afternoon and by 1615 were less capable of supporting deep convection. Apparently as a result, Storm II began to weaken after 1630 and by 1700 had completely dissipated. To the southeast of Storm I a dry outflow region appeared at the surface, thought to be the result of a dissipating cell, and it seems quite likely that Storm I, being deprived of any suitable moisture source, was also dissipating after 1615. It was not being scanned by radar at this time. Thus, by 1615 the region of air capable of sustaining the storms (mixing ratios greater than about 9 g kg^{-1}, or equivalent potential temperatures greater than about 343 K) had decreased in size to a narrow sector that extended from the southeastern corner of the mesonet northwestward toward Storm III. The northwesternmost extent of this potentially unstable air was the source region that had been feeding the southwestern branch of Storm III, Cells C, D, E, and H, as discussed in Chapter 16. However, the southward-moving outflow from Storm III and to an extent also the eastward-moving outflow from Storm I were rapidly displacing this moist air, which was being used up by Storm III and not replaced by the low-level flow. Thus, the failure of any new cells to form on the southwest side of the storm after 1630 can reasonably be attributed to the storm's encounter with the drier air located over the south-central portions of the network.

As described in Chapter 16, new cells began forming on the southeastern flank of Storm III between 1600 and 1615. It

Figure 17.4 *Photograph of the subcloud region of Storm II, taken at 1550 MDT. The view is toward the southeast from a position approximately 20 km northwest of the storm. Location is near Grover coordinates (55 km, 31 km) in Fig. 17.3. The distance from the ground to cloud base is approximately 2 km, and the horizontal distance from the precipitation wall to the right edge of the photo is approximately 10 km. (Photo courtesy of Larry Neubauer, University of Wyoming.)*

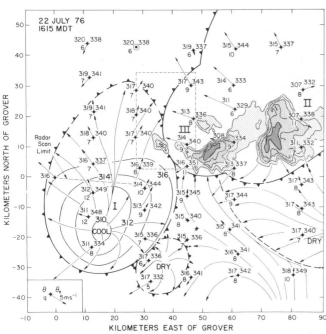

Figure 17.5 *Same as Fig. 17.2, except at 1615 MDT.*

the point where it was intersecting the inflow-outflow interface from Storm III on that storm's southeastern flank, thereby providing further support for the general pattern of convergence that existed along the gust front of Storm III. The confluence line extending southeastward from Storm III probably also played a role by acting to deepen the layer of moisture ahead of the storm, thereby reducing the amount of further lifting required to realize the potential instability.

In order to estimate the surface divergence that developed in association with these features, the surface flow at 1630 was interpolated to a 28 × 26 array with a 2-km grid spacing. A subjective time-space conversion of the mesonet data was used to enhance the details. The resulting vector field is shown on the left side of Fig. 17.6; the divergence field, computed with a second-order approximation, is shown on the right. Two regions of maximum divergence (magnitude of about $5 \times 10^{-3} \, s^{-1}$) are seen positioned beneath the high-reflectivity cores of the storms. Values of convergence in the range of 1 to $2 \times 10^{-3} \, s^{-1}$ were located along Storm III's outflow boundary and increased to nearly $3 \times 10^{-3} \, s^{-1}$ on the southeast flank of Storm III at the location where the two gust fronts intersect. Along the confluence line extending southeastward from Storm III, maximum convergence values were somewhat smaller, though still on the order of $1 \times 10^{-3} \, s^{-1}$. The combination of a favorable moisture source to the southeast (see the hatched area in Fig. 17.6, indicating $\theta_e \geq 343$ K) along with a zone of low-level convergence seems to explain plausibly the storm's continued growth in that direction.

appears from Fig. 17.5 that in addition to the fact that the convectively unstable air was lying to the southeast, a local dynamic mechanism existed there that could aid in releasing the instability. This mechanism was the outflow boundary from Storm II, which had spread southwestward by 1615 to

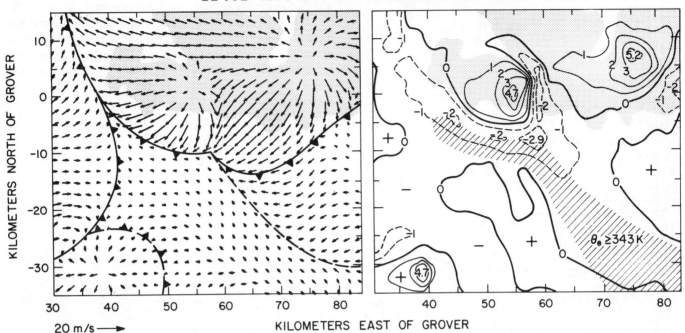

Figure 17.6 Surface wind and divergence fields for 1630 MDT. Wind data from 1615 and 1645 MDT were interpolated to a 28 × 26 grid using a subjective time-space conversion scheme. The position of the low-level radar echo at 1630 MDT is shown in the left panel with light stipple while the cross-hatching in the right panel shows the location of surface air with $\theta_e \geq 343$ K.

The flow pattern that leads to relatively greater convergence being found near the gust fronts than along the line extending to the southeast from Storm III (Fig. 17.6) can be visualized more clearly by examining the surface streamlines drawn relative to the moving storm. Figure 17.7 shows the same data as Fig. 17.6, except that the storm motion vector of 8 m s^{-1} from 330° has been subtracted from all the winds. The degree of confluence seen in the relative flow to the southeast of Storm III is markedly reduced, while the contrast in relative flows across the interface separating outflows from Storms II and III is increased, more in line with the calculated convergence pattern. Interestingly, the relative flow behind the gust front of both storms has only a weak component perpendicular to the front. This is consistent with the fact that the gust front of Storm III was moving only slowly away from the storm at this time.

The most important mesonet station for constructing the flow pattern in Fig. 17.6 was station 6192 at Grover coordinates (61 km, −8 km). The station was situated to the southeast of Storm III at 1615, and was in a position to observe the early pre-storm flow as well as the outflow air from both Storms II and III. As shown in Fig. 17.8, between 1530 and 1615 the station experienced relatively steady temperature and humidity conditions with an easterly wind at 3-4 m s^{-1}. At about 1621 the outflow from Storm II arrived from the northeast. The winds shifted to a northeasterly direction and gradually increased in speed to 10 m s^{-1}. The pressure increased slightly and both potential temperature and mixing ratio decreased. By 1639 the outflow from Storm

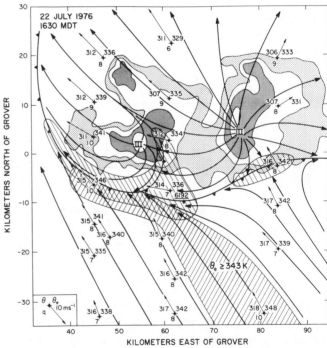

Figure 17.7 Surface conditions at 1630 MDT as in Fig. 17.2, except that the winds are drawn relative to the moving storm. The region of θ_e greater than 343 K and regions of convergence greater than 10^{-3} s^{-1} are cross-hatched.

III had arrived from the northwest and, perhaps because of the proximity of the storm, produced a more dramatic sequence of events. The sharp shift in wind direction to the

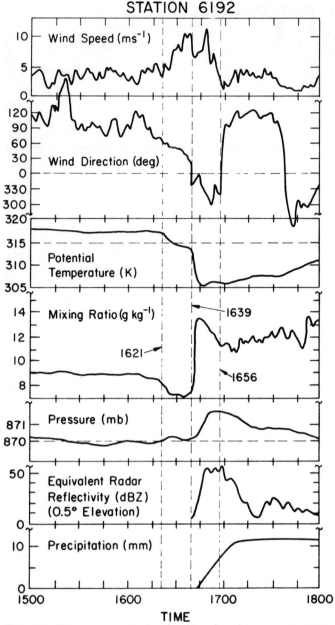

STATION 6192

Figure 17.8 Time sequence of meteorological conditions observed at mesonet station 6192 located at Grover coordinates (61 km, − 8 km). Outflow from Storm II arrived at the station at about 1621 MDT, with the primary outflow surge from Storm III arriving around 1639 MDT. By 1656 MDT the precipitation core from Storm III was passing just northeast of the station.

By 1645 (Fig. 17.9) the combined outflows from Storms II and III had spread some 10-12 km to the south of Storm III, and the storm was propagating to the southeast along the moisture tongue shown in Fig. 17.5. On the southwest corner of Storm III, Cell H dominated the PPI at this time but was rapidly dissipating. Storm II had been ingesting dry air and was now merely raining out, its reflectivity decreasing markedly.

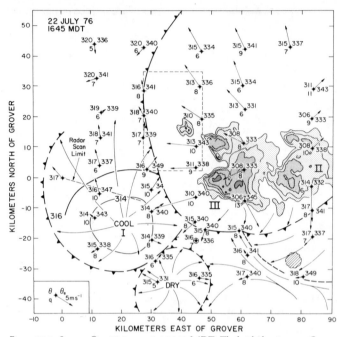

Figure 17.9 Same as Fig. 17.2, except at 1645 MDT. The hatched region near Grover coordinates (80 km, − 30 km) represents the location of a newly developing convective storm, which was only partially being scanned by the CP-2 radar at this time.

It can be noted in Fig. 17.9 and in previous figures that Storms II and III do not appear in the center of the cold dome. This asymmetry is believed to be a result of their southward motion, rather than the tendency for the inflow to slow down the southward advance of the colder air. In fact, according to Clarke (1961) and Goff (1976) the momentum in the inflow can be ignored in discussing the advance of the gust front, though Miller and Betts (1977) do note a weak dependence on inflow speed.

To the south of the storms, photographs from aircraft show that a new cloud was beginning to develop near the southeastern corner of the network. The position of this new cloud at 1645 is indicated schematically in Fig. 17.9. The formation of this cell well in advance of the outflow boundaries from Storms II and III suggests that gust fronts were not a factor. Its initiating mechanism may have been the persistent line of boundary layer convergence that existed to the southeast of Storm III, coupled with the high moisture content of the air entering the network from the southeast. Since the moisture feeding Storm III was confined to a relatively narrow sector, the development of this new cell in an

northwest was at first accompanied by a decrease in speed, which then increased back to 11 m s^{-1} by 1648. The potential temperature dropped rapidly to nearly 305 K and was accompanied by a strong increase in mixing ratio to more than 13 g kg^{-1} (85 % relative humidity). The surface pressure also began to increase, rising about 1.6 mb as the storm passed overhead. Rain started falling at 1644 and continued until about 1710 with a total of 11 mm recorded. By 1656 the winds had decreased below 2 m s^{-1} and the direction was shifting sharply back to southeasterly, indicating the passage of the downdraft core nearly directly over the site.

upstream position began to rob Storm III of its moisture supply. By 1715, as shown in Fig. 17.10, the combined outflow air from Storms II and III had spread to nearly all portions of the network and was acting to suppress further convective development there. Also, the new storm was developing in the path of Storm III, and by 1730 Storm III was dissipating. The next section gives further details of the evolution of the inflow structure using aircraft data collected near cloud base.

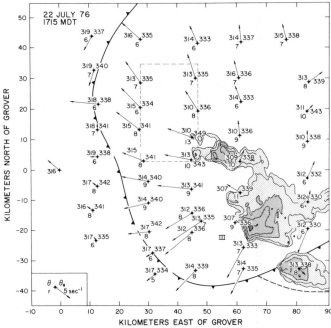

Figure 17.10 Same as Fig. 17.2, except at 1715 MDT. The new storm, near (80 km, − 33 km), is now clearly visible to the southeast of Storm III, and Storm II has dissipated.

17.5 AIRCRAFT MEASUREMENTS NEAR CLOUD BASE

Between 1615 and 1730 the Queen Air aircraft made 15 passes below cloud base, flying at altitudes between about 2.8 and 3.6 km, generally in the upper third of the subcloud layer. Figure 17.11 shows the track of the Queen Air during this period as it moved to the southeast, keeping just ahead of the storm. The dashed line on the figure represents the locations where updrafts greater than 2 m s⁻¹ were encountered. By positioning the track relative to the storm one can combine the data from different passes to infer the spatial characteristics. The method is valid as long as the structure does not change too much over the period of measurement.

Two time periods centered roughly on 1620 and 1640 have been chosen to illustrate in detail the evolving structure of the subcloud region. For two later periods, at 1650 and 1710, the further evolution is illustrated schematically. The data for the first period are presented in Fig. 17.12. The upper diagram shows a plan view of the aircraft track relative to the 1618 radar echo (the track has been adjusted to remove

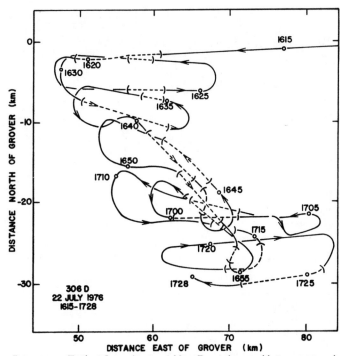

Figure 17.11 Track of Queen Air aircraft N306D over the ground between 1615 and 1728 MDT. Dashed lines represent the locations of regions of updraft greater than 2 m s⁻¹.

the motion of the storm). The wind vectors along the aircraft track are shown in their normal sense relative to the ground. The lower diagram shows time histories of five meteorological parameters observed by the aircraft during this period. Reference values of potential temperature (θ = 315.5 k), mixing ratio (q = 9.0 g kg⁻¹) and equivalent potential temperature (θ_e = 343 K) are shown on the traces to help identify the various characteristics.

During the time period shown, from 1617 to 1630, the aircraft made three passes in the inflow sector. These passes were made at slightly different altitudes as well as varying distances from the storm. Pass 1 was flown very near cloud base, 2-3 km south of the precipitation curtain. A broad updraft region extending about 13 km in the east-west direction was encountered. Two updraft maxima, with peak velocities of 8 and 10 m s⁻¹, respectively, are evident in the vertical velocity trace and are associated with Cells G and H, described in Chapter 16.

Pass 2 was flown at a slightly lower altitude near the southern edge of the updraft region. The updraft velocities were considerably weaker on this pass, and only one peak, at 1623, exceeded 4 m s⁻¹.

Pass 3 was flown directly along the leading edge of the precipitation curtain, about 0.8 km below cloud base. The sudden drop in potential temperature at 1627:10, accompanied by a northerly wind shift, is an indication that the aircraft had entered outflow air. The approximate location of the leading edge of this cooler air at 3 km MSL is indicated in the figure by the cold front symbol. The surface gust front at

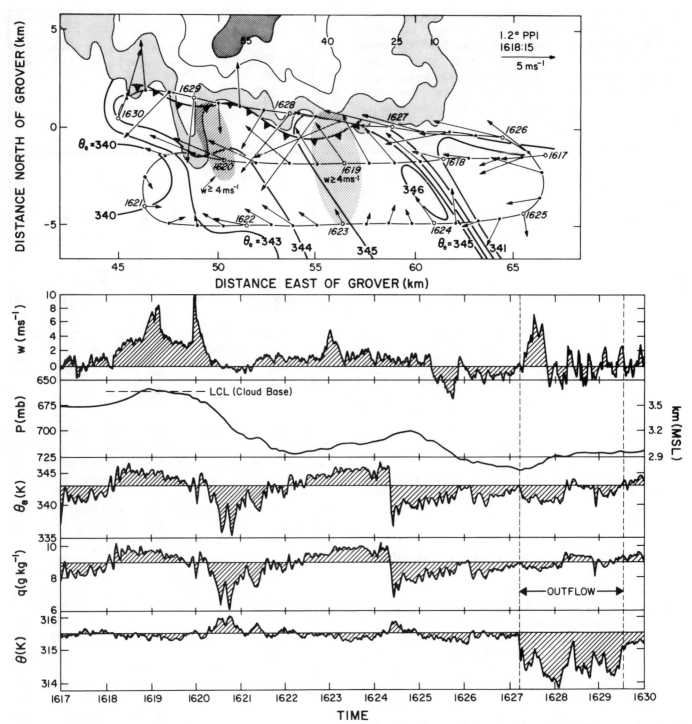

Figure 17.12 Subcloud aircraft data for the period 1617-1630 MDT. The upper panel shows the aircraft track, adjusted for storm motion and drawn relative to the radar PPI at 1618:15. Ground-relative wind vectors are plotted along the track at 15-s intervals. The shaded elliptical regions represent the locations of updrafts greater than 4 m s⁻¹. A composite of the field of equivalent potential temperature near cloud base is shown with the bold solid lines. The cold front symbol near the southern edge of the PPI represents the estimated position of the outflow boundary near cloud base. The surface position of the gust front at this time lies aproximately beneath the aircraft track from 1619 to 1620. The lower panel shows plots of five quantities observed by the aircraft during the three passes.

this time is estimated to be just south of the aircraft track for pass 1 (1619 to 1620).

The shaded elliptical regions represent a composite of the updraft cores ($w \geq 4$ m s⁻¹) near cloud base for Cells G and H based on the three passes. The cores appear to be elongated

in the north-south direction approximately along the direction of relative inflow into the storm, although their shape is not defined very well by the data.

The winds in the updraft region were generally out of the southeast, similar to the surface flow, with speeds ranging up to 5 m s⁻¹. Velocities relative to the storm were roughly twice this magnitude. Between passes 1 and 2 (some 3 to 4 km apart) the horizontal component of the wind is seen to increase from an average of 2 m s⁻¹ to 5 m s⁻¹. A velocity change of this magnitude could be produced by a pressure difference as small as a few tenths of a millibar.

Equivalent potential temperatures in the updraft region were mostly greater than 343 K, with the highest values (near 346 K) tending to be found on the east side of the updraft. Such values are consistent with those found near the surface, as discussed in the previous section. To the east and west of the updraft region the environmental air was characterized by mixing ratios less than 8 g kg⁻¹ and equivalent potential temperature generally less than 341 K.

The structure of the updraft region near cloud base was different in the 1640 time period shown in Fig. 17.13, though the pattern of θ_e was similar. Since the aircraft track extended farther to the southeast during this second period, the extent of the region of high θ_e is better defined at this time. Indeed, as implied also in Fig. 17.12, the region of $\theta_e > 345$ K in Fig. 17.13 extends well to the southeast into the inflow sector, and is much more elongated than the area of strong updraft. In a study of a similar storm, Fankhauser (1976) showed a comparable region of enhanced θ_e extending into the inflow at cloud base altitude. In both cases this high-θ_e air observed in front of the updraft is apparently a result of deepening of the moist surface layer by convergence in the inflow sector and weak ascent. Ragette (1973) has also reported observations of such ascent ahead of a storm. A region of weak surface convergence in the inflow region was described in the previous section.

It is evident from Fig. 17.13 that the broad maxima in w and θ_e do not coincide, the highest θ_e's being located to the southeast of the updraft core. This is different from the cases in Fankhauser (1976) and others, and it is natural to ask under what conditions there should be a strong positive correlation between updraft velocity at cloud base and θ_e in the inflowing air. In a steady state situation one might possibly expect the inflow air with the highest θ_e to have the greatest vertical velocity at cloud base since that air becomes the most unstable aloft. In this case, however, the gust front in the subcloud region probably forces the vertical motion below cloud base, decreasing the correlation between updraft and θ_e that might otherwise be expected. Dry air located just east of the updraft core is entering Storm III just ahead of the precipitation curtain, accounting for the gradient in θ_e across the updraft core at cloud base. Much of the air being forced upward at cloud base is not quite as potentially unstable as that which lies just ahead of the storm. The role of the gust front is discussed further in the next section.

It is interesting to note two implications of this mismatch between updraft velocity at cloud base and potential instability. Radar chaff introduced into the region of maximum updraft at cloud base with the aim of tracking it aloft to deduce storm structure (e.g., Marwitz, 1973), may well not enter the updraft core aloft. The same complication could influence the results of cloud seeding in the location of the updraft maximum at cloud base (e.g., Foote and Knight, 1979). This may be an explanation for the result reported by Linkletter and Warburton (1977).

According to the sounding in Chapter 16, Fig. 16.4, air with θ_e less than about 343 K has potentially about 2 °C of temperature excess, and is thus only marginally unstable. This includes the air in roughly the northern third of the updraft. In the north edge of the updraft where $\theta_e \leq 341$ K, the air will not have much buoyancy at any altitude, and the vertical motion there may be primarily a result of lifting by the outflow boundary.

Comparison of Figures 17.12 and 17.13 yields estimates of the changes in the position and structure of the updraft and the outflow boundary. Maximum updrafts were 6-8 m s⁻¹, about the same as at 1620. However, the position of the updraft core, shown stippled in the upper part of the figure, has moved several kilometers farther south from the echo. It may also be somewhat more elongated than in Fig. 17.12, though as previously stated the updraft shape is not well defined by the data.

The position of the outflow interface is also farther from the echo at 1640. For example, on the southeastward leg shown in Fig. 17.13, the aircraft encountered outflow air between 1639:30 and 1640:40 at an altitude of approximately 3.1 km (position of the outflow boundary is shown by a darkened cold front symbol). Since the aircraft was some 200 m higher at this time than the earlier pass at 1628, we may conclude that the outflow boundary had moved several kilometers south with respect to the storm at the aircraft altitude. In fact, the surface position of the gust front (indicated in Fig. 17.13) had also moved 3-4 km farther away from the storm, a fact that can also be seen by comparing the gust front positions in Figs. 17.6 and 17.10. The shift in position of the gust front and updraft are probably casually related, as discussed in more detail in the next section.

On the return pass to the northwest when the aircraft encountered the outflow boundary at 1646:45 (second darkened cold front symbol in Fig. 17.13) it was some 200 m higher than at 16:40. The distance between these last two positions of the outflow boundary as determined by the aircraft is about 3 km, which would suggest an interface slope of about 4°, though if the interface were irregular, as seems likely, a slope determined in this way would have less meaning.

To the east of the updraft core in Fig. 17.13 a region of drier air ($\theta_e \leq 341$ K) is shown schematically. It was sampled on other aircraft traverses, though not on the passes included in Fig. 17.13. The source of this drier air, with a water vapor mixing ratio 1-2 g kg⁻¹ less than the ambient southeasterly inflow, may be inferred from the relative surface flow in Fig. 17.7, where it is apparent that air with the appropriate equivalent potential temperature and momentum existed to the south of the dissipating Storm II. With time this drier air spread across the entire subcloud updraft region of Storm III.

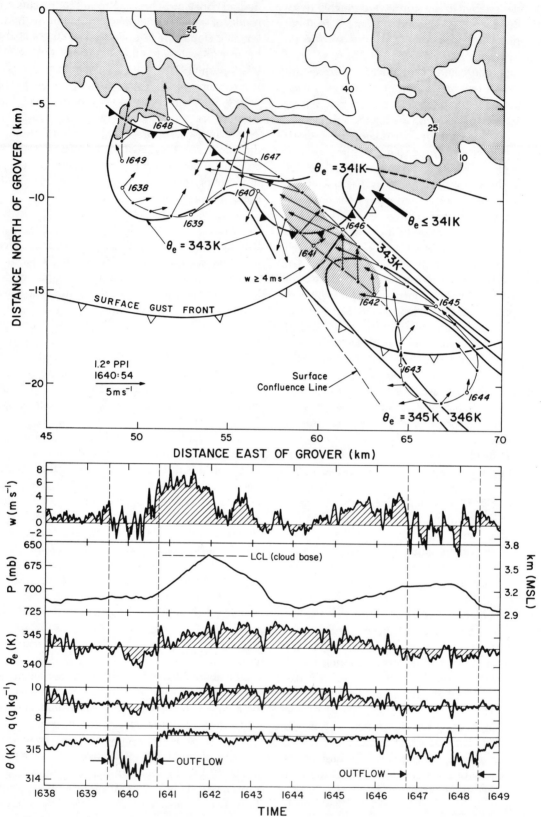

Figure 17.13 Same as Fig. 17.12, for the period 1638-1649. The darkened cold front symbols represent the position of the leading edge of the storm outflow near cloud base for the two aircraft passes as discussed in the text. The difference in their location is believed to be a reflection of the shallow slope of the gust front in this region. The surface position of the gust fronts from Storm II and Storm III is shown using the open frontal barbs. Note that the updraft core near cloud base lies just to the northeast of the intersection of these two surface boundaries. The θ_e composite shows that higher θ_e values are located southeast of the main updraft region, and that lower θ_e's are intruding into the subcloud region from the east.

This is illustrated in Fig. 17.14 with composites centered on 1650 and 1710. The bold arrow represents the likely trajectory of this dry air lifted over the combined outflow boundary from both storms. At 1650 the leading edge of the drier air was located through the eastern one-third of the updraft region, with moist southeasterly flow still present on the west. Visual evidence of this dry intrusion was noted by the observer on board the Queen Air who recorded that the cloud base had become considerably higher on the north and northeast sides of the updraft.

By 1710 the relative inflow was entirely from the east and there was no longer any evidence of the moist southeasterly air that had been feeding the storm up to this point. From Fig. 17.14 it appears that air entering the updraft region was forced to pass around outflow from the new cell that had developed to the southeast of Storm III. The combination of the blocking of the southeasterly inflow by the new cell and the switch to the drier easterly inflow (factors not likely to have been indpendent) then lead to the gradual decline and dissipation of Storm III during the next half hour.

In this section we have documented the structure of the updraft near cloud base. In spite of the persistence of the overall radar echo pattern at this time, discussed in detail in Chapter

16, a gradual progression in updraft structure is noted, including a relative shift in the position of the updraft core, apparently as a result of a similar movement of the gust front, and a drying trend as air from the east gradually replaced the moist southeasterly inflow that had been entering the storm. Certain aspects of these and other observations are considered further in the next section.

17.6 DISCUSSION

Few if any studies of convective storms have been successful in understanding the factors that influence the details of their structure, motion, and cellular development. The motion of a storm as a whole is of course influenced by both (1) the translation of individual cells comprising the storm, and (2) the formation of new cells in preferred locations with respect to the parent storm. The former can be a combination of simple advection and continuous propagation. With regard to the latter, the forcing of new cells by storm outflows is often cited as the dominant mechanism, following Humphreys (1940), Byers and Braham (1949), and others, though there are few field studies that actually demonstrate such a connection.

In this chapter, we have considered the evolution of a storm system in the context of observations made at the surface and by aircraft flying near cloud base. The storm evolution had certain dominant features, described in Chapter 16, and it also had a well-developed surface outflow, documented here. It seems worthwhile, then, to consider further the possible effects of the gust front, and, more generally, to summarize which of the observed storm characteristics seem explainable by the subcloud measurements presented in the previous sections.

17.6.1 Structure and Primary Effects of the Gust Front

It is interesting to note from Figs. 17.12 and 17.13 that at both 1620 and 1640 the updraft core is found approximately centered on the position of the gust front at the surface. The relative positioning is seen somewhat more clearly in Fig. 17.15, which shows a vertical section oriented along the inflow direction and including data from the 1640 aircraft pass from Fig. 17.13. The wind vectors in the plane of the figure are shown in storm-relative coordinates, with strong updrafts in the general region of the gust front, and anomalous momentum and weakening updrafts as one moves behind the outflow boundary. The surface position of the gust front is known from mesonet observations, but the shape of the interface in cross section was not measured. The shape in-

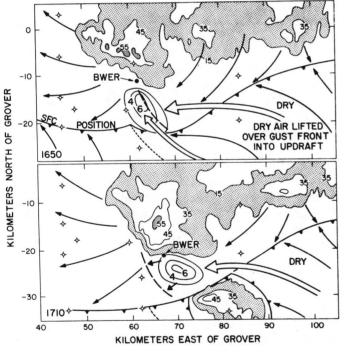

Figure 17.14 *Composites summarizing the evolving kinematic structure of the subcloud region of Storm III for 1650 and 1710 MDT. Each composite represents a synthesis of aircraft and mesonet data for a 20-min period centered on each time shown. Solid arrows show the storm-relative surface flow, while the larger open arrows represent the path of the air being lifted above the gust front and entering the updraft region of the storm. Updraft regions greater than 2 m s^{-1} in magnitude are represented by the concentric solid lines with values of 4 and 6 m s^{-1} labeled. The dashed line, oriented approximately northwest to southeast, represents the leading edge of the drier air near cloud base which is spreading westward across the updraft region. Mixing ratios behind this interface are approximately 2 g kg^{-1} lower than the air ahead of it. The presence of drier air in the inflow region of the storm after 1700 manifested itself in higher cloud bases, observed from the subcloud aircraft.*

dicated in Fig. 17.15 is based on work by Simpson (1969), Charba (1974), and Goff (1976), and on similar descriptions by others. The maximum speeds in the outflow (occurring perhaps a few hundred meters above the surface) tend to be greater than the speed of advance of the gust front itself, so that some air is deflected upward at the front to form the leading "head." The head is typically 1000-1500 m in depth (Ragette, 1973; Charba, 1974; Goff, 1976; Ellrod and Marwitz, 1976). Though the high-speed outflow is apparently confined to a relatively shallow layer, perhaps 500 m or so thick, there is a turbulent zone behind the head where mixing between inflow and outflow air occurs. Benjamin (1968) attributes this turbulence to a breaking "head wave," which appears as a shearing instability in the tank experiments of Simpson (1969, 1972). On other legs of the flight the Queen Air passed in and out of cold pockets of air with anomalous momentum, suggesting that it was flying just above the interface, occasionally encountering eddies that extended somewhat higher than normal. Simpson (1969) emphasized that the shape of the interface at both the leading edge and the mixing zone are quite unsteady, a point also supported by the variety of gust front profiles observed by Colmer (1971) and Goff (1976).

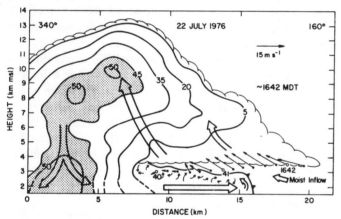

Figure 17.15 *Vertical section through Storm III at 1642 MDT, showing the relationship of the gust front, the winds measured by the aircraft, and the storm echo. The vertical section is oriented along the direction of storm motion (toward 160°), and the aircraft winds are shown in storm-relative coordinates. The structure of the gust front is estimated from its position over the mesonet as well as its position derived from the subcloud aircraft data.*

The importance of the gust front to storm evolution is linked to its ability to force air upward. The induced updraft to be expected from the passage of inflow air over the gust front can be estimated from the work of Simpson (1969) on density currents. In the important forward region of the front he finds that the flow is in good agreement with that of an ideal fluid over an obstacle of such a shape. This means that under typical conditions the induced updraft may be as large as 5-10 m s^{-1}, very close to the leading edge of the gust front, but that at heights of more than twice the obstacle height the induced updraft will be small, on the order of 1 m s^{-1} or less. The latter does not agree with the numerical results of

Mitchell and Hovermale (1977), who found induced updrafts extending to greater heights. However, it is clear by inspection of their streamline patterns that their general solutions are strongly influenced by the rigid lateral boundaries assumed in the model, and their results, at least in regard to updraft extent, may not be realistic.

According to the results of Simpson (1969), it might be reasonable to use two-dimensional solutions for potential flow over obstacles to estimate how far air can be lifted by passage over a gust front. The amount of lifting to be expected depends on the initial height of the air parcel at some large distance upstream, and the size and to some extent shape of the obstacle. Under conditions of potential flow, the lifted distance, L, is equal to the obstacle height, b, for air initially at the surface, and decreases uniformly for air located initially at higher altitudes. If the obstacle can be represented by the upper half of a cylinder of circular cross section then one can compute that L decreases to 0.62 b for air initially at the obstacle height b upwind, and L drops further to 0.41 b for air initially at height 2 b. If one chooses for the obstacle geometry an elliptical cross-section with vertical to horizontal axial ratio of 1/2, then a streamline at initial height b is lifted by the amount 0.72 b, and if the axial ratio is instead 1/3, it is lifted a distance 0.78 b. The facts that real gust fronts are three-dimensional and that the lower troposphere is slightly stable both tend to make the above numbers overestimates. Nevertheless it seems reasonable to expect that lifting by the gust front can deflect air initially at an altitude corresponding to the gust front depth, b, by approximately half to three-quarters of that amount.

Thus, if one considers a gust front 1700 m deep, as seems reasonable for the present case based on the aircraft measurements, the direct effect of the gust front is to lift surface air some 1700 m, and lift air initially at 1700 m AGL by perhaps 850 to 1300 m. Interestingly, this degree of mechanical lifting is about the same as the amount, H, required to bring an air parcel initially at 1700 m AGL to its level of free convection (LFC). For example, if one considers air with $\theta_e = 346$ K, then $H = 1100$ m according to the sounding in Fig. 16.4.

One would expect a more important role for the gust front when the frontal lifting raises an air parcel above its LFC, for then a traveling gust front would be able to initiate new convection even when it outruns its parent storm. When the lifting is not sufficient to bring an air parcel to its LFC the degree of control the gust front can exert would be more restricted, though probably not negligible. If the gust front is appropriately positioned, it can play a significant role in helping to push inflow air through the slightly stable boundary layer, a process usually thought of as being accomplished by non-hydrostatic pressure gradients (Marwitz, 1973; Foote and Fankhauser, 1973; Ellrod and Marwitz, 1976). Hane (1973), for example, observed from his numerical simulations that the most suitable position of the gust front from the standpoint of supporting the storm circulation is likely to be right under the updraft core, or perhaps slightly forward. These considerations then provide a plausible basis for arguing that storms like the present one "attempt," within limits, to organize themselves so as to maintain the updraft at cloud base over or very near the surface gust front. The movements of both updraft and gust front noted in the previous section seem to be an example of this, with a relative shift in position of the gust front apparently causing a similar shift in the updraft location.

17.6.2 Effect of Gust Front on Storm Intensification and Vaults

It is important to ask whether there were observable changes in the echo structure of the storm attributable to the presence of the gust front or to changes in its position. To help answer this, Fig. 17.16 was drawn to compare the timing of the gust front advance with certain echo characteristics. The approximate times that the various cells identified in Chapter 16 were present are also indicated. The major intensification of the storm during its early history occurred with Cell D at about 1535, as shown in the middle panel of Fig. 17.16. This coincides well with the earliest detection of spreading downdrafts in the mesonet. It is believed that these two observations are related. Indeed, it may be argued that a relationship between the earliest storm-scale downdraft and thunderstorm intensification is commonly important in the High Plains. The impression gained by investigators during

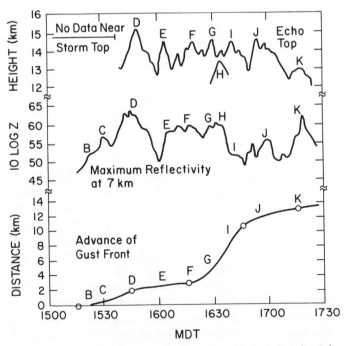

Figure 17.16 Plot of the advance of the gust front ahead of the leading edge of the precipitation curtain of Storm III. Reflectivity histories of Storm III in the upper two curves are repeated from Chapter 16. The letters along the curves represent various cells identified in Chapter 16.

cloud flights in NHRE (including those reported here, and by Foote and Fankhauser, 1973, Fankhauser, 1976, and Browning and Foote, 1976) is that a storm never becomes very vigorous (for example, as gauged by the presence of an organized and vigorous updraft at cloud base and a subjective judgement of the intensity of precipitation) until precipitation has fallen for a long enough time to generate a downdraft and surface outflow. The gust front then provides a steady forcing mechanism that organizes the low-level updraft. If the updraft has been on the back side of the storm, as described, for example, by Dennis et al. (1970), and Auer et al. (1970), then it is at this stage that it moves to the front side. If these ideas, which are similar to those proposed by Newton (1963) and others, are correct, then they imply an important role for the low-level outflow in both organizing and invigorating the young storm. It is also implied that surface measurements may have some predictive value in this regard, and further investigations using mesonet data seem in order.

The rapid movement of the gust front discussed in Section 17.5 in the context of the aircraft measurements is shown in Fig. 17.16 to start at about 1620. Cells I, J, and K occurred after this time, and Cells F and G of the southeastern branch formed when the gust front was closer in to the precipitation curtain. With reference to the analysis in Chapter 16, the only significant difference in the echo morphology and evolution between these cells was that I, J, and K all displayed transient weak-echo vaults, while F and G did not. In terms of cell size and echo top height the various cells were all fairly similar. The maximum reflectivities of I and J were somewhat smaller than F and G, though that of K was comparable. It is possible that the motion of the gust front away from the storm during I-J-K caused the low-level updrafts to be somewhat stronger, thereby giving rise to the occasional vaults. However, the general conclusion seems to be that the storm was able to adapt without major structural response to changes of as much as 10 km in the relative position of the gust front.

The problem of the compatibility of storm speed and gust front speed has been emphasized by Moncrieff and Miller (1976), and a number of numerical cloud simulations have also illustrated the importance of the gust front as described here (e.g., Hane, 1973; Orville and Kopp, 1977; Thorpe and Miller, 1978; Klemp and Wilhelmson, 1978).

17.6.3 Variations in Inflow Moisture

Certain conditions are necessary if storms are to maintain themselves. Of the various factors that influence storm behavior, the most critical are the availability of air that is convectively unstable and a suitable environmental wind profile. If either of these two factors disappears, a storm will dissipate. In the High Plains the instability tends to depend mostly on the low-level humidity. Surface mixing ratios can easily change by several grams per kilogram in only a few tens of kilometers, as seen in the present study, and this is often enough to make the difference between convectively unstable air and stable air.

The development of various regions of drier air in the inflow sectors of Storms I and II and the southwest branch of Storm III (Cell H, Fig. 16.10) was documented in Section 17.4. Perhaps the dominant feature was the dissipation of the storms. The time correspondence is good enough that the dissipation of Storms I and II and the southwest branch of III seems adequately explained by the decrease in low-level moisture below values necessary for convective instability. If this result should turn out to be general, then surface mesonet observations should be useful in predicting such behavior.

The situation with regard to Storm III is slightly more complicated. By 1715 (as shown in Fig. 17.10) a new cell had formed some 20 km to the southeast of Storm III and its outflow was blocking the previous southeasterly inflow. As inferred from the aircraft data (Fig. 17.14), the inflow to Storm III then passed around the east side of the developing cell and entered the updraft with easterly and later northeasterly momentum relative to the ground. This air was much drier, with a θ_e of only 341 K or so, and had lessened convective instability. Because of its change in direction the relative mass influx to the storm decreased by almost half, though this was probably a secondary effect in the present case. The dissipation of Storm III can thus be attributed to a combination of a reduced inflow followed by the ingestion of dry air.

The reason for the appearance of the dry air in the network is not clear. One possibility is that compensating downdrafts in the storm environment may have brought down drier air, as hypothesized for example in the study of Feteris (1961). The physical explanation given by Feteris, that of subsidence caused by evaporation under the anvil, cannot be correct for the present case, though, as the position of the dry air to the south and southeast of the storms does not correspond to the position of the anvil, which extended to the northeast. Evidence supporting the existence of compensating downdrafts in the vicinity of storms has been presented by Fritsch (1975) and interpreted by Hoxit et al. (1976) as a mechanism for organizing and invigorating storms. However, these authors do not visualize such downdrafts as extending to the surface, as would appear necessary in the present case. Apart from the rapid drying that occurred just southeast of Storm I, the general drying trend in the inflow sector may have simply been part of the normal diurnal cycle with lower absolute humidities by late afternoon in association with the mixing taking place in the boundary layer.

The upper panel of Fig. 17.16 shows that the maximum echo heights for two cells are significantly lower than for the others. It is perhaps significant that these two cells, H and K, grew in regions and at times when the inflow air at the surface was much drier, resulting in weaker instability. Cell H was the last of the southwestern branch of Storm III, and Cell

K the last of the southeastern branch. In a similar vein, the most intense cell of the storm, in terms both of its size and maximum reflectivity, was Cell D. It existed at a time when the mesonet observations indicated that inflow moisture was at a maximum.

17.6.4 Storm Motion

The factors that control the motion of storms are not well understood, though many theories exist (e.g., Byers and Braham, 1949; Newton and Newton, 1959; Newton and Fankhauser, 1964; Browning, 1964; Fujita and Grandoso, 1968; Charba and Sasaki, 1971; Moncrieff and Green, 1972; Raymond, 1975; Moncrieff and Miller, 1976; Browning, 1977). During the early phase of Storm III the cells of the southwest branch tended to move with the mid-level wind toward the southeast, but because of discrete propagation on the west flank the storm as a whole moved to the southwest and then south (see Chapter 16). During its steady phase (1615-1715) the storm propagated more or less continuously to the southeast. The reflectivity perturbations during this time tended to move slowly toward the north. This is to be interpreted not as cell motion but as the motion of a batch of precipitation being processed in an updraft having southerly momentum.

Of course, the two obvious factors influencing propagation on the south side of Storm III were the presence of the gust front imbedded in an environment with southeasterly low-level flow, and suitable moisture to the south. The lifting effected by the spreading cold air would have been most vigorous on the south side of the storm, and one would expect propagation, whether continuous or discrete, to occur there (Thorpe and Miller, 1978, also stress this). Support for the importance of such a process would seem to be strengthened by noting that as the inflow direction changed from southwest to southeast the direction of storm motion changed accordingly, always moving into the inflow. On the other hand, as already noted, this storm motion was a combined result of propagation and translation. Often the new cells formed on the west and even northwest of the parent storm, rather than on the southwest and south as the simple convergence argument would predict. In addition, as we have stressed, the moisture distribution probably played a dominant role in determining which of many three-dimensional disturbances induced along the periphery of the gust front could grow to large amplitude.

The above mechanism does account for the common (but not universal) observation that storms tend to propagate into the inflow. Though it may turn out that factors other than convergence in front of the outflow boundary are responsible, such a simple forecast rule deserves further checking, particularly in light of the relative ease with which measurements in the surface layer can be made.

17.7 CONCLUDING REMARKS

In this chapter we have continued the description of the storm as begun in Chapter 16, emphasizing the characteristics of the airflow in the surface layer and near cloud base. Though the general importance of gust fronts has been emphasized by a number of previous investigators, very little detailed discussion of mechanisms has been presented. An examination of the physical effects of gust frontal lifting indicates` that it probably does play a significant role in the dynamical organization of storms even though the amount of direct lifting is rather limited (1 to 2 km). During a rapid movement of the gust front in the present case it was observed that the updraft at cloud base was also able to move, apparently in response, without causing major structural changes in the storm's reflectivity pattern. The initial major intensification of the storm took place just after the first detection of spreading downdrafts at the surface, and it seems likely that the organized, storm-scale lifting over this dome of cold air leads to the storm invigoration. Much more needs to be learned about the effects of gust fronts. Though investigators have claimed relationships between such things as lifting over the gust front and the formation of new cells, the arguments remain plausible but largely undocumented. Also, the role of the gust front in current numerical cloud models has not been clarified, partly because of the poor resolution of the models in the vertical relative to the depth of the surface outflow.

The general factors influencing storm longevity should be a matter of some interest, because of such things as the forecast problem and the relationship between total storm precipitation and storm lifetime, but the subject has received little attention. Two phenomena were noted here as being commonly important for the dissipation of long-lived storms: (1) the lack of suitably moist, unstable air, and (2) disruption of the storm inflow. The latter often arises from the growth of a new cell in a position that is upstream with respect to the low-level flow, with the result that the inflow is blocked or diverted (a secondary effect for the southeast branch of Storm III discussed here). If the gust front outruns the parent storm the inflow can also be effectively obstructed. In a similar vein, Brandes (1978) has documented how the occlusion of the mesolow in some Oklahoma storms can disrupt the inflow. The dominant factor causing the dissipation of Storms I and II and both branches of Storm III, however, was the decrease in moisture of the inflowing air. It is worth emphasizing that rather small variations in local moisture (as little as 1 to 2 g kg^{-1} changes in mixing ratio) can have a profound effect on a storm's ability to persist.

Perhaps the more striking result here is not that such subtle moisture variations seem to be important and can occur over relatively small distances, but rather that these and other factors appear to be observable with a mesometeorological network. It is likely that with more common usage of such observational facilities, particularly in real time, many seemingly random aspects of storm behavior will become understandable and perhaps predictable.

References

Auer, A. H., D. L. Veal, and J. D. Marwitz, 1970: The identification of organized cloud base updrafts. *J. Rech. Atmos.* 6, 1-6.

Barnes, S. L., 1978: Oklahoma thunderstorms on 29-30 April 1970. Part I: Morphology of a tornadic storm. *Mon. Weather Rev.* 106, 673-684.

Benjamin, T. B., 1968: Gravity currents and related phenomena. *J. Fluid Mech.* 31, 209-248.

Brandes, E. A., 1978: Mesocyclone evolution and tornadogenesis: some observations. *Mon. Weather Rev.* 106, 995-1011.

Browning, K. A., 1964: Airflow and precipitation trajectories within several local storms which travel to the right of the winds. *J. Atmos. Sci.* 21, 634-639.

-----, 1977: The structure and mechanisms of hailstorms. *Meteorol Monogr.* 16 (38), 1-43.

-----, and G. B. Foote, 1976: Airflow and hail growth in supercell storms and some implications for hail suppression. *Q. J. R. Meteorol. Soc.* 102, 499-533.

Byers, H. R., and R. R. Braham, 1949: *The Thunderstorm.* U. S. Government Printing Office, Washington, D. C., 287 pp.

Charba, J., 1974: Application of gravity current model to analysis of squall-line gust front. *Mon. Weather Rev.* 102, 140-156.

-----, and Y. Sasaki, 1971: Structure and movement of the severe thunderstorms of 3 April 1964 as revealed from radar and surface mesonetwork data analysis. *J. Meteorol. Soc. Japan* 49, 191-214.

Clarke, R. H., 1961: Mesostructure of dry cold fronts over featureless terrain. *J. Meteorol.* 18, 715-735.

Colmer, M. J., 1971: On the character of thunderstorm gust fronts. Royal Aircraft Establishment, Bedford, England, 11 pp.

Dennis, A. S., C. A. Schock, and A. Koscielski, 1970: Characteristics of hailstorms of western South Dakota. *j. Appl. Meteorol.* 9, 127-135.

Ellrod, G. P., and J. D. Marwitz, 1976: Structure and interaction in the subcloud region of thunderstorms. *J. Appl. Meteorol.* 10, 1083-1091.

Fankhauser, J. C., 1976: Structure of an evolving hailstorm: Part II. Thermodynamic structure and airflow in the near environment. *Mon. Weather Rev.* 104, 576-587.

Feteris, P. J., 1961: The influence of the circulation around cumulonimbus clouds on the surface humidity pattern. *Swiss Aero-Rev.* 36(11), 626-630.

Foote, G. B., and J. C. Fankhauser, 1973: Airflow and moisture budget beneath a northeast Colorado hailstorm. *J. Appl. Meteorol.* 12, 1330-1353.

-----, and C. A. Knight, 1979: Results of a randomized hail suppression experiment in northeast Colorado: Part I. Design and conduct of the experiment. *J. Appl. Meteorol.* 18, 1526-1537.

Fritsch, J. M., 1975: Cumulus dynamics: local compensating subsidence and its implications for cumulus parameterization. *Pageoph* 113, 851-867.

Fujita, T., 1959a: Study of mesosystems associated with stationary radar echoes. *J. Meteorol.* 16, 38-52.

-----, 1959b: Precipitation and cold air production in mesoscale thunderstorm systems. *J. Meteorol.* 16, 454-466.

-----, 1963: Analytical mesometeorology: a review. *Meteorol. Monogr.* 5 (27), 77-125.

-----, and H. Grandoso, 1968: Split of a thunderstorm into anticyclonic and cyclonic storms and their motion from numerical model experiments. *J. Atmos. Sci.* 25, 416-439.

Goff, R. C., 1976: Vertical structure of thunderstorm outflows. *Mon. Weather Rev.* 104, 1429-1440.

Gurka, J. J., 1976: Satellite and surface observations of strong wind zones accompanying thunderstorms. *Mon. Weather Rev.* 104, 1484-1493.

Hane, C. E., 1973: The squall line thunderstorm: numerical experimentation. *J. Atmos. Sci.* 30, 1672-1690.

Hoxit, L. R., C. F. Chappell, and J. M. Fritsch, 1976: Formation of mesolows pressure troughs in advance of cumulonimbus clouds. *Mon. Weather Rev.* 104, 1419-1428.

Humphreys, W. J., 1940: *Physics of the Air.* McGraw-Hill, New York, N.Y., 676 pp. (Also reprinted by Dover Publications, New York, N.Y., 1964).

Keulegan, G. H., 1958: The motion of saline fronts in still water (12th progress report on model laws for density currents), U. S. National Bureau of Standards, Dept. of Commerce, Washington, D. C., 29 pp.

Klemp, J. B., and R. B. Wilhelmson, 1978: Simulations of right- and left-moving storms produced through storm splitting. *J. Atmos. Sci.* 35, 1097-1110.

Kropfli, R. A., and L. J. Miller, 1976: Kinematic structure and flux quantities in a convective storm from dual-Doppler radar observations. *J. Atmos. Sci.* 33, 520-529.

Lenschow, D. H., 1976: Estimating updraft velocity from an airplane response. *Mon. Weather Rev.* 104, 618-627.

-----, B. B. Stankov, and L. Mahrt, 1979: The rapid morning boundary-layer transition. *J. Atmos. Sci.* 36, 2108-2124.

Linkletter, G. O., and J. A. Warburton, 1977: An assessment of NHRE hail suppression seeding technology based on silver analysis. *J. Appl. Meteorol.* 16, 1332-1348.

Marwitz, J. D., 1973: Trajectories within the weak echo regions of hailstorms *J. Appl. Meteorol.* 12, 1174-1182.

-----, A. H. Auer, and D. L. Veal, 1972: Locating the organized updraft on severe thunderstorms. *J. Appl. Meteorol.* 11, 236-238.

Miller, M. J., 1978: The Hampstead storm: a numerical simulation of a quasi-stationary cumulonimbus system. *Q. J. R. Meteorol. Soc.* 104, 413-427.

-----, and A. K. Betts, 1977: Traveling convective storms over Venezuela. *Mon. Weather Rev.* 105, 833-848.

Mitchell, K. E., and J. B. Hovermale, 1977: A numerical investigation of the severe thunderstorm gust front. *Mon. Weather Rev.* 105, 657-675.

Moncrieff, M. W., and J. S. A. Green, 1972: The propagation and transfer properties of steady convective overturning in shear. *Q. J. R. Meteorol. Soc.* 98, 336-352.

-----, and M. J. Miller, 1976: The dynamics and simulation of tropical cumulonimbus and squall lines. *Q. J. R. Meteorol. Soc.* 102, 373-394.

Newton, C. W., 1963: Dynamics of severe convective storms. *Meteorol. Monogr.* 5 (27), 33-58.

-----, and J. C. Fankhauser, 1964: On the movements of convective storms, with emphasis on size discrimination in relation to water-budget requirements. *J. Appl. Meteorol.* 3, 651-668.

-----, and H. R. Newton, 1959: Dynamical interactions between large convective clouds and environment with vertical shear. *J. Meteorol.* 16, 483-496.

Orville, H. D., and F. J. Kopp, 1977: Numerical simulation of the life history of a hailstorm. *J. Atmos. Sci.* 34, 1596-1618.

Purdom, J. F. W., 1973: Meso-highs and satellite imagery. *Mon. Weather Rev.* 101, 180-181.

-----, 1976: Some uses of high-resolution GOES imagery in the mesoscale forecasting of convection and its behavior. *Mon. Weather Rev.* 104, 1474-1483.

Ragette, G., 1973: Mesoscale circulations associated with Alberta hailstorms. *Mon. Weather Rev.* 101, 150-159.

Raymond, D. J., 1975: A model for predicting the movement of continuously propagating convective storms. *J. Atmos. Sci.* 32, 1308-1317.

Simpson, J. E., 1969: A comparison between laboratory and atmospheric density currents. *Q. J. R. Meteorol. Soc.* 95, 758-765.

-----, 1972: Effects of the lower boundary on the head of a gravity current. *J. Fluid Mech.* 53, 759-768.

Thorpe, A. J., and M. J. Miller, 1978: Numerical simulations showing the role of the downdraught in cumulonimbus motion and splitting. *Q. J. R. Meteorol. Soc.* 104, 873-893.

Tepper, M., 1950: A proposed mechanism of squall lines: the pressure jump line. *J. Meteorol.* 7, 21-29.

Ulanski, S. L., and M. Garstang, 1978: The role of surface divergence and vorticity in the life cycle of convective rainfall: Part I. Observations and analysis. *J. Atmos. Sci.* 35, 1047-1062.

Zipser, E. J., 1977: Mesoscale and convective-scale downdrafts as distinct components of squall-line structure. *Mon. Weather Rev.* 105, 1568-1589.

CHAPTER 18

The 22 July 1976 Case Study: Storm Airflow, Updraft Structure, and Mass Flux from Triple-Doppler Measurements

Harold W. Frank and G. Brant Foote

18.1 THE NATURE OF THE DATA

On 22 July, successive scans from the four Doppler radars were begun at 5-min intervals, each requiring about 4 min. Except for occasional missed scans, this sampling schedule was maintained while observing the storm from 1551 to 1724. The discussion here is restricted to three consecutive volume scans between 1623 and 1641 MDT and three between 1704 and 1719. The two periods correspond to the development of Cell I and of the K series (see Fig. 18.1). Data from the CP-4 radar were not used in this analysis (see Appendix), except qualitatively. Using software of Kohn et al. (1978), we interpolated measurements obtained in each volume scan by the other three radars to a common grid, shifted slightly to account for storm motion during the scan times and then combined to determine horizontal (u, v) components of air velocity. The vertical air velocity, w, was estimated by assuming $w = 0$ near the top of the data region and applying the mass continuity equation in a downward integration. The main emphasis here is on the structure in the middle parts of the cloud, so the volume scans have been assigned times appropriate for the mid-level w fields, which rely on data from the upper half of the scanned volume.

Errors in the air velocity components, discussed in the Appendix, are thought to have root-mean-square values (over the entire data set discussed) of 2 to 3 m s⁻¹ for u and v and about 5 m s⁻¹ for w. Some sources of error are highly dependent on position, and maximum errors in w could be in excess of 10 m s⁻¹. The following section and the Appendix include comparisons of updraft measurements obtained from the Doppler radars and from the penetrating T-28 aircraft. These are in reasonable agreement, lending support to the major velocity features evident in the Doppler data, though for reasons discussed in the Appendix the comparison is not entirely straightforward.

Spatial resolution of the field of motion should be adequate to represent three-dimensional features (e.g., updrafts) larger than 6 km in diameter at better than 90% of their original amplitude (see Appendix). Features of smaller scale are attenuated more; for example, features less than 2 km across are damped by more than 50%. Distortion of the shorter-lived and generally smaller features of the velocity field is also likely as a result of the 4-min period needed to scan the whole storm.

Doppler measurements were made in most of the radar echo volume of Storm III, but much of the discussion focuses on a cubical region, 15 km on a side, located in the forward (south) part of the storm. We refer to this as the "active region," as it contains the intense updrafts and downdrafts. The positions of the active region over the ground for each of the six volume scans are shown in Fig. 18.1. All velocities are displayed relative to the storm, which moves from 330° at 8 m s⁻¹. Where sequences are shown the display area has been moved along with the storm.

18.2 THE GENERAL AIRFLOW PATTERN

Other parts of this case study illustrate some basic features of the storm that remained essentially unchanged for long periods relative to the lifetimes of individual impulses or internal variations of the storm structure. Disregarding variations that were evident only in the high-reflectivity regions (say, for reflectivity greater than about 30 dBZ$_e$), the basic radar echo pattern (Chapter 16) was persistent, consisting of an echo core with an overhang extending from it on the south and southwest sides. Penetrations of the updraft region by the T-28 aircraft (Chapter 19) indicate that the same basic

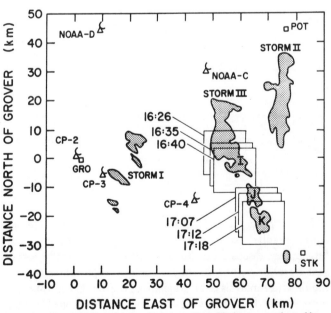

Figure 18.1 *Layout of the radar network relative to Storm III, the storm emphasized here. Doppler radars NOAA-C, NOAA-D, and CP-3 are used for triple-Doppler analysis. Envelopes of reflectivity in excess of 55 dBZ$_e$ are shown for Storm III and neighboring Storms I and II for the period 1500-1800 MDT on 22 July 1976. Positions of the so-called "active region" for six Doppler scans discussed are shown. The S-band CP-2 radar is used for reflectivity data. Rawinsonde launch sites are shown at Grover (GRO) and Sterling (STK), Colorado, and Potter (POT), Nebraska.*

updraft/downdraft configuration was present in association with three separate reflectivity impulses. Furthermore, in Chapter 17 it has been noted that within the limits of available measurements the area and magnitude of the updraft near cloud base were fairly constant, at least in the last half of the storm lifetime. In this section we address the persistent aspects of the three-dimensional airflow that these observations reflect.

Consider first the hodograph of the environmental wind in Fig. 18.2, taken from the 1627 Grover sounding; V_s is the storm motion in the period discussed here (approximately 1625-1720). In a coordinate system attached to the moving storm there is considerable veering with height in the lower levels (the 1605 Sterling sounding, dashed, is most representative here). The relative wind is south-southeasterly at about 10 m s^{-1} near the surface, westerly at less than 5 m s^{-1} in the middle troposphere, and southwesterly at 10 to 15 m s^{-1} in the upper troposphere.

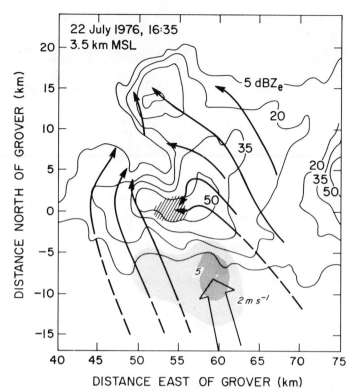

Figure 18.3 Storm-relative airflow and reflectivity structure at 1635 MDT at 3.5 km MSL, near cloud base. Streamlines (dashed where data are not present) and downdraft (hatched for w < −5 m s^{-1}) are from Doppler observations. The area of updraft near cloud base (stippled) is mapped from aircraft data obtained during several traverses near this time. Bold arrow on the south represents the southeasterly inflow below cloud base.

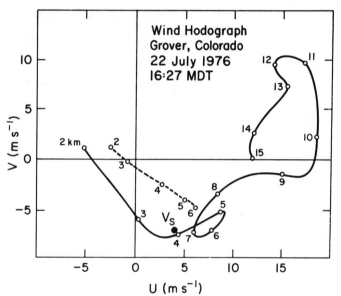

Figure 18.2 Hodograph of environmental winds from the GRO sounding at 1627 MDT (solid line), heights given in kilometers (MSL). The dashed line shows data from the 1605 STK sounding, which is more representative at low heights. Mean storm motion is represented by V$_S$.

To illustrate the basic storm airflow pattern we refer to some results from the 1635 Doppler volume scan. Streamlines discussed in this section do not generally represent trajectories, nor do they accurately depict the speed of the flow (an analysis of air trajectories follows in the next section). Figure 18.3 shows a streamline analysis from Doppler data at 3.5 km MSL for Storm III, as it has been termed in the preceding two chapters. The shaded region signifies updraft, as determined by Queen Air N306D flying below cloud base (see Chapter 17). Most of the moist air entering the storm is thought to pass through this region, but there is no echo there

at 3.5 km, and thus no Doppler data. The south-southeasterly motion in the inflow sector as measured by an aircraft is shown schematically by the bold arrow. The southeasterly relative flow seen in Fig. 18.3 tends to be dominated by the motion of the storm itself toward that direction.

An area of downdraft (hatched for $w < -5$ m s^{-1}) is situated in the region of highest reflectivity at this height, overlying an area of fairly strong horizontal divergence at the surface, nearly 5×10^{-3} s^{-1} (see Chapter 17). Confluent streamlines in the vicinity of the downdraft reflect convergence of air into it at this height, as will be discussed later. Part of the echo from the neighboring Storm II is seen on the right side of Fig. 18.3.

The updraft shown at cloud base is actually feeding two cells at the time of these data, as seen more clearly in Fig. 18.4 for 6.5 km altitude. Here the updraft has been determined from Doppler measurements. The two large areas of updraft, corresponding to Cells I and H as discussed in Chapter 16, are directly linked to the updraft at cloud base. The eastern updraft (Cell I) is much the stronger of the two at this time, with a maximum speed of 29 m s^{-1}. According to reflectivity analyses the western Cell H has already passed its maximum intensity. Several other small updraft cores shown in the

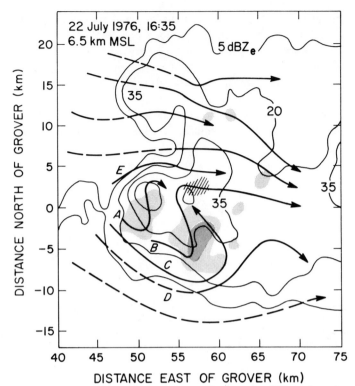

Figure 18.4 Storm-relative airflow as in Fig. 18.3 but at 6.5 km. Updrafts shown for this altitude, which is near the height of maximum updraft, are from Doppler data. The streamlines designated by letters are referenced in the text.

Heymsfield (1978), Brandes (1978), and Lemon and Doswell (1979). In a veering environment such a pattern should be expected when the storm updraft is large enough that inflow momentum tends to be conserved in the updraft, and the environmental winds aloft are strong enough to produce the blocking effect that leads to cyclonic curvature of streamlines on one side of the storm and anticyclonic curvature on the other. Rotunno (1981) has recently considered certain aspects of this configuration from a theoretical viewpoint.

The region of general confluence downwind of the major updrafts contains a weak anticyclonic eddy. Such an eddy in the lee of an updraft has been previously reported by Lemon (1976). In the present case this feature has a vertical extent of 1-2 km.

The horizontal section in Fig. 18.4, at an altitude of 6.5 km, is only 1 km below the level of maximum updraft, where horizontal divergence in the updraft is small. In contrast, Fig. 18.5 shows the streamlines at 9.5 km, about 2 km above the updraft maximum and 4 to 6 km below the storm top. The strongly divergent airflow in Fig. 18.5 illustrates the basic structure in the upper part of the storm. The streamline pattern is dominated by a source region on the south side of the storm, reflecting outflow from the major updraft. This source, superimposed on the southwesterly environmental flow at this altitude, produces a barrier-like flow pattern similar to that seen at 6.5 km. The anvil region is seen as an area of relatively weak reflectivity extending downwind from the updraft and reflectivity core.

figure are not persistent, but the larger-scale pattern is similar from one volume scan to the next.

The streamlines of the horizontal flow at 6.5 km show large perturbations in the vicinity of the updrafts, with two features most evident: (1) the tendency for air coming from the west to pass around the updrafts, as in streaming flow past a barrier, and (2) the southerly to southeasterly momentum apparent in the regions of updraft. These features have been previously reported in both field observations and numerical modeling results (e.g., Fankhauser, 1971; Kropfli and Miller, 1976; Heymsfield, 1978; Brandes, 1978; Moncrieff and Miller, 1976; Schlesinger, 1978, 1980; Wilhelmson and Klemp, 1981; Klemp et al., 1981). The latter seems to reflect conservation of horizontal momentum as boundary layer air is lifted into the storm.

When viewing streamlines A, B, and C, which pass around the south side of the storm and then turn into the updraft, it should be remembered that they are not air trajectories, because of the strong vertical component of vertical motion. Nevertheless, in Chapter 20 it will be shown that certain trajectories of hailstones do look something like this because of their large terminal velocity in the updraft. The S-shaped streamlines like A and B in Fig. 18.4 seem to be characteristic of large storms that occur in an environment for which the wind profile relative to the storm veers with height. Similar patterns can be seen in the data of Eagleman and Lin (1977),

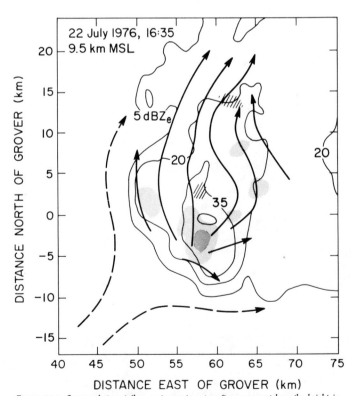

Figure 18.5 Storm-relative airflow as in previous two figures, except here the height is 9.5 km, near the height of maximum upper-level mass outflow.

Figure 18.6 shows the details of the Doppler wind field for an area 15 km on a side that focuses on the major updraft. In the vertical motion field (center panels) the pattern is dominated by a single large updraft of dimensions 5 to 10 km across, extending throughout the depth of the storm. A weak-echo region, which is most obvious at the 6.5 and 8.0 km levels, is centered on the updraft maximum, as originally proposed by Browning and Ludlam (1962). As

discussed in Chapter 16, though, the vault is never strongly developed in this storm and is not a persistent feature.

Two general regions of downdraft are observed in the middle levels, one to the north of the strong updraft and one just to the east. The northern downdraft is behind the sloping updraft (opposite the inflow side) in a region of high reflectivity where precipitation loading may be a factor. A second persistent downdraft is located just a few kilometers to the east of

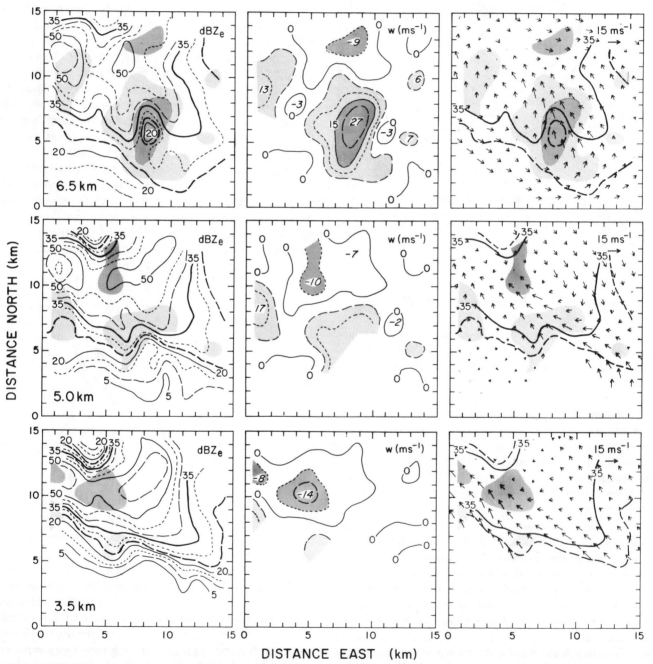

Figure 18.6 Horizontal sections showing detailed storm-relative airflow and reflectivity in the active region at 1635 MDT, with data for six altitudes. The origin of the coordinate system is at (49 km, −10 km) relative to Grover. At each altitude the three columns indicate: (left) radar reflectivity factor, Z_e, contoured at 5-dB intervals; (center) Doppler-derived vertical air velocity, w, contoured at 5 m s^{-1} intervals (extremes labeled in italics); (right) horizontal wind vectors. Selected contours of Z_e and w are superimposed on other data for reference.

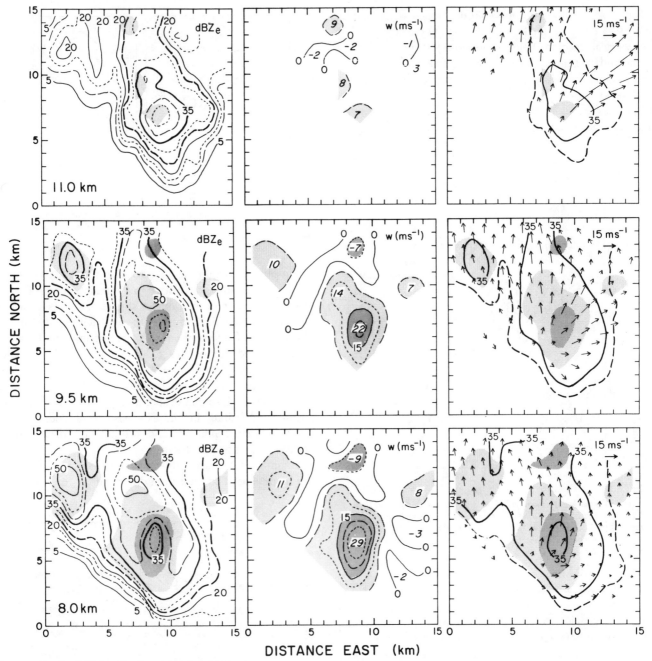

Figure 18.6 (continued) Horizontal sections at 8, 9.5, and 11 km.

the updraft core. Although Fig. 18.6 indicates that it is weaker than the northern downdraft, at other times it is of the same magnitude or even slightly stronger than the northern downdraft. T-28 aircraft observations near the time shown indicate that it was in fact stronger than estimated from the Doppler data. Figure 18.7 compares vertical air motion from the T-28 data with Doppler data collected during the 1640 scan sequence. Although the updraft magnitudes seem to be comparable, data from the T-28 lead to estimates as large as 14 m s^{-1} for the eastern downdraft, while the Doppler estimate is only about 4 m s^{-1}. Further comparisons of

vertical motion from the T-28 and Doppler systems are made in the Appendix.

The eastern downdraft is imbedded in the cyclonically streaming flow that has passed around the south side of the storm, and it occurs on the lee side of the updraft with respect to the environmental wind at mid-cloud level. It is probably caused mainly by evaporative chilling of the dry mid-level air during its traverse in a region of weak reflectivity around the south side of the storm, though other factors may also be important. Downdrafts occurring within this cyclonically streaming flow have been reported for Oklahoma storms

THE 22 JULY 1976 CASE STUDY

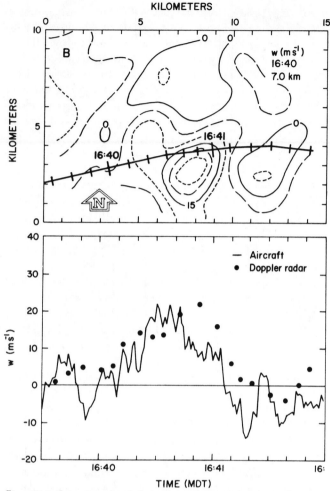

Figure 18.7 *Comparison of vertical air velocity estimates near 1640 MDT from the T-28 aircraft and triple-Doppler radars. Top: The storm-relative aircraft track is superimposed on the Doppler updraft field at 7 km MSL, near the flight altitude. Bottom: Aircraft estimates are represented by solid line; Doppler estimates obtained by bi-linear interpolation are shown as dots.*

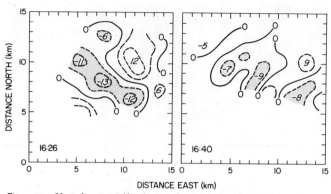

Figure 18.8 *Vertical motion field in horizontal sections at 3.5 km MSL for 1626 and 1640 MDT. The downdraft area at this altitude corresponds to the two distinct downdrafts noted at higher elevations. Contours are shown at intervals of 5 m s⁻¹ with downdrafts stronger than 5 m s⁻¹ stippled.*

An obvious feature of the downdrafts in Fig. 18.6 is their small size and strength relative to the main updraft. This has been observed in other Doppler studies of storms (for example, Miller, 1975, and Kropfli and Miller, 1976), but downdrafts of intensity equal to that of updrafts have also been reported (Brandes, 1978). One reason for the weakness of the northern downdraft may be that it is poorly ventilated by the weak environmental winds relative to the storm, as seen in Fig. 18.4, so that evaporative cooling is perhaps not a strong contributing factor. The eastern downdraft is apparently well ventilated, and precipitation drag may be relatively unimportant there.

Figure 18.9 shows a north-south cross section through both the main updraft and northern downdraft at 1635. The flow

(e.g., Heymsfield, 1978; Brandes, 1978). In the Oklahoma studies downdrafts are observed farther upstream than in the present storm, i.e., to the south of the updraft along streamline C in Fig. 18.4, rather than to the east. This position is distinct from that of the so-called "rear-flank downdraft" inferred in some studies of Oklahoma storms (e.g., Nelson, 1977; Barnes, 1978; Lemon and Doswell, 1979) to exist near the upwind stagnation point in the middle levels of the cloud.

At the lower levels, below about 5 km, the eastern and northern downdrafts tend to merge into a single region of downdraft elongated in a northwest-southeast direction. This is more obvious in the 1626 and 1640 data, shown in Fig. 18.8, than in the data for 1635. Air converges into this region from both the north and south sides of the storm, as evident in the 5-km winds in Fig. 18.6, to feed the surface outflow pool. This is shown more clearly by trajectory analyses in the following section.

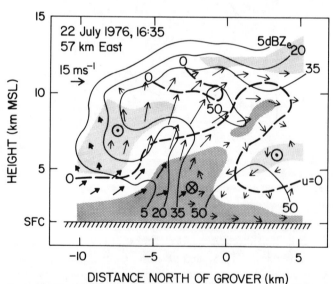

Figure 18.9 *Vertical section oriented south (left) to north along x = 57 km, showing the relative locations of the weak-echo vault, updraft, downdraft, and echo core at 1635. Reflectivity is contoured (thin lines) at 15-dBZₑ increments. The scale for the storm-relative wind vectors is shown at the upper left. Wind vectors with triangular arrowheads identify subjectively extrapolated velocity data. Dark stippling denotes horizontal flow greater than 5 m s⁻¹ into the page, and light stippling denotes similar flow out of the page. The thick dashed line is the zero contour of flow normal to page, and the symbols x and ⊙ are placed near maxima of the components into and out of the page, respectively.*

in the updraft is approximately in the plane of the cross section, as seen from Fig. 18.6 and the zero contour of horizontal velocity in Fig. 18.9. The juxtaposition of the weak vault and the updraft is seen more clearly here than in the plan view, and the position of the northern downdraft to the rear of the updraft is also clear.

The position of the strong updraft in weak echo on the inflow side of the storm is consistent with many earlier studies based on reflectivity measurements and aircraft observations below cloud base (e.g., Marwitz, 1972a and b; Chisholm, 1973; Foote and Fankhauser, 1973). However, it might be noted that several Doppler radar studies have found the updraft to be contained within the high-reflectivity region even at low elevations (Miller, 1975; Ray, 1976; Brandes, 1978), so apparently either structure is possible. Battan (1980) also notes no high correlation between reflectivity factor and updraft strength, though the latter study of vertically-pointing Doppler observations is in a slightly different category because there is no assurance that the strongest updrafts of the storm were sampled.

18.3 AIR PARCEL TRAJECTORIES

Trajectories of air parcels were calculated forward in time from initial positions located throughout the active region, using wind data at 1626, 1635, and 1640 for 7 min each to represent approximately the 1622-1643 time period. They were extended until they reached a boundary of the data volume or until 21 min had passed. There appears to be sufficient time continuity between the successive wind data to represent storm-scale aspects of the airflow configuration, and we choose to consider these trajectories before discussing in the following section the question of temporal evolution. Doppler measurements were lacking below about 6 km in much of the updraft region on the south side of the storm, where reflectivity was relatively weak. Because this region is very important in understanding inflow trajectories, we interpolated and extrapolated the velocity data there by eye to fill data voids as shown, for example, in Fig. 18.9.

Figure 18.10 shows selected air parcel trajectories that illustrate the airflow pattern in the vicinity of the strong updraft of Cell I. Trajectories reaching the upper part of the storm are shown on the left, those confined mainly to middle heights are shown in the center, and downdraft trajectories are on the right.

Trajectories H1 and H2 are representative of air parcels that experience the strongest updraft. In 5 min they ascend from cloud base to nearly 10 km, with a mean ascent rate of about 20 m s^{-1}. While passing through the updraft core they turn gradually to the right, and, upon reaching the upper levels of the storm, they exit toward the northeast (H2 must make a leftward turn in the upper levels after leaving the data region). Parcels following H3 and H4 ascend in the "shoulder" of the strong updraft, and their trajectories are

more tilted away from the vertical. They reach positions above the mid-level reflectivity core about 10 min after passing through the cloud base and then level out to leave the active region toward the north at an altitude of about 10 km. Parcels following H5 and H6 ascend on the west side of the updraft core at about the same rate as H3 and H4, but their trajectories show little tilt until they reach the upper levels. Above about 8 km trajectories H5 and H6 diverge toward the north and south, respectively, as they approach the divergent upper part of the main updraft core from the west side.

On the whole, air parcels comprising the updraft branch of the flow follow a rather simple path in the storm-relative framework. They approach from the south-southeast and retain southerly momentum as they rise, such that the updraft tilts toward the back of the storm. They veer gradually (by 30° to 90°) as they ascend and exit the storm toward the north and northeast. Trajectories in the south and southwest parts of the updraft are more erect, and tilt toward the east under the influence of westerly flow around the south perimeter. While these represent a rather small proportion of the updraft flux they nevertheless may be important in terms of particle transport and microphysical structure.

It might be noted that the simple veering with height of the principal updraft trajectory, H1, occurs despite the fact that the updraft possesses predominantly cyclonic vorticity of magnitude typically 4 to 10 × 10^{-3} s^{-1} in the middle levels of the storm. Similar updraft trajectories have been proposed by Marwitz (1972a) and Browning and Foote (1976), in contrast to the cyclonically rotating (strongly backing) trajectories inferred by Browning (1964). However, trajectories similar in sense of rotation to those of Browning (1964) have since been reported in the multiple-Doppler analysis of Brandes (1981), so apparently both types can occur. Heymsfield (1978) also infers a cyclonically turning updraft trajectory, but as shown here the existence of positive vorticity in the updraft is not sufficient to guarantee this. The storm model of Eagleman and Lin (1977) showing a cyclonically spiraling updraft is erroneously based on so-called "perturbation winds" and is not consistent with the basic data.

Parcels ascending in weak updrafts on the front and right front flanks are represented by trajectories M1 and M2, which enjoy rather long residence times in the middle heights as they follow cyclonically curved paths around the periphery of the updraft region. Special emphasis has been placed on trajectories like M1, which rises on the west and passes over the principal inflow region. As discussed by Browning and Foote (1976) particles attaining large enough terminal velocities along such a path could descend into the southeasterly inflow and be carried into the strong updraft where further growth into hail is possible. This is discussed further in Chapter 20. Trajectories like M1 and M2 probably often originate in young cumulus turrets on the flanks of the mature storm.

Trajectory M3 illustrates the nature of mid-level flow

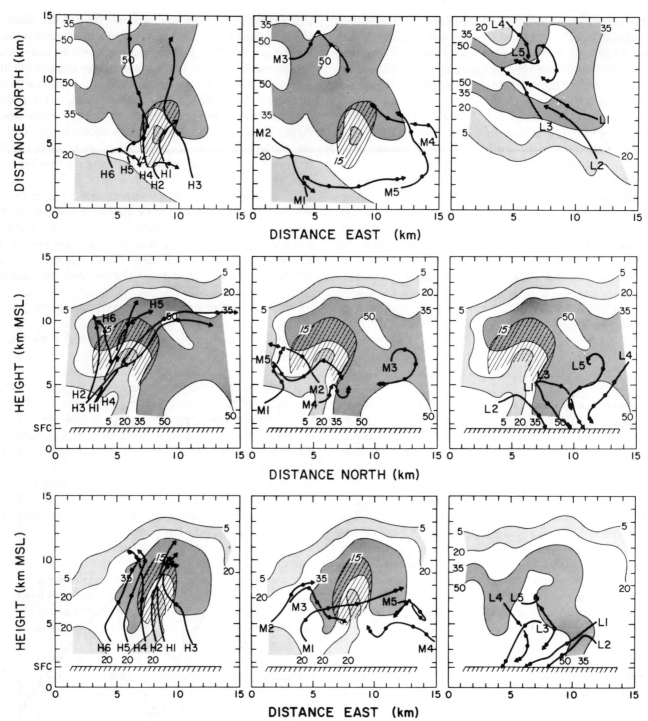

Figure 18.10 Storm-relative air parcel trajectories calculated from Doppler wind fields for 1626, 1635, and 1640 MDT. At 1635 the origin of the coordinate system is at (49 km, −10 km) relative to Grover. The top row shows the trajectories in plan view; the middle and bottom rows show projections on vertical planes oriented north-south and east-west as indicated. The left column depicts updraft trajectories that are initiated at 3.5 km MSL, about the altitude of cloud base. The middle column shows trajectories that reside mainly in middle levels. The right column illustrates the dominant downdraft trajec-tories. Dots along the trajectories indicate 5-min time intervals. Reflectivity contours and up-draft location (hatched for w > 15 m s⁻¹) are from the 1635 scan sequence. They are shown for an altitude of 6.5 km MSL for the left and middle frames of the top row, and for 3.5 km on the right. In the center row the contours are for the plane passing through the updraft maximum, x = 8 km in the coordinates of this figure. In the bottom row the con-tours are for y = 6 km (left and center frames), and y = 11 (right frame).

around the north side of the updraft region. Originating in an area of weak updraft on the back side of Cell H, it follows a helical path as it subsequently descends and then approaches the main updraft from the back side. Other trajectories (not shown) in this region of the storm suggest that some air parcels could undergo at least two successive up-and-down excursions along trajectories similar to M3. Inflow to the storm at low to middle heights on the southeast side of the updraft is represented by trajectories M4 and M5. M4 begins at cloud base height in the relatively inactive region to the east of the updraft, and M5 traverses the leeward wake of the updraft in the middle levels. Both of these arrive after about 20 min at positions on the east side of the downdraft region (see Fig. 18.8). It appears that parcels following trajectories like these might subsequently descend into the low-level outflow layer along trajectories like L1, which is initiated near the endpoints of M4 and M5. (Because the velocity field evolves in time this interpretation is not strictly valid, but the general pattern is thought to be essentially correct.) Trajectory L2 enters the radar echo at cloud base near the eastern edge of the updraft and L3 is initiated on the north edge, so that they rise slightly before entering the downdraft. Trajectories (not shown) originating around cloud base height in the region to the east and northeast of the updraft also descend to the surface.

It appears that the group of trajectories just discussed, including M4, M5, L1, L2, L3, and others originating at lower heights, account for a major proportion of the air that reaches the low-level outflow layer. It is not possible from this analysis to determine the exact origin in the storm's far environment of air contributing to this branch of the circulation, but these trajectories suggest that the low-level outflow is comprised of a mixture of air from middle levels down to the surface. Figure 18.11 shows schematically some of the possible trajectories by which air can reach the low-level outflow area. By reference to the larger-scale flow shown in Fig. 18.4 and trajectory M5, it appears that some of this air originates in the mid-level environment (5 to 7 km) to the west of the storm, passes around the updraft approximately along streamline D in Fig. 18.4, and then descends in the leeward wake region (path B in Fig. 18.11). Along such a path it would presumably be mixed with cloudy air on the south

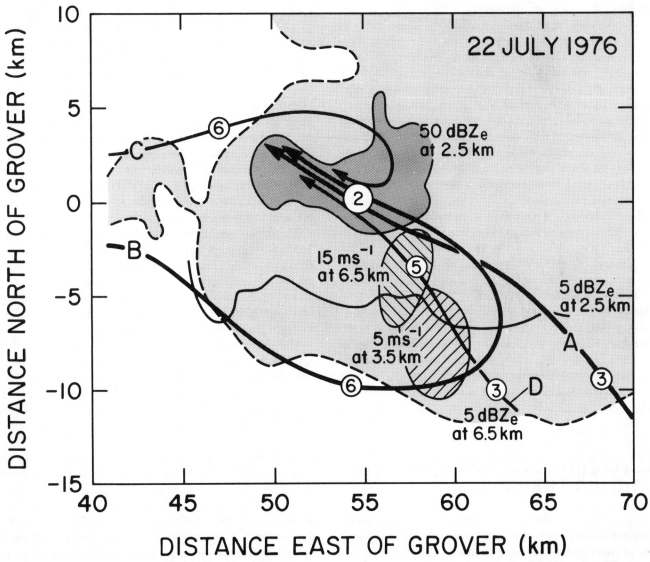

Figure 18.11 Plan view indicating the principal downdraft trajectories. Paths A, B, C, and D converge into the region of high reflectivity as they descend to near the surface. Circled numbers denote approximate heights in kilometers. The solid contours represent reflec- tivities of 5 and 50 dBZ$_e$ at 2.5 km MSL. The dashed line is the 5-dBZ$_e$ contour at 6.5 km. The updraft cores at 3.5 and 6.5 km are shown as hatched regions.

and east sides of the updraft, and would be chilled by evaporation of cloud material as a result. Trajectories M4 and L2 seem to reflect a direct intrusion from the southeast of air originating near the cloud base height (path A). Trajectory L3 begins in the north shoulder of the updraft at cloud base where, according to the observations in Chapter 17, air is not convectively unstable. Subjective extrapolation of trajectory L3 backward suggests it may have originated in cold outflow spreading westward from Storm II. Surface data indicate that this air might be lifted over the outflow from Storm III (the storm discussed here) and into the back side of the updraft along path D in Fig. 18.11. Some air following trajectories like L2 could also have a similar history.

Downdraft trajectories L4 and L5 indicate that some of the low-level outflow also comes from the northwest side of the storm along path C in Fig. 18.11, entering at heights below about 7 km (again, many trajectories reaching the outflow layer originate at lower levels than those shown). Extrapolating these trajectories backward, they appear to enter the storm system approximately following streamline E in Fig. 18.4, the anticyclonic branch of the barrier flow. Considering that the rate of descent along these trajectories is only about one-half that of the ones coming from the southeast, and that fewer trajectories in the northwest region are found to reach the surface layer, it seems that this branch provides a smaller contribution to the downdraft and outflow in the active region than does the branch on the southeast side of the storm. However, additional downdraft to the north of the active region, not considered here, would also probably ingest air from the northern (anticyclonic) branch of the mid-level barrier flow.

The convergence of paths B and C in the downdraft region is a reflection of the velocity field shown for the 5-km level in Fig. 18.6. One notes from Fig. 18.11 the interesting result that some of the mid-level streaming flow that is forced to part on the upwind side of the storm is apparently reunited in the vicinity of the reflectivity core, with both branches contributing to the low-level downdraft and surface outflow. The dominant southern branch of this pair is the same as the downdraft trajectory proposed in the schematic model of Browning (1964), being wrapped around the strong updraft. Analysis of multiple-Doppler radar observations of a storm in Oklahoma led Heymsfield (1978) to deduce two downdraft trajectories qualitatively similar to branches B and C in Fig. 18.11.

18.4 EVOLUTION OF THE MOTION FIELD IN THE ACTIVE REGION

To examine the evolution we consider two sets of triple-Doppler data, each including three consecutive volume scans. The first set, consisting of scans at 1626, 1635, and 1640, has been emphasized in previous sections because the data are more complete and represent a more intense portion of the

storm lifetime. It spans the time period corresponding to Cell I as discussed in Chapter 16. The second set is comprised of scans at 1707, 1712, and 1718, and relates similarly to the K series of Chapter 16.

18.4.1 The 1626-1640 Period

We illustrate the evolution here with horizontal sections at 6.5 km (Fig. 18.12), and vertical cross sections through the strong updraft along the north-south (Fig. 18.13) and east-west (Fig. 18.14) directions. Reflectivity factor, Z_e, is shown on the left, vertical velocity, w, in the middle, and projected storm-relative wind vectors on the right. Selected contours of vertical velocity and reflectivity are repeated in other plots for reference. In the vertical sections, the shading represents the cross-plane component of motion, rather than w.

We consider first the horizontal sections at 6.5 km in Fig. 18.12; the middle row for 1635 has already been discussed. There are two major cells in the reflectivity plots, Cell H on the left, and the G-I sequence on the right, as discussed in detail in Chapter 16. At 1626 updraft maxima are shown just south of the reflectivity cores for both of these cells. The updraft near Cell I is less compact than at 1635, and a continuous band of updraft connects the two cells, in contrast with the discrete updrafts observed later at 1635. A fairly large updraft is also delineated on the northeast side of the echo core at 1626 (marked in the center figure with a 19 m s⁻¹ maximum). It cannot be determined whether this is a remnant of some previous cell (perhaps part of Cell F in Fig. 16.15) or merely a short-lived feature that did not persist long enough to produce a characteristic cellular echo. Only weak updrafts are seen in this position at other times.

The updraft structure at 1640 is little changed from that at 1635, with a single updraft core for Cell I dominating the field in a position coincident with the weak-echo region, seen on the left. At all three times the dominant updraft is 5 to 10 km across, with a maximum speed of 25 to 30 m s⁻¹. Downdrafts are always present on the east and north sides of the strong updraft, as discussed previously.

The horizontal wind fields are all characterized by southeasterly (low-level inflow) momentum in the vicinity of the updrafts and cyclonic flow (westerly on the south sides turning to southerly on the east sides) around the major updrafts, as noted in the previous discussion for 1635, though the cyclonic turning of the flow is somewhat less pronounced at 1620.

The strong updraft appears with about the same size and intensity and in the same relative position in the storm at the three times, and the horizontal wind field shows only minor variations between the volume scans. Thus, it appears that the macroscale organization of the storm is basically steady in this period. Some deviations from the mean structure are observed in both the updraft and reflectivity structure, however, and more detailed consideration of these below will be useful.

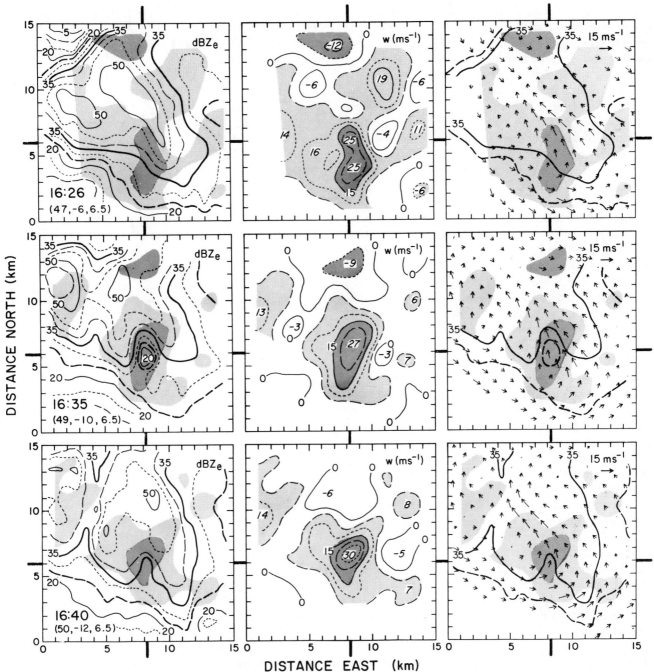

Figure 18.12 Horizontal sections showing reflectivity and wind fields in the active region at 6.5 km at 1626 MDT (top row), 1635 (middle row), and 1640 (bottom row). The format is the same as in Fig. 18.6. The coordinates of the local origin with respect to Grover and elevation (MSL) are given in kilometers at the lower left of each row. Heavy tick marks on the perimeter of the grid indicate the locations of vertical sections shown in Figs. 18.13 and 18.14.

The vertical structure of the storm in the vicinity of the updraft is shown in Figs. 18.13 and 18.14, in sections that pass through the updraft maximum. Since the horizontal flow in the updraft is dominated by southerly momentum, Fig. 18.13 is the more revealing, as the wind vectors in the panel on the right are fairly representative of streamlines. The shaded areas on this panel represent flow of more than 5 m s⁻¹ into or out of the plane, primarily below or above the updraft core.

Figure 18.14 aids the interpretation of the three-dimensional flow.

The most striking time-dependent features in the vertical sections are seen at 1626 and 1635. Changes of the updraft structure between these two times and the development of the vault in the region of the updraft core between 6 and 8 km altitude elucidate the nature of the cellular propagation discussed in Chapter 16. On the basis of detailed reflectivity

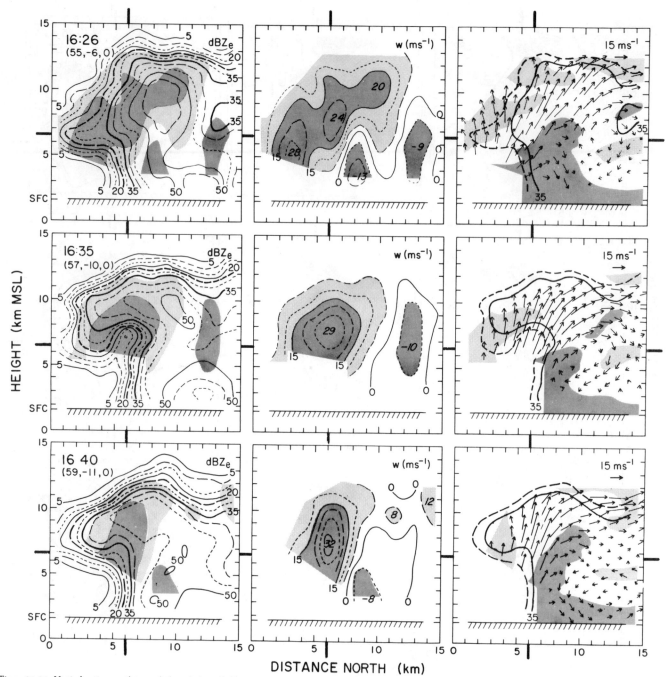

Figure 18.13 Vertical sections south to north through the updraft core in a format similar to that of Fig. 18.12. The dark shading in the frames on the right indicates where the com-ponent of the flow into the page is greater than 5 m s⁻¹; light shading indicates flow out of the page of more than 5 m s⁻¹.

analyses there, it was concluded that the evolution of the updraft can be understood in terms of a steady upward flow with superimposed fluctuations. The fluctuations give rise to the variations of radar reflectivity in the storm, as seen for example in Fig. 16.9, and the steady component of the flow accounts for the persistence of the basic mean structure. The Doppler updraft analyses support and clarify this picture, with the data for 1626, in particular, being the key. The 1626 view represents a time just before the reflectivity perturbations associated with Cell I are evident (compare Figs. 16.21 and 16.24). Subsequent development of the reflectivity structure, discussed in Chapter 16, is as follows. A small vault forms in the underside of the radar echo overhang, implying an updraft intensification starting in the lower levels, and then a local reflectivity maximum forms over the vault. The height of the echo top and the visual cloud top both increase as this occurs. These observations seem to be consistent with the updraft evolution seen in Fig. 18.13 and interpreted further in Fig. 18.15, where intermediate and adjoining times are added subjectively.

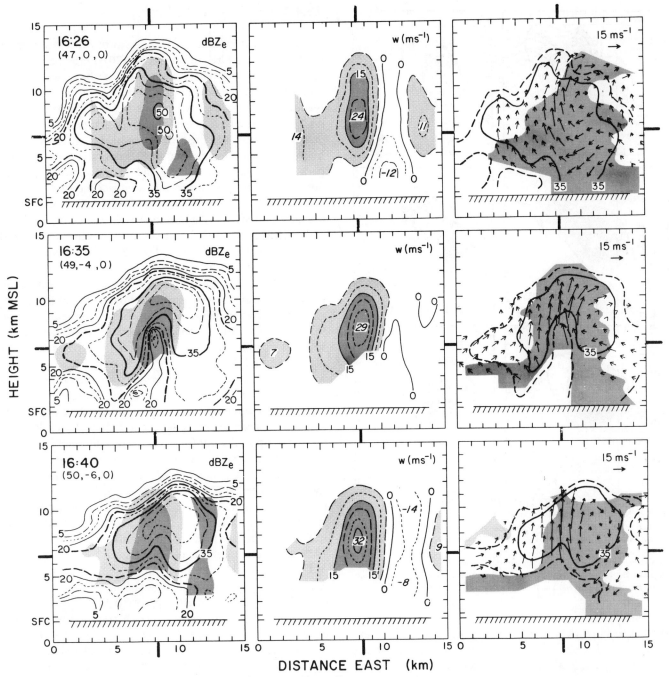

Figure 18.14 Vertical sections west to east through the updraft core, similar to Fig. 18.13.

In Figure 18.15 four updraft impulses are labeled α through δ. Three of these relative maxima, α, β, and γ, are seen in the data at 1626. It appears that γ is the impulse that corresponds to Cell I in the reflectivity analysis, and that α and β have a correspondence with an earlier Cell G. The exact correspondence of G with α and β can not be determined from these data, but the appearance of γ with a maximum at a relatively low level seems to explain the weak-echo vault seen clearly at 1635. As the impulse γ rises it disturbs the trajectories of growing precipitation particles, pushing them up-

ward locally and changing the reflectivity profiles according-ly. The vault starts developing about 1630, and the time history of γ as sketched in Fig. 18.15 seems about right for this. By 1630 α probably would have dissipated, to be replaced by β.

At 1635 γ is at maximum development, with perhaps a trace of β still evident in the Doppler data. The movement of α and β to the north in the figures is consistent with the fairly strong component (10-15 m s^{-1}) of the relative environmental wind in this direction at the upper levels. At 1640 the size of

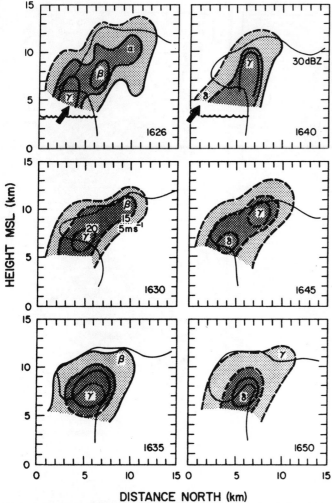

Figure 18.15 Vertical sections (south-to-north) showing the proposed updraft evolution from 1626 to 1650 MDT. Coordinates correspond to those in Fig. 18.13 at 1635. The thick solid lines (1626, 1635, and 1640) are contours of vertical air velocity, w, based on the Doppler measurements of Fig. 18.13. Dashed lines are hypothetical contours of w. Thin lines are the measured 30-dBZ$_e$ contour of reflectivity. Bold arrows at 1626 and 1640 show the positions of the updraft core at cloud base as ascertained from aircraft measurements. Four updraft maxima are identified as α, β, γ, and δ.

γ seems to have diminished in its upper part, and by extrapolation of the updraft contour downward one can imagine a new impulse γ taking shape. In fact, the existence of δ is strongly supported by the aircraft measurements taken at cloud-base and discussed in Chapter 17 (see Fig. 17.14). At this time the updraft core at cloud base has moved further south from the precipitation as shown by the bold arrow for the 1640 plot in Fig. 18.15, and does not appear to support γ directly. Earlier aircraft analyses show the updraft closer to the echo at cloud base, for example as indicated by the bold arrow in the 1626 plot. Though no Doppler analyses are available for 1645 and 1650 we visualize the subsequent structure of δ to be similar to that hypothesized at 1630 and measured at 1635 for γ. The development of impulse δ would then be responsible for the vault associated with Cell J, first noted at about 1645 (Chapter 16).

18.4.2 The 1707-1718 Period

As discussed in Chapter 16, the storm echo shows more small-scale structure during the 1707-1718 period corresponding to the intensification of Cell K than it does during the 1626-1640 period just considered. Actually the smaller features are noticeable by the time of Cell J, 1650 or so, but no Doppler data have been reduced for that period. Since the increased structure in the reflectivity field is probably caused by the presence of additional smaller updrafts, we would expect the vertical motion field also to show more structure during the later period, and this is indeed the case.

Figure 18.16 shows the Doppler data at 6.5 km altitude for the later period. The data on the southern side of the storm are generally less complete than in the earlier period, and additional plots are shown in Fig. 18.17 for 8.0 km, where more of the updraft is detected.

At 1707 the Doppler results show a broad region of updraft across the whole south to southwest side of the storm, containing two general updraft cores. The one on the southwest lies within an arm of reflectivity that is developing southward from the echo core. This is the echo feature that was termed J$_1$/J$_2$ in Chapter 16.

Fig. 18.17 shows that the updraft on the south side is deeper than that on the southwest and considerably stronger aloft. The southern updraft is in the same relative position as the large updraft in the earlier period (1626-1640). It is apparently responsible for the developing vault seen in its location at 1712. The subsequent intensification of the echo core termed Cell K is evident at 1718. The two maxima in the southern updraft at 6.5 km (the one on the left is only partially revealed) seem to correspond with reflectivity minima seen in the developing vault at 1707 and 1712. An arm of the southern updraft extends to the northeast at 1707, and at 1712 a discrete updraft is observed there.

The main southern updraft and the southwest updraft are less distinct at 1712, though separate maxima are evident at other heights. The updrafts appear to have moved closer together and slightly westward with respect to the storm echo. The vault is more developed by this time, and the southern updraft at 8.0 km has a more regular shape, more like that at 1635. Unfortunately the data are severely truncated on the south at 6.5 km. The position of the southern updraft is displaced slightly from that of the vault, by perhaps 2 km, in contrast to their juxtaposition noted in the earlier period. Because this region lies near the edge of the data, the offset may in fact be an artifact caused by errors (see the Appendix). However, the subsequent view at 1718 suggests it may be realistic.

At 1718 the vault has become filled with echo. The large updraft is now observed more on the southwest corner of the storm within an indentation of the reflectivity pattern that would appear to be consistent with early stages of vault development. A new updraft has appeared to the east of the large updraft, and a small weak-echo region is associated with

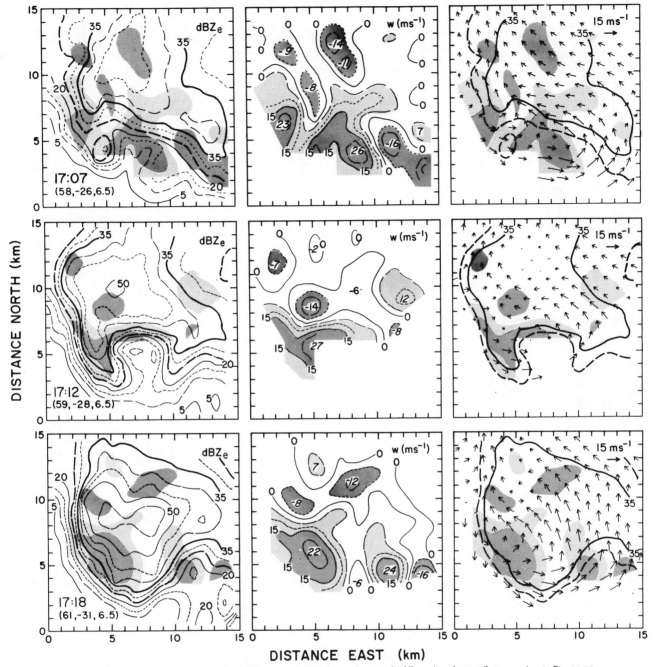

Figure 18.16 Horizontal sections of the active region at 6.5 km MSL at 1707 MDT (top row), 1712 (middle row), and 1718 (bottom row) as in Fig. 18.12.

it. This updraft gives rise to the echo of Cell K₄ mentioned in Chapter 16. The existence and, in fact, the general strength of both updrafts is confirmed by the T-28 aircraft data collected between 1714 and 1716 (see Appendix). Cell K₄ never becomes as large as Cell K, and in fact within 10 min both K and K₄ are dissipating rapidly, as described in Chapters 16 and 17.

The cyclonic turning of horizontal winds near the updraft region is generally more pronounced in the 1707-1718 period than in the earlier one and the direction of flow on the north side of the updrafts is more easterly. This seems consis-

tent with a stronger component of easterly momentum in the low-level inflow at this time noted in Chapter 17. In addition, westerly flow on the south sides is stronger and penetrates well into the strong updrafts. These westerly speeds are in fact about four times greater than those in the environment at this altitude, as judged from the hodograph in Fig. 18.2, and they can not be explained solely in terms of the barrier effect unless the environmental wind has increased substantially. Downdrafts are also more intense in this period, especially to the east of the updraft region.

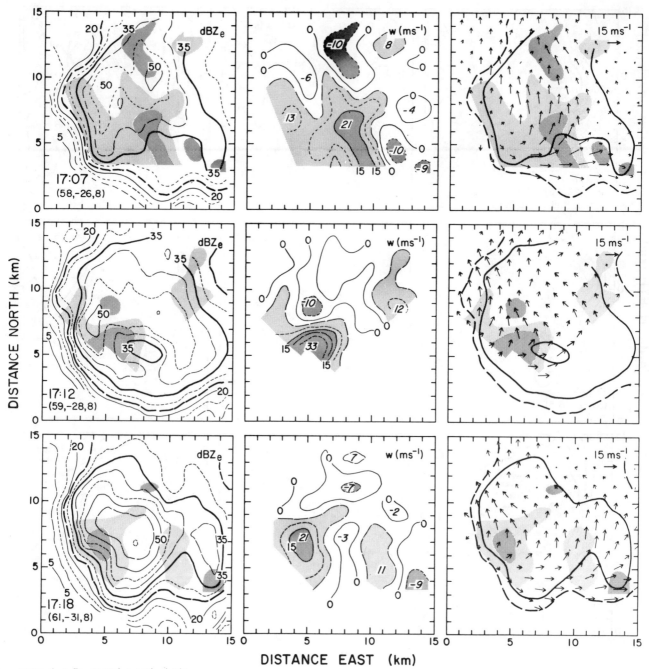

Figure 18.17 As in Fig. 18.16 but at 8 km height.

18.5 PROFILES OF UPDRAFT AND DOWNDRAFT VELOCITIES

In this section we examine profiles of vertical velocity along vertical and horizontal lines through local maxima and minima in the vertical velocity field. Figures 18.18 and 18.19 show, on the left, the positions in plan view of updraft and downdraft entities selected for this discussion. The 20-, 35-, and 50-dBZ$_e$ reflectivity contours for 6.5 km height are included for reference to Figs. 18.12 and 18.16. All updrafts and downdrafts that could be clearly distinguished with a

5 m s^{-1} contour interval, and for which reasonably complete profiles could be constructed, have been included.

In the center frames are profiles of maximum and minimum w as a function of height. These are taken along slightly tilted axes in some cases (ten degrees or less from the vertical) but will be called vertical profiles for convenience. The vertical profiles generally cover the entire depth for which data are available, but in a few cases they are truncated because local maxima or minima exist only in a limited height interval. On the right, profiles of w along horizontal lines passing west to east and south to north through the updraft

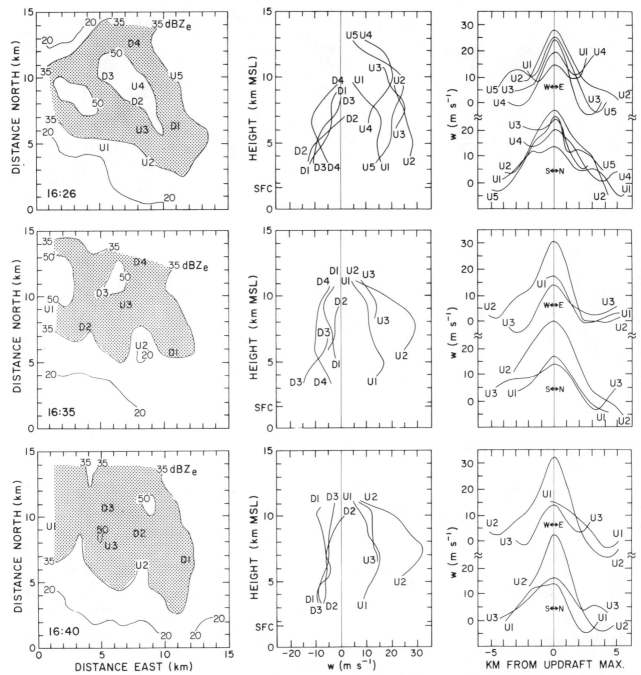

Figure 18.18 Profiles of vertical air velocity, w, in updrafts and downdrafts at 1626 MDT (top row), 1635 (middle row), and 1640 (bottom row). Left: Positions of updrafts (U) and downdrafts (D) are identified in plan view; reflectivity contours in dBZ_e at 6.5 km MSL are included for reference to Fig. 18.12. Center: Plots of maximum up-draft and downdraft as a function of height. Right: Horizontal profiles of updraft in west-to-east (top set of curves) and south-to-north (bottom set) directions through the altitude of maximum w. Labels U1, U2, etc. refer to one time only; thus the same label generally refers to different entities at different times.

maxima are plotted against distance from the respective maxima.

The broadest updrafts observed (U2 at 1635 and 1640, and U1 at 1712), are about 8 km in diameter. Maximum updraft speeds of about 30 m s⁻¹ occur in them at heights between 7 and 8 km, about 4 km above cloud base. Other updrafts of various sizes are observed at heights ranging from near cloud base upward to 10 km. Those located at higher levels tend to be in the central or back portions of the general updraft region, while those at lower heights are toward the front side. Surprisingly large vertical velocities of 20 to 30 m s⁻¹ occur at heights as low as 4 to 5 km in some cases. Although their magnitude may be exaggerated, it is fairly certain that these low-level updraft maxima are indeed present (see Appendix).

147

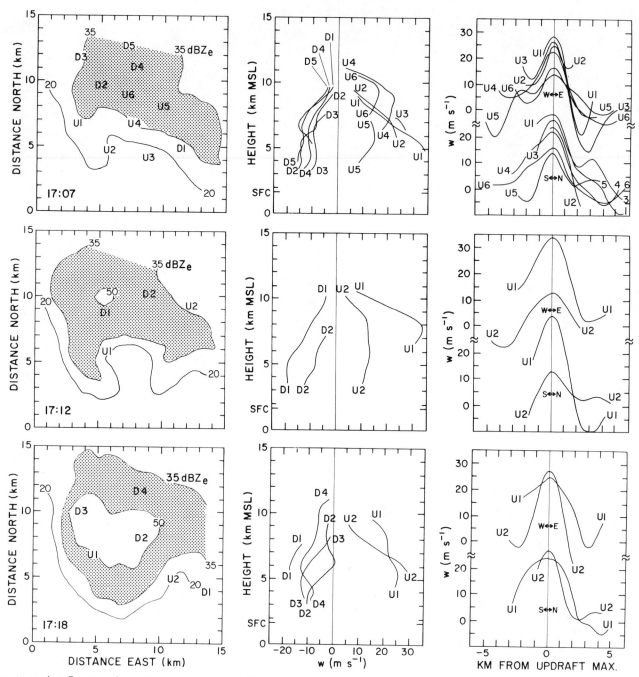

Figure 18.19 As in Fig. 18.18, but for 1707, 1712, and 1718 MDT.

Figure 18.20a shows some quasi-vertical profiles, determined by Marwitz (1973) from trajectories of radar-reflective "chaff" parcels in northeastern Colorado hailstorm updrafts, presented as a function of height above cloud base. Chaff parcels were released near cloud base in areas of relatively strong, smooth updraft, and were tracked for periods of 4 to 8 min until they were lost in precipitation echo. For 16 cases where the height of maximum vertical velocity could be determined, the heights range from 1.7 to 4.2 km above cloud base. These heights are consistent with those observed

toward the front side of the 22 July 1976 storm, and Marwitz's profile shapes provide some insight regarding the possible shape of profiles for the present storm in their lower portions where Doppler measurements are lacking. Though Marwitz's method differs essentially from that used here, the profile shapes are thought to be comparable to the extent that the updrafts were steady over the time periods involved.

Near the core, the horizontal updraft profiles (on the right in Figs. 18.18 and 18.19) are nearly Gaussian in shape, quite similar to profiles found in axially symmetric jets. Kyle et al.

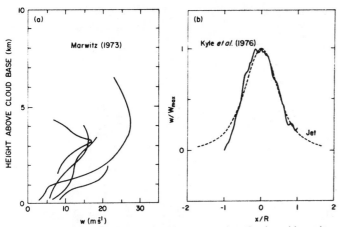

Figure 18.20 Measured updraft profiles: (a) quasi-vertical profiles obtained by tracking chaff parcels released near cloud base, from Marwitz (1973), and (b) horizontal profile obtained by averaging twelve normalized profiles measured from aircraft (Kyle et al., 1976). An empirical jet profile (Schlicting, 1960) is shown dashed.

(1976) found that quasi-horizontal profiles of w obtained from aircraft penetrations were also well described by Gaussian or jet-like profiles. The profile obtained by Kyle et al. by averaging 12 normalized individual profiles is shown in Fig. 18.20b, compared with a jet profile due to Schlichting (1960).

Perhaps the most important point here with regard to updraft structure is that the shape of the "tails" of the horizontal profiles, away from the central peaks, suggests that each updraft maximum is at the center of a ring vortex. This concept of updraft structure has been applied to cumulus convection by numerous investigators (see, for example, Ludlam and Scorer, 1973) who have referred to the convective elements as buoyant bubbles or thermals. Laboratory experiments (for example, Turner, 1964; Woodward, 1959) have established that buoyant parcels of fluid in neutrally stratified surroundings take a form similar to the spherical Hill's vortex (Lamb, 1945). Vertical and horizontal profiles through the center of Hill's vortex are shown by solid lines in Figs. 18.21a and b,

scaled to the radius of the sphere that encloses the region of vorticity. Comparable profiles observed in the laboratory by Woodward are shown by dashed lines. The laboratory experiments of Woodward and of Turner, performed in neutrally stratified fluids, indicate that roughly spherical buoyant vortices translate upward at a rate of about one-half the maximum updraft speed at the center, expanding in size at a constant rate as they move away from their virtual source. Velocity is scaled to this translation rate in the figures.

The most significant difference between these "thermal" velocity profiles and the nearly Gaussian profile for a jet is that the thermals exhibit rather sharp local minima of vertical velocity on their lateral edges. This feature is most pronounced at the height of the vortex center, and horizontal profiles toward the top or bottom of the spherical vortex region would approach a Gaussian shape (Turner, 1962) more consistent with the observations of Kyle et al. Turner (1963) and Woodward (1959) demonstrate that there is mixing between a thermal element and its surroundings along the top and lateral edges, and that the torroidal flow pattern transports mixed fluid downward around the periphery and then inward toward the central updraft. Such downward mixing across the top might be even more pronounced in the cloud environment than in the laboratory thermals, as latent heat is available to help drive penetrative downdrafts (Squires, 1958). Paluch (1979) discusses "vertical" entrainment mechanisms of this kind, and presents evidence that they are in fact important in some cumulus clouds.

The thermal flow pattern also accommodates recirculation of precipitation particles, as discussed by Woodward (1959), such that certain particles can pass more than once through the central updraft portion of a thermal. However in a motion field containing many thermal-like elements, actual particle trajectories depend on the persistence and translation properties of the thermals and on possible interactions between thermals. These properties are not resolved in this analysis.

18.6 AIR MASS FLUX

To help quantify the air mass budget of the storm, we now consider the net vertical exchange of air accomplished in major updraft and downdraft regions of the storm. While the vertical flux properties are of interest with regard to the storm's dynamical structure, the value of this exercise is more obvious when one considers that, given sufficient information about the humidity field, this kind of analysis would yield the water vapor budget as well. We do not address the water vapor flux here, but this may become feasible in future studies.

The approach used here is to define the closed region occupied by a particular updraft or downdraft entity (these regions are nominally bounded by iso-surfaces where $w = 0$), and to examine as a function of height the total or integrated

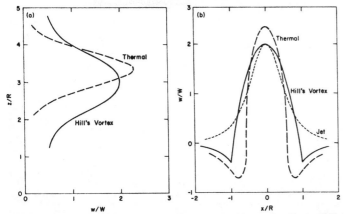

Figure 18.21 Theoretical and experimental updraft profiles: (a) vertical profiles for Hill's vortex (Lamb, 1945) and for laboratory thermals (Woodward, 1959), and (b) horizontal profiles for Hill's vortex, Woodward's thermal, and the jet of Fig. 18.20b. Scales are normalized to the radius and rise rate of a spherical bubble.

vertical flux over the horizontal area, $A(Z)$, of the defined region. The total flux through a particular level has units of kg s^{-1}, being the areal integral of ϱw. To describe the net horizontal flux through the bounding surface that is required by mass continuity, we differentiate the flux profile with respect to height to obtain the profile of horizontally integrated flux divergence, which has units of kg s^{-1} km^{-1}. First we consider the "main" updraft of the storm at 1635 (i.e., the relatively large updraft cell identified with Cell I, as discussed in Chapter 16).

18.6.1 Flux in the Main Updraft

To obtain the total vertical air mass flux in the main updraft, $F_m(z)$, at a given height, z, we have delineated the main updraft at that height and calculated the flux according to

$$F_m(z) = \rho(z)\int_{A_m}w(z)dA$$
$$= \rho(z)\bar{w}_m(z)A_m(z),$$

where ϱ is the air density, A_m is the area of the main updraft, w is vertical velocity and $\bar{w}_m$ is the mean vertical velocity for the area A_m. Figure 18.22 shows how the updraft area was delineated at $z = 6.5$ km, and Fig. 18.23 shows $A_m(z)$ and $\bar{w}_m(z)$. Values of $F_m(z)$ determined from Doppler data are shown by small circles in Fig. 18.24a. Figure 18.24b shows the associated profile of $D_m(z)$, the integrated horizontal flux divergence, which was obtained by numerically differentiating $F_m(z)$ in the height range above 6 km where Doppler data were available, as indicated by diagonal hatching. The region below 6 km is discussed below.

The error in F_m at 7 km is thought to be less than 40% (see Appendix). Comparing measurements of w obtained by the penetrating T-28 aircraft near 1640 with Doppler estimates along the aircraft track (which, as shown in Fig. 18.7, passed through the main updraft region), we find that $\int w\,dl$, where l is the distance along the aircraft track and the integral is over the width of the main updraft, is 69 and 91 m s^{-1} km for the aircraft and Doppler data, respectively. As is pointed out in the Appendix, much of the 30% discrepancy could result from errors in the aircraft position and errors in the aircraft estimates of w.

Doppler coverage of the main updraft is not achieved below 6 km because of insufficient radar echo. The total flux through the 3.5-km level in the main updraft region was estimated from a map of the cloud base updraft structure, similar to those presented in Chapter 17, prepared by using a composite of aircraft data from several traverses of the updraft near the time of interest. The position and structure of the cloud-base updraft thus obtained correspond well with the main updraft at middle heights. Part of the updraft area in the vicinity of Cell H, further to the west, was not included. We obtained the value $F_m(3.5$ km$) = 2 \times 10^8$ kg s^{-1} by this procedure and feel that this is correct within $\pm 25\%$. Except for the vertical flux at cloud base, just discussed, there are no

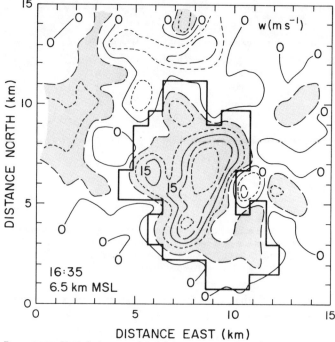

Figure 18.22 Vertical air motion (stippled for $w > 5$ m s^{-1}) in the active region at 6.5 km MSL at 1635 MDT (without Hanning filter). Heavy line encloses what is called the main updraft at this height.

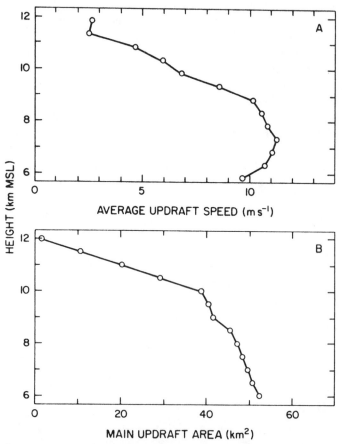

Figure 18.23 Average updraft speed (a) and area (b) for the main updraft at 1635 MDT, as a function of height.

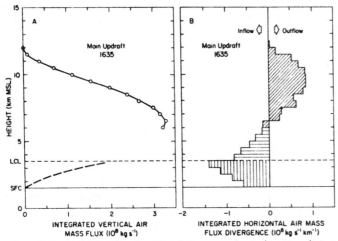

Figure 18.24 (a) Horizontally integrated vertical air mass flux in the main updraft at 1635 MDT. Solid line connects Doppler estimates. The value at cloud base (LCL) is derived from aircraft measurements there. (b) Integrated horizontal air mass flux divergence associated with (a). The vertically, hatched portion below cloud base is based on data of Fankhauser (1974) shown in Fig. 18.25. Horizontal hatching indicates subjective interpolation. Errors and further details are discussed in the text.

measurements of flux or flux divergence below 6 km in the main updraft. We speculate, however, about the shape of the D_m profile in the lower levels as follows. Figure 18.25 shows estimates of integrated horizontal air mass flux divergence obtained by Fankhauser (1974) for the subcloud layer of a hailstorm observed in the NHRE area on 9 July 1973. This storm was of comparable size and intensity to the 22 July 1976 storm, and has been discussed at some length in the literature (see, for example, the series of papers summarized by Browning et al., 1976). Measurements of horizontal winds and thermodynamic variables from surface instruments and from three stacked aircraft that circumnavigated the storm were used by Fankhauser to estimate the horizontal flux divergences associated with updraft and downdraft branches of the storm, with air being assigned to updraft and

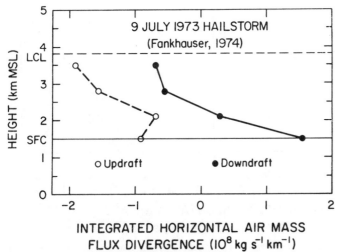

Figure 18.25 Subcloud profiles of integrated horizontal air mass flux divergence. The measurements were obtained during aircraft circumnavigations of a different hailstorm by Fankhauser (1974).

downdraft branches on the basis of its equivalent potential temperature. One can see from Fig. 18.25 that the horizontal inflow to the updraft shows a general increase with height in the layer below cloud base. It is assumed here that the slope of the profile D_m (Z) is similar to the slope of Fankhauser's profile of flux divergence for the updraft in this layer.

If we take the estimates of F_m at cloud base and middle heights to be correct, mass continuity requires additional inflow to the main updraft in the layer between 3.5 and 6 km. The horizontally hatched area in Fig. 18.24b represents this inflow, the slope of the D_m profile in this layer having been subjectively chosen to obtain a relatively smooth profile.

The maximum upward flux of 3.3×10^8 kg s^{-1} occurs at 6.5 km height and most of the upper-level outflow from the main updraft is in the layer between 8 and 11 km (see Fig. 18.24). The estimated 2×10^8 kg s^{-1} flux through the cloud base level is approximately 65% of the maximum flux obtained from the Doppler analysis. While this suggests that a significant amount of air enters the main updraft at heights above cloud base, the approach used here does not permit us to address the origin (in the "undisturbed" environment) of air that contributes to the updraft mass flux at any given level. The boundary of the updraft area defined here is not an interface between the updraft and the storm environment, and we do not account for vertical flux outside the boundary. As we consider only the net inflow or outflow in each height interval we do not necessarily account for the total exchange of air across the updraft boundary.

18.6.2 Fluxes in the Active Region

We now consider the regions of updraft and downdraft in the active region of the storm, defined as the 15-km-by-15-km area including the largest and most intense regions of updraft and downdraft at middle heights, plus the area on the south occupied by the main cloud-base updraft. Two time periods centered around 1635 and 1712, respectively, are addressed. By restricting the calculations to the active region we avoid problems associated with different areal coverage by the Doppler radars in different volume scans, and also confine the calculations to a region where the meteorological "signal" is relatively strong, compared with errors in the measurements. The active region has been defined such that it encompasses comparable airflow features at the different times, but in both periods there was other convective activity in the immediate vicinity, as described in Chapters 16 and 17, which was not included. The flux calculations used data that were not subjected to the Hanning filter. These data extend one to two grid intervals outside the data areas shown in preceding sections and appear to represent nearly all the flux in the active region, with the possible exception of data for 1712, when part of the updraft may be omitted (see Section 18.4.2).

We define F_u (z) and F_d (z) as the integrated updraft and

151

downdraft fluxes, respectively, in the active region. We obtained $F_u(z)$ by essentially the same method as for the main updraft. Profiles obtained from consecutive volume scans at 1626, 1635, and 1640, and at 1707, 1712, and 1718, respectively, were averaged together and then smoothed over height by taking three-point running averages. For each time period Doppler data provided profiles covering heights down to 7 km (solid lines and diagonally hatched areas in Fig. 18.26). As with the main updraft, the error in $F_m(Z)$ is estimated to be less than ± 40%. To obtain cloud base values of F_u, the flux of 2×10^8 kg s^{-1} for the main updraft at cloud base (which lies outside the Doppler data region) was added to the fluxes obtained from available Doppler measurements at this height in the active region. The aircraft data indicated that the amount of flux not accounted for by Doppler measurements at cloud base was about the same in both periods. Profiles of F_u and D_u were then completed as for the main updraft.

Figure 18.26a shows $F_u(Z)$ for the 1635 (thick line) and 1712 (thin line) time periods; the profile for the main updraft at 1635 (dotted line) is included for reference to Fig. 18.24. Although the maxima of F_u are not seen in the figure, the profile shapes above and below suggest that they lie between about 4 and 6 km. For the main updraft, on the other hand, the maximum is around 6.5 km, and the main updraft accounts for only about half the upward flux in the active region around 1635. The remainder reflects the presence of weaker and shallower updrafts outside the main updraft. The height of the storm top (and the updraft) is about 1.5 km less

in the later period, and the maximum flux derived is about 10% smaller. At an altitude of 10 km, however, the upward flux during the later period is less than half that during the earlier period. The profiles of D_u in Figs. 18.26b and c illustrate that the maximum upper-level outflow occurs just above 10 km in the 1635 period and just below 9 km in the 1712 period.

The integrated downdraft flux, F_d, and integrated flux divergence, D_d, were calculated as for the updraft, except Doppler data are used down to 3.5 km. Because Doppler estimates are not reliable in the subcloud layer (see Appendix) D_d profiles in the subcloud layer are again given shapes suggested by Fankhauser's (1974) data. The flux divergence just below cloud base has been matched to that derived from Doppler measurements just above, and all the low-level outflow from the downdraft is assigned to a 500-m layer just above the surface. The rationale for this is discussed further below.

The procedure used here leads to a much larger estimate of flux divergence near the surface than was derived by Fankhauser (1974). To address this discrepancy we consider that the outflow (except in the immediate vicinity of strong downdrafts) probably takes the general form of a density current, as suggested, for example, by studies of Charba (1974), Goff (1976), Greene et al. (1977), and Simpson (1969). These authors indicate that thunderstorm outflows are typically on the order of 1-2 km deep, although Greene et al. discuss shallower currents of uncertain origin. For most cases where data permit a comparison, the basic flow structure is similar

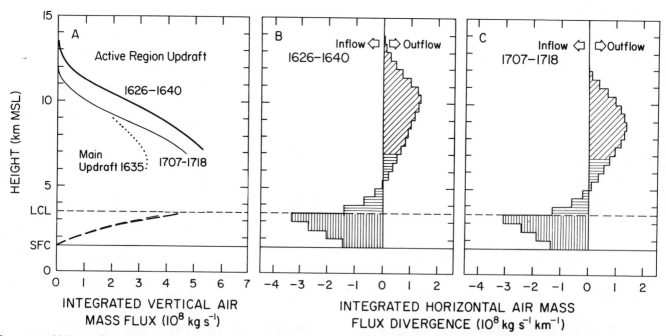

Figure 18.26 (a) Horizontally integrated vertical air mass flux in the portion of the active region where w > 0. Solid lines represent Doppler estimates averaged over three volume scans in each indicated time period. The dotted line is for the main updraft only (from Fig. 18.24a). The cloud base flux is obtained by adding the flux in the main updraft (obtained from aircraft measurements and not covered by Doppler data) to the Doppler-derived flux in the rest of the active region. (b,c) Horizontal air mass flux divergence associated with curves in (a) for the two time periods. The vertically hatched portion below cloud base is based on data of Fankhauser (1974). Horizontal hatching indicates subjective interpolation.

to that inferred from laboratory observations (e.g., Keulegan, 1957, 1958; Middleton, 1966) and theory (e.g., Benjamin, 1968). The coldest temperatures and most of the horizontal flux occur in approximately the lower half of the current, but above a relatively thin "friction layer;" Sinclair et al. (1973) suggests a logarithmic profile below 100 m. The "core" of the current at low heights is surmounted by a mixed layer that contains most of the vertical shear of the horizontal wind.

Fankhauser's data (Fig. 18.25) indicate inflow to the downdraft in the upper half of the subcloud layer, slight outflow at the lowest aircraft altitude, about 600 m above the surface, and more substantial outflow at the surface; the total depth of outflow is evidently no more than about 1 km. If the structure is as suggested above, it seems that his observations may not resolve the rather thin layer in which most of the outflow occurs. The surface measurements probably underestimate the speed of the flow because of frictional retardation near the ground, and the aircraft measurements 600 m above the ground probably lie in the sheared upper part of the outflow layer, above the layer of maximum horizontal flux.

We gain some insight as to the outflow depth in the present case from the CP-4 Doppler radar, which was positioned such that at 1632 it was roughly 5 km away from the surface gust front on the southwest side of Storm III. Figure 18.27 shows radial velocity measurements obtained by CP-4 near this time in a vertical plane oriented along a radial at 40° azimuth. Contours were drawn by hand from original measurements spaced 800 m apart in range and at 3° increments of elevation angle (dashed lines), and the radial component of overall storm motion was removed. These measurements are probably contaminated to some extent by

ground clutter, and no attempt has been made to correct for this. The effect would be to bias low-level measurements toward zero radial velocity, leading in this case to underestimation of the vertical shear of radial velocity near the ground. The nominal radar beamwidth is about 1.2°.

The cross section of Fig. 18.27 approximately intersects the center of surface divergence determined in Chapter 17, which is closely coincident with the maximum downdraft and the reflectivity core at low levels as determined from radar measurements. The point of intersection is indicated. Negative radial velocity for CP-4 (i.e., toward the radar) is consistent with the direction of flow at the ground under the storm (see Chapter 17) as well as with the apparent trajectories of downdraft air discussed earlier. The shaded area for $x > 12$ km designates where (storm-relative) radial velocities are less than -2 m s^{-1}.

Everywhere beyond $x = 12$ km in the figure there is pronounced shear of the radial wind between the lowest radar beam (at 0.5° elevation angle) and the second beam (3.5°). On the near side of the downdraft at a distance of 13 km, the axis of the second beam is 800 m above the surface. At 20 km, on the far side of the downdraft, it is at 1220 m. These heights are upper limits of the top of the shear layer at the positions observed. If the flow is basically as described above, most of the horizontal flux of cool downdraft air occurs well below these heights.

The shaded area centered about 4 km away from the radar indicates a region where the radar signal was found to be above the "noise" value, but for which reliable radial velocities could not be determined. The weak radar signal in this region is attributed to unknown targets (probably dust, insects, and other particulates) that are commonly associated

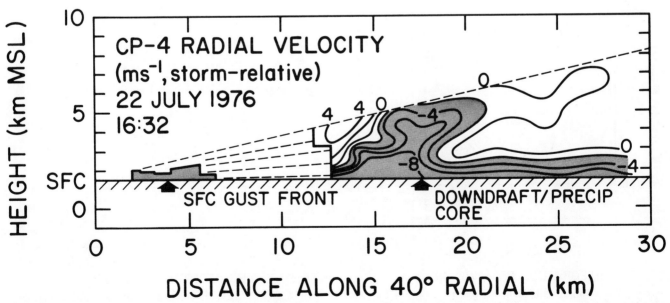

Figure 18.27 Vertical section of CP-4 Doppler radial velocity field at 1632, for azimuth of 40°. The component of storm motion along this direction has been subtracted. The plane of this section intersects the surface gust front and the downdraft core as indicated. Stippled area at distances > 10 km indicates radial velocity < −2 m s⁻¹. Stippled area centered at 4 km is an area where weak radar signal is detected but velocity estimates are unreliable.

with gust fronts. The approximate intersection of the surface front (see Chapter 17) with the plane of the figure coincides with the radar feature as indicated in the figure. The maximum depth of the region of discernible radar signal is not considered to be a useful indicator of the outflow depth because targets could have been present over a greater depth, but in low concentrations not detectable by the radar, and because in any case, the depth of the region containing such targets could differ substantially from the depth of the main outflow current.

Although aircraft observations reveal the presence of outflow air on the back side of the main updraft near cloud base, at a height of 1700 m above the ground and several kilometers behind the gust front, this air does not appear to be part of the main outflow current from Storm III. It in fact possesses a component of motion into the storm, and probably represents a mixture of Storm III outflow with outflow from Storm II (see the earlier discussion of trajectory L3) and with environmental air.

Profiles of $F_d(Z)$ and $D_d(Z)$ for the active region downdraft are shown in Figure 18.28 (note that the scale of the abscissa for D_d is changed from Fig. 18.26). The error is thought to be less than 70% at 3.5 km. Convergence of air into the downdraft is indicated throughout the depth of the storm, except for the outflow layer at the bottom.

It is interesting to note that the estimated downdraft mass flux in the 1712 period is approximately twice that in the 1635 period. Reflectivity magnitudes are comparable during the two periods, so that differences in precipitation loading are probably not responsible. The most apparent difference in the airflow structure is the greater ventilation of the storm

by environmental air in the middle levels in the later period, as seen for example in the greater storm-relative wind speeds at middle heights on the exposed south side of the updraft region in Fig. 18.16, as compared with Fig. 18.12. This suggests that the intrusion of potentially cold air into the cloud and the attendant evaporation of cloud material are more pronounced in the later period.

The trend in the flux estimates from all six volume scans is shown in Fig. 18.29. It can be seen that the variation among the three estimates in each of the two periods is not extreme, and is in fact well within the stated error limits. The slight trend toward decreasing updraft flux with time may not be significant, as there is some overlap of the individual estimates. However, all three of the downdraft flux estimates for the later period are less than those for the earlier period, thus strengthening the assertion that the downdraft flux was stronger at the later time. The relatively small variation of results from different times, compared to the rather large error limits given above, is consistent with the expectation that the most serious errors are similar at all times (see Appendix). Thus even if all estimates reflect a large absolute error, errors in differences from one time to another may be relatively small.

Figure 18.30 shows estimated profiles of net integrated vertical air mass flux (the sum of updraft and downdraft fluxes) for the two periods. According to these the net vertical momentum of air in the active region (proportional to the area enclosed on the right-hand side of the ordinate minus the area on the left-hand side) decreases from about 2.5×10^{14} kg m s^{-1} in the 1635 period to about 1.0×10^{14} kg m s^{-1} for the later period. This change reflects a

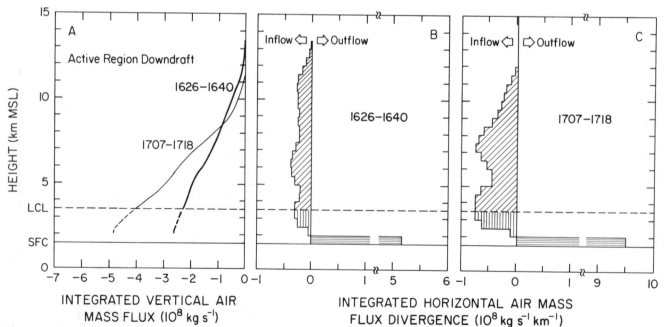

Figure 18.28 (a) Horizontally integrated downdraft mass flux, and (b,c) the associated integrated flux divergence in the active region for the two time periods. Solid lines and diagonal hatching represent Doppler-estimated portions. The vertically hatched portions are based on data of Fankhauser (1974). Horizontal hatching below 0.5 km represents low-level outflow required by mass continuity.

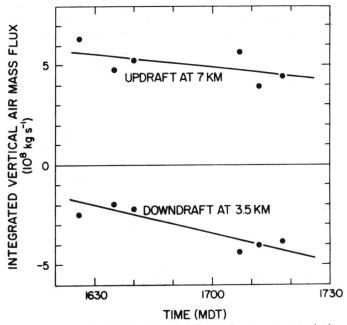

Figure 18.29 Integrated updraft and downdraft fluxes in the active region at two heights, from Doppler data of individual volume scans. Lines connect mean values for sets of three scans.

rather small (about 10%) decrease of the updraft flux, and a rather large (about 83%) increase of the downdraft flux between the two times. The net horizontal inflow to the active region, which occupies the layer from 2 to 6 km, is in fact greater at the later time (about 7.0×10^8 compared to 6.2×10^8 kg s^{-1}),but the net outflow above 6 km drops from 4.3×10^8 to 2.9×10^8 kg s^{-1}, while net outflow beneath the storm increases from 1.9×10^8 to 4.1×10^8 kg s^{-1}.

18.6.3 Comparisons With Other Observations

Braham (1952) summarized flux properties of "air-mass" thunderstorms (typically smaller than the one discussed here) observed in the Thunderstorm Project (Byers and Braham, 1949) and derived representative integrated vertical air mass flux values for various height ranges. He obtained these values by first considering measurements of draft speeds and widths of individual updrafts and downdrafts obtained in various height intervals, and then scaling the values (by the same factor at all heights) to be consistent with horizontal flux divergence estimates obtained from balloon tracks in the vicinity of mature thunderstorm cells. Data taken from Table 2 of Braham (1952) are plotted in Fig. 18.31. Although the magnitudes are nearly one order of magnitude less than have been estimated for the storm considered here, the shapes of his profiles are broadly consistent with the present results. His maximum updraft flux occurs between 3 and 4 km above the mean cloud base and in this aspect his updraft profile

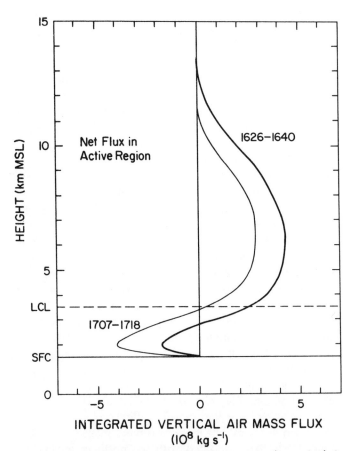

Figure 18.30 Integrated net vertical air mass flux in the active region, the sum of updraft flux (Fig. 18.26a) and downdraft flux (Fig. 18.27a).

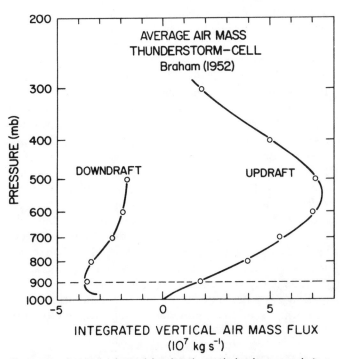

Figure 18.31 Integrated updraft and downdraft flux profiles based on average draft properties observed by aircraft in the Thunderstorm Project. Pressure at cloud base averaged 900 mb. The data are taken from Table 2 of Braham (1952).

matches that for the main updraft in this analysis fairly well. Our profiles for the active region indicate that the maximum storm-scale updraft occurs at a lower height than the maximum for the main updraft, and our data suggest that the flux through cloud base is larger, relative to the maximum flux, than indicated by Braham's profile. Braham's downdraft profile indicates that there is inflow to the downdraft throughout most of the cloud depth and that the maximum downward flux is near cloud-base height, consistent with results for the present storm.

The integrated updraft air mass flux at cloud base has been estimated for a number of convective storms, and results are generally consistent with estimates given here. Some examples follow, all stated in units of 10^8 kg s^{-1}. Auer and Marwitz (1968) found average values of 3.1 for eight northeastern Colorado storms, 1.3 for nine South Dakota storms, and 4.4 for one Oklahoma storm. These values were obtained from aircraft measurements by a method similar to that used in this analysis. Fankhauser (1974) obtained a value of 2.8 from the horizontal flux divergence measurements mentioned earlier, where the area considered was about 400 km^2. We have estimated the cloud base updraft flux to be about 4.3 in the 225 km^2 active region. Foote and Fankhauser (1973) derived a value of 20 using the same technique for a storm in which the area of updraft alone was estimated at 350 km^2. They estimated that about half the flux (9.6) was concentrated in an area of 120 km^2. All the above estimates pertain to storms that evidently produced hail. Braham's (1952) estimate of 0.18 pertains to smaller storms, as mentioned above.

Measurements of net integrated horizontal flux divergence as a function of height obtained from dual-Doppler analyses of small south Florida thunderstorms have been reported by Lhermitte and Gilet (1975) and by Frank and Lhermitte (1976). As in Braham's data, magnitudes are roughly one order of magnitude smaller than found in the present case. Both of these studies indicate net outflow in a layer just above the ground; the thickness of this layer appears to be about 2 km (except for a period when the storm was dissipating) in the former case and less than 1 km in the latter. In both cases it seems likely that a significant amount of inflow to the updraft at low-to-middle levels, and outflow from the downdraft at very low heights, was not seen by the radars. Thus actual depths of the layer of net outflow are probably less than these data indicate. Lhermitte and Gilet found that during the period of storm dissipation the layer of net outflow evidently deepened to at least 4 km. Both studies showed net inflow above the outflow layer up to 5 km to 6 km. In the present case our assumption that the downdraft outflow has a depth of 500 m determines the depth of net outflow. We find net inflow above this layer up to about 6 km.

18.7 DISCUSSION AND SUMMARY

The early evolution of Storm III, as discussed in Chapter

16, was dominated by the regular appearance on its right flank of new cells that grew to dominate the storm as older cells dissipated. Such a pattern of evolution is typical of what were called "organized multicells" in Chapter 16. During the last hour or so of its lifetime, however, including the period covered here by the Doppler analysis, the large-scale features of the echo pattern did not undergo significant change. During this period the storm had certain features in common with previous descriptions of supercell storms, including a quasi-steady reflectivity structure with a pronounced forward overhang, a persistent updraft at cloud base, and weak echo vaults (though the vaults were transient in the present case). It has been shown here, particularly for the 1626-1640 period, that the airflow from the triple-Doppler measurements was also similar in important ways to the supercell model of Browning and Foote (1976). Most significant in this regard were the existence of a large and long-lived updraft in the middle levels of the storm (a region of importance for the growth of hail), and the presence of a cyclonically streaming flow around the forward side of the updraft that was capable of transporting hail embryos into the vicinity of the updraft. The latter feature, and other microphysical implications of the storm airflow will be amplified in the following two chapters.

Trajectory analyses accomplished for the 1626-1640 period indicated that air contributing to the low-level downdraft had a variety of origins. Apparently air in both the cyclonic and anticyclonic branches of the flow passing around the sides of the storm in the middle levels contributed to the downdraft flux, the path of the cyclonic branch being the same as that inferred in Bronwing's (1964) model. Air initially in the low levels on the southeast and east sides of the storm also entered the downdraft after first being lifted several kilometers on the front side of the storm. The participation of this air in the storm circulation is qualitatively similar to that proposed in the original supercell model of Browning and Ludlam (1962).

While the general features of the flow during the later period, 1707-1718, were similar to those of the earlier period, including the presence of large updrafts and the ventilation of the updraft by the cyclonically streaming flow, the detailed aspects were perhaps more complicated. While low-level southeasterly momentum was observed in the updrafts at 6.5 km during the early period, westerly momentum tended to be seen there in the later period. It is possible that this change was related to the significantly weaker component of southerly inflow momentum in the later period. The presence of large updrafts in close proximity to one another (as shown, for example, in Fig. 18.17), is also a significant change, and may have led to important microphysical effects as proposed by Heymsfield et al. (1980). The downdraft branch originating with the cyclonic flow on the south side of the storm in the middle levels was more intense at the later time, while that associated with the northern anticyclonic branch had disappeared.

While the storm had attributes of previous supercell models during the 1615-1730 period, it also had certain unsteady features reminiscent of the cellular evolution of the multicell model (e.g., Browning et al., 1976). These included periodic echo intensifications and variations in the height of the cloud top. This behavior was interpreted in Chapter 16 in terms of fluctuations in an otherwise steady updraft. The Doppler analyses of 1626-1640 supported this view and, furthermore, suggested the nature of the updraft fluctuations. During this period the storm evolved not simply by the formation of discrete adjacent updrafts, as in the multicell model, nor by the pulsations of a single large updraft, which preserved exactly its form and position relative to the storm. Rather, the evolution was intermediate. As shown in Fig. 18.15, a new updraft surge started in the low levels at a position slightly ahead of the old updraft position but close enough so that the new updraft surge represented more a modulation in the shape of the existing updraft than the formation of a new updraft. In Chapter 20 this is referred to as an example of "weak evolution," as opposed to the "strong evolution" of the multicell model. Further discussion of the ramifications of such behavior is included in Chapter 20 after the presentation of the microphysical data in Chapter 19.

The profiles of updrafts deduced from the Doppler analysis were examined in both the horizontal and vertical dimensions. Maximum updraft values were in the range of 25 to 30 m s^{-1} and tended to occur between 7 and 8 km MSL, or about 4 km above cloud base. This is similar to some of the measurements reported by Marwitz (1973). In horizontal profile the updrafts had a nearly Gaussian shape near their centers, similar to axisymmetric jets. However, near the edges of updrafts the profiles tended to show pronounced minima, suggesting a structure more like a ring vortex or thermal.

The mass flux in the vertical direction was computed for various regions using the Doppler results along with aircraft data. Confining attention to the single large updraft at 1635, the upward flux attained a maximum value of 3.3 × 10^8 kg s^{-1} at an altitude of 6.5 km. Comparing this with the value of 2.0 × 10^8 kg s^{-1} determined from Queen Air data at cloud base one sees that about one-third of the flux into the convergent lower region of the main updraft may have taken place above cloud base.

If one considers a somewhat larger region 15 km on a side that contained the main updraft, termed here the active region, the flux was about 6 × 10^8 kg s^{-1}, or about double that in the main updraft. Thus, an approximately equal contribution to the upward flux in the active region came from the many smaller updrafts that surrounded the main updraft. They tended to be shallower and have their maxima at a lower altitude, so that the maximum in the total upward flux in the active region occurred at 5 km or so, somewhat lower than for the main updraft itself.

While convergence into the updraft was found to occur only in the lower part of the storm, convergence into the

downdraft in the active region occurred at all heights sampled by the Doppler radars. The outflow from the downdraft appeared to be confined to a layer near the surface, perhaps only 500 m thick. The maximum flux estimated for the downdraft was about 2 to 4 × 10^8 kg s^{-1}, somewhat less than that for the updraft.

Appendix

Errors and Resolution of Triple-Doppler Results

In this appendix we describe procedures used in the triple-Doppler radar data collection and analysis, consider what scales of motion are resolved, discuss the major sources of velocity errors and compare vertical air velocity estimates with measurements obtained from the T-28 aircraft and other sources.

A.1 DATA COLLECTION AND PROCESSING PROCEDURES

Velocity fields discussed in this paper were estimated by use of radial velocity measurements from the CP-3, NOAA-C, and NOAA-D Doppler radars (Fig. 18.1). Each radar sampled numerous range intervals while scanning back and forth in the azimuth sector containing the storm, starting at a low elevation angle and incrementing the elevation angle upward between sector scans until the top of the storm was sampled. Each volume scan took about 4 min. Sampling volumes were spaced 0.26 to 1.5 km apart and were nominally 1° across in azimuth and elevation and 150 m across in range.

Estimates of reflectivity-weighted radial velocity, V_r, were obtained using the covariance or pulse-pair method (see, for example, Berger and Groginski, 1973). Except for CP-3 the zero component of the Doppler spectrum was removed to reduce the influence of ground targets. Software of Kohn et al. (1978) was used for subsequent processing. Radar measurements were edited and then interpolated, using an isotropic inverse-distance weighting function with 1 km radius of influence, onto a rectangular grid arranged at 750-m horizontal and 500-m vertical intervals. At edges of the data volume extrapolation was permitted when two or more measurements were within the volume of influence. The locations of radial velocity data were adjusted by as much as 6 km to place them in a coordinate system moving with the radar echo (from 330° at 8 m s^{-1}, as deduced in Chapter 16.)

Interpolated data from the three radars were combined to estimate the components of precipitation and air motion, using transformations discussed by Armijo (1969) and detailed in the appendix to Chapter 14. A five-point scheme was used to estimate the horizontal divergence, $D = \partial u/\partial x + \partial v/\partial y$, and the boundary condition $w_b = 0$ was applied at the one grid interval (500 m) above the maximum height at which D was evaluated. Integration of the mass continuity equation was done by the trapezoid rule. Finally, the wind fields were smoothed with a two-dimensional Hanning filter applied in horizontal planes.

In this study we focus on the vertical velocity structure at middle heights. Because mid-level vertical velocities were derived by downward integration through the upper half of the storm, the approximate mean time of data in that region (to the nearest minute) is assigned as the effective mean time of a volume scan. For example, the 1626 scan began at the bottom of the storm at 1623 and ended at the top at 1627.

A.2 SCALES OF MOTION

Two low-pass filtering operations, namely the interpolation and the Hanning filter, were employed to reduce the effect of randomly distributed measurement errors and to help equalize the spacing of independent data in different directions. Smoothing caused by averaging of radial velocity over radar beamwidths has relatively little effect for scales passed by these operations, and is not considered here. The interpolation was done with inverse-distance weighting of radial velocity measurements in spherical volumes of 1 km radius. The number of measurements contributing to a particular interpolated value varied from about 20 to more than 100, depending on the radar and the location of the data. The Hanning filter had weights of 1/4, 1/8, and 1/16 for center, side and corner points of three-by-three point arrays.

To describe the response of these filters we consider the three-dimensional, point-symmetric radial velocity feature

$$V_r(R) = e^{-R^2/L^2}$$

where R is distance from an arbitrary center, and L is regarded as the "radius" of the feature. The amplitude at $R = O$, after interpolation only, is plotted as a function of diameter in Fig. 18.32 (dashed line); the corresponding curve for the Hanning filter only (not shown) is nearly the same. The solid line in Fig. 18.32 represents the product of the two response functions. For diameters larger than about 6 km the peak amplitude of velocity perturbations is diminished by less than 10%, while for diameters less than 2 or 3 km it is reduced to less than half.

Inspection of the wind fields shows that time continuity of the storm-scale velocity structure is fairly good between volume scans separated by as much as 27 min, while imbedded features of a few kilometers diameter are often difficult to identify between scans only 5 min apart. The smaller spatial scales of motion evident in the data appear to have evolved too rapidly to be described well by the observations.

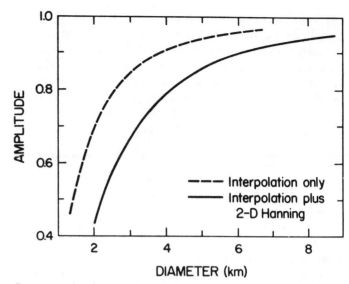

Figure 18.32 Smoothing properties of filtering operations used in triple-Doppler data processing. Dashed curve shows the effect of interpolation on the amplitude of a three-dimensional radial velocity feature defined by $V_r(R) = exp(-R^2/L^2)$. Solid curve shows approximately the effect of interpolation followed by 2-D Hanning filter.

A.3 VELOCITY ERRORS

Three types of velocity errors are discussed here: randomly distributed errors in the pulse-pair estimates of Doppler spectrum means, bias errors in radial velocity estimates that reflect differences betweem spectrum means and the true velocities we wish to represent, and errors in the boundary condition used to compute vertical air velocity, The following notation will be used: $a = a_T + a_R + a_B + a_b$, where a is a velocity or divergence estimate determined by the Doppler analysis and subscript T denotes the true value, R the random error, B the bias error, and b the boundary condition error.

The magnitude of random errors is illustrated by assuming a radar signal-to-noise ratio appropriate to 20 dBZ_e reflectivity at a representative location, and a Doppler spectrum standard deviation of 5 m s^{-1}. According to the formula of Berger and Groginsky (1973), the standard deviation of this error (over an ensemble of realizations) is approximately 1 m s^{-1} for each radar. Corresponding standard deviations, $\sigma(u_R)$ and $\sigma(v_R)$, range from about 0.5 m s^{-1} to 2.5 m s^{-1} for different positions in the storm, and $\sigma(D_R)$ varies from 1.3 × 10^{-3} s^{-1} to 2.2 × 10^{-3} s^{-1}. The effect of random measurement errors has been discussed more generally by Bohne and Srivastava (1976). Random errors accumulate when the divergence is integrated such that

$$\sigma[w_R(z)] = \frac{\sigma(D_R)}{\rho(z)}\left[\int_z^{z_0}\rho^2(\zeta)d\zeta\right]^{1/2}$$

For example if the boundary condition is set at $z_o = 10$ km and $\sigma(D_R) = 2 \times 10^{-3}$ s^{-1} at all heights, then $\sigma(W_R)$ is 3.8 m s^{-1} at $z = 3.5$ km. Average errors, $\overline{w}_R$, over areas used in mass flux calculations are smaller by a factor of nine or more due to cancellation of random errors.

Bias errors are thought to arise mainly from beam sidelobes that cause the radar signal to be influenced by off-axis targets, including but not limited to ground clutter. Data in the lowest 2 km above the ground are subject to large errors due to beam blocking and ground clutter. Other regions of particular concern are low-reflectivity regions near the edges of the storm and regions of large reflectivity and/or velocity gradients in the updraft region. Differences in the measurement times and the volumes sampled by different radars, pointing and ranging errors, and other sources also contribute to the bias errors. Unlike the random errors, these depend strongly on location in the storm, as they are related to reflectivity, velocity, and ground clutter distributions, as well as to the illumination patterns of the radar beams.

The standard deviation of divergence estimates, $\sigma(D)$, which includes the effects of true variations and random errors as well as bias errors, is approximately $3 \times 10^{-3}\text{s}^{-1}$. We propose that D_B is less than $\sigma(D)$ for most of the data. As an illustration suppose that $D_B = 3 \times 10^{-3}$ s^{-1} for all heights in a vertical column at a given (x, y) position and that $z_o = 10$ km. The error caused in w, given by

$$w_B(z) = \frac{D_B}{\rho(z)} \int_z^{z_o} \rho(\zeta)d\zeta.$$

is 9 m s^{-1} at 6.5 km and 14 m s^{-1} at 3.5 km.

The boundary condition on w was typically applied between 10 km and 12 km, in regions of 10-dBZ$_e$ to 15-dBZ$_e$ reflectivity near the top of the storm. Examination of north-south vertical sections through the regions of strongest updraft and downdraft has revealed that the 15-dBZ$_e$ reflectivity contour moves vertically by as much as ± 6 m s^{-1} over 90-s periods; this is in close agreement with the motion of the 5-dBZ$_e$ contour depicted in Fig. 17.9.

If the motion of the 15-dBZ$_e$ contour reflects mainly the advection of particles, the true vertical air motion, $w_T(z_o)$, equals the motion of the contour plus the fall speed. To illustrate, we assume $w_b = 10$ m s^{-1} and $z_o = 10$ km. The resulting error at lower heights is given by

$$w_b(z) = \frac{\rho(z_o)}{\rho(z)} w_b(z_o).$$

At 6.5 km $w_b = 7$ m s^{-1} and at 3.5 km $w_b = 5$ m s^{-1}.

Consider a square area 7 km on a side, which encloses an area comparable to the main updraft. Suppose that everywhere on the perimeter the bias error in velocity normal to the perimeter is 1 m s^{-1} and directed outward. The error in the average divergence over the enclosed area is then about 0.6×10^{-3} s^{-1}. If the error is in the same sense at all heights, then $\overline{w}_B$ is about 2 m s^{-1} at 6.5 km and 3 m s^{-1} at 3.5 km.

Average values of w, designated $\overline{w}$, associated with maximum updraft and downdraft fluxes are about 11 m s^{-1} at 6.5 km in the main updraft, 8 m s^{-1} at 7 km in the active region updrafts, and -5.5 m s^{-1} at 3.5 km in the active region downdrafts. Relative to these values, the errors

depicted in illustrations above are quite large, but average errors involved in the flux calculations are thought to be much less. Random errors are nearly eliminated by averaging, and average bias and boundary condition errors are assessed subjectively as follows.

Regarding boundary condition errors, the largest motions of the reflectivity contours appear to occur at scales of 3 km to 6 km, while average motions over distances comparable to the dimensions of the main updraft appear to be no more than ± 3 m s^{-1}. Over periods of about 15 min the average motion of the contours in the selected cross section was about -2 m s^{-1}. This is comparable to the expected particle fallspeeds, suggesting that average boundary condition error is fairly small. We suggest representative values of $\overline{w}_b(z_o) = 4$ m s^{-1} for the main updraft, and $\overline{w}_b(z_o) = 2$ m s^{-1} for the active region updrafts and downdrafts. At the heights of maximum fluxes these yield values of w_b less than 3 m s^{-1} and 1.5 m s^{-1}, respectively.

Assuming that bias and boundary condition errors have magnitudes as suggested above and that they are additive, errors on the order of $\pm 40\%$ in the updraft fluxes and $\pm 70\%$ in the downdraft fluxes are indicated. Smaller errors associated with random errors, errors in updraft and downdraft areas, and errors caused by extrapolation have been neglected, but the error limits given are nevertheless thought to be conservative.

Errors in differences between updraft and downdraft fluxes in the two periods considered are probably substantially less than would be indicated by these limits. Because the basic reflectivity and velocity structures are similar at all times, the bias errors, whatever they may be, are also likely to be similar at different times. Based on the motion of the 15-dBZ$_e$ reflectivity contour near the storm top, it appears that the difference in the average boundary condition error between the two time periods might be as small as 1 m s^{-1}. It seems reasonable to expect that error limits for the differences are less than one-half of the limits stated above.

A.4 COMPARISON OF VERTICAL VELOCITY ESTIMATES

Measurements obtained during four penetrations by the T-28 aircraft are available for comparison with the triple-Doppler radar results discussed here. Though such a comparison between independent measurements is valuable, a number of difficulties arise to complicate the interpretation. These include the fact that the aircraft and Doppler measurements represent quite different time and space scales, and that they are not exactly coincident in time (though the correspondence is probably as good as one could expect to obtain). For example, it took about 2 min to acquire the Doppler data required for determining vertical velocity at the T-28 flight level near 6.5 km MSL, and the intercomparisons involve T-28 data collected as much as 3 min before or after the mean time of Doppler observations. Also, vertical accelerations due to buoyancy could be as large as 10^{-1} m s^{-2}, so that the evolution of non-steady updrafts alone could account for differences as large as the errors just discussed. Finally, it should be noted that the T-28 estimates of w are also subject to uncertainty (see Chapter 19). Despite this situation, agreement between the two independent measurements serves to strengthen confidence in both.

Figures 18.33 through 18.35 compare T-28 estimates with Doppler estimates of w. The T-28 tracks (adjusted to storm-relative coordinates) are shown in the upper part of the figures, superimposed on the Doppler updraft fields for horizontal planes near the flight altitude. Doppler estimates used for the comparison shown in the lower part of the figures are determined by two-dimensional linear interpolation to points where the T-28 track intersects vertical planes of the analysis grid that are approximately normal to the track. These figures, along with the similar Fig. 18.7 already presented for the 1640-1642 period, show that the agreement between the independent radar and aircraft measurements, though hardly impressive, is perhaps as good as one should expect. While the two estimates tend to agree about the sign of w, differences in magnitude of 5 to 10 m s^{-1} are typical. The best agreement is seen for the broad and intense updraft in Fig. 18.7, discussed previously, where the time differences involved are also the least. Also the existence of two large updrafts in Fig. 18.34 is supported by both measurements. The biggest difference in the estimates, at 1715:20, is almost 30 m s^{-1}. However, it is apparent at this time, and in fact throughout, that considerable improvement in the comparison would result from making a small shift in the position or time of one of the measurements. The nature of such an adjustment for the various times does not indicate a systematic time or positioning error, however. Given the difficulties mentioned above, the two data sets can be considered as supporting one another as regards the general existence, position, and areal extent of the updrafts. For Figs. 18.7 and 18.34 the agreement in the magnitude of the peak updrafts, to within a few meters per second, is really quite good.

Measurements at cloud base by the NCAR Queen Air and surface mesonet data (both described in Chapter 17) are also available for comparison, though they are not spatially coincident with the Doppler observations. The aircraft measurements of the location of the updraft at cloud base are basically consistent with the Doppler results. For the larger updrafts, the magnitude is also consistent. However, in certain isolated cases (for example on the south edge of the echo at 1626, see Fig. 18.13) the Doppler results indicate very strong updraft cells centered at rather low heights. In this extreme case w is about 30 m s^{-1} at 5 km, about 3.5 km above the ground. This implies extremely strong convergence ($D \cong -10^{-2}$ s^{-1}) over the 3.5-km interval, which is not confirmed by independent measurements. Maximum convergence calculated from surface wind mesurements is about 3 10^{-3} s^{-1}, and maximum cloud base updraft values of 6 to 8 m s^{-1} at 2 km above the ground indicate a similar magnitude of the average convergence throughout the

subcloud layer. Isolated areas of much stronger convergence could have escaped observation, but it appears that the Doppler results may overestimate some updrafts by 10 m s^{-1} or so.

Downdrafts in the Doppler results reach peak speeds of 10 to 20 m s^{-1} at cloud base (as mentioned previously, reliable measurements are not available below cloud base). The aircraft at cloud base did not observe downdrafts stronger than about 5 m s^{-1} but it did not sample in the region of strong radar echo where downdrafts were identified in the Doppler analysis. Maximum surface divergence at 1630 was about 5×10^{-3} s^{-1} and positioned beneath the dominant downdraft indicated in 1635 Doppler results. If the surface divergence estimate is applied throughout the 2-km subcloud layer, a downdraft of 10 m s^{-1} is indicated; the Doppler results show a peak speed of 14 m s^{-1} in this downdraft. The strongest downdraft in the Doppler data (19 m s^{-1} at cloud base at 1712) would imply a mean divergence of 10^{-2} s^{-1} in the subcloud layer, which cannot be ruled out for limited areas not resolved in the mesonet observations.

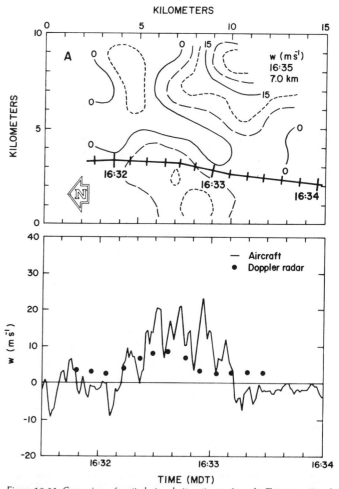

Figure 18.33 Comparison of vertical air velocity estimates from the T-28 aircraft and triple-Doppler radars near 1633 MDT. Top: The storm-relative aircraft track is superimposed on updraft contours from the Doppler radar analysis at a height near the flight altitude. Bottom: Aircraft estimates are represented by solid line, and Doppler estimates obtained by bilinear interpolation to the aircraft track are shown by dots.

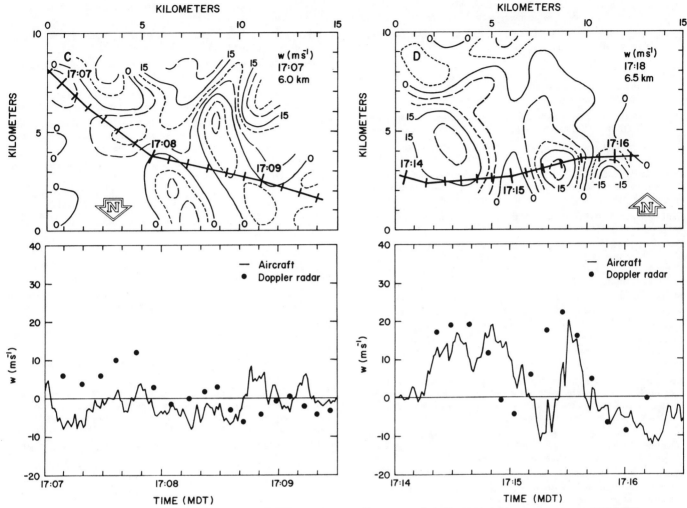

Figure 18.34 As in Fig. 18.33, but for measurements near 1708 MDT.

Figure 18.35 As in Fig. 18.33, but for measurements near 1715 MDT.

References

Armijo, L., 1969: A theory for the determination of wind and precipitation velocities with Doppler radars. *J. Atmos. Sci.* 26, 570-573.

Auer, A. H., Jr., and J. D. Marwitz, 1968: Estimates of air and moisture flux into hailstorms on the high plains. *J. Appl. Meteorol.* 2, 196-198.

Barnes, S. L., 1978: Oklahoma thunderstorms on 29-30 April 1970: Part I. Morphology of a tornadic storm. *Mon. Weather Rev.* 106, 673-684.

Battan, L. J., 1980: Observations of two Colorado thunderstorms by means of a zenith-pointing Doppler radar. *J. Appl. Meteorol.* 19, 580-592.

Benjamin, T. B., 1968: Gravity currents and related phenomena. *J. Fluid Mech.* 31, 290-248.

Berger, T., and H. L. Groginsky, 1973: Estimates of spectral moments of pulse trains. IEEE Conf. on Information Theory, Tel Aviv, Israel, 1973, Inst. of Electrical and Electronics Engineers, New York, N.Y.

Bohne, A. R., and R. C. Srivastava, 1976: Random errors in wind and precipitation fall speed measurement by a triple Doppler radar system. Prepr. 17th Conf. on Radar Meteorology, Seattle, Wash., 1976, Am. Meteorol. Soc., Boston, Mass., 7-14.

Braham, R. R., Jr., 1952: The water and energy budgets of the thunderstorm and their relation to thunderstorm development. *J. Meteorol.* 9, 227-242.

Brandes, E. A., 1978: Mesocyclone evolution and tornadogenesis: some observations. *Mon. Weather Rev.* 106, 995-1011.

-----, 1981: Fine structure of the Del City-Edmond tornadic mesocirculation. *Mon. Weather Rev.* 109, 635-647.

Browning, K. A., 1964: Airflow and precipitation trajectories within severe local storms which travel to the right of the winds. *J. Atmos. Sci.* 21, 634-639.

-----, and G. B. Foote, 1976: Airflow and hail growth in supercell storms and some implications for hail suppression. *Q. J. R. Meteorol. Soc.* 102, 499-533.

-----, and F. H. Ludlam, 1962: Airflow in convective storms. *Q. J. R. Meteorol. Soc.* 88, 117-135..

-----, J. C. Fankhauser, J-P. Chalon, P. J. Eccles, R. G. Strauch, F. H. Merrem, D. J. Musil, E. L. May, and W. R. Sand, 1976: Structure of an evolving hailstorm: Part V. Synthesis and implications for hail growth and hail suppression. *Mon. Weather Rev.* 104, 603-610.

Byers, H. R., and R. R. Braham, Jr., 1949: *The Thunderstorm.* U. S. Govt. Printing Office, Washington, D. C., 287 pp.

Charba, J., 1974: Application of gravity current model to analysis of squall-line gust front. *Mon. Weather Rev.* 102, 140-156.

Chisholm, A. J., 1973: Alberta hailstorms: Part I. Radar studies and airflow models. *Meteorol. Monogr.* 14 (36), 1-36.

Eagleman, J. R. and W. C. Lin, 1977: Severe thunderstorm internal structure from dual-Doppler radar measurements. *J. Appl. Meteorol.* 16, 1036-1048.

Fankhauser, J. C., 1971: Thunderstorm-environment interactions determined from aircraft and radar observations. *Mon. Weather Rev.* 99, 171-192.

----, 1974: Subcloud air mass flux and moisture flux attending a northeast Colorado thunderstorm complex. Prepr. Conf. on Cloud Physics, Tucson, Ariz., 1974, Am. Meteorol. Soc., Boston, Mass., 271-275.

-----, 1976: Structure of an evolving hailstorm: Part II. Thermodynamic structure and airflow in the near environment. *Mon. Weather Rev.* 104, 576-587.

Foote, G. B., and J. C. Fankhauser, 1973: Airflow and moisture budget beneath a northeast Colorado hailstorm. *J. Appl. Meteorol.* 12, 1330-1353.

Frank, H. W., and R. M. Lhermitte, 1976: Cell interaction and merger in a south Florida thunderstorm. Prepr. 17th Conf. on Radar Meteorology, Seattle, Wash., 1976, Am. Meteorol. Soc., Boston, Mass., 151-156.

Goff, R. C., 1976: Vertical structure of thunderstorm outflows. *Mon. Weather Rev.* 104, 1429-1440.

Greene, G. E., H. W. Frank, A. J. Bedard, Jr., J. A. Correll, M. M. Cairns, and P. A. Mandics, 1977: Wind shear characterization. U.S. Dept. of Commerce, Federal Aviation Admin., Rep. No. FAA-RD-77-33, 120 pp. Available through NTIS, Springfield, Virginia, 22161.

Heymsfield, A. J., A. R. Jameson and H. W. Frank, 1980: Hail growth mechanisms in a Colorado storm: Part II. Hail formation processes. *J. Atmos. Sci.* 37, 1779-1807.

Heymsfield, G. M., 1978: Kinematic and dynamic aspects of the Harrah tornadic storm analyzed from dual-Doppler radar data. *Mon. Weather Rev.* 106, 233-254.

Keulegan, G. H., 1957: Thirteenth progress report on model laws for density currents: an experimental study of the motion of saline water from locks into fresh water channels. National Bureau of Standards Rep. No. 5168, U.S. Dept. of Commerce, Washington, D. C., 21 pp.

-----, 1958: Twelfth progress report on model laws for density currents: the motion of saline fronts in still water. National Bureau of Standards Rep. No. 5831. U.S. Dept. of Commerce, Washington, D. C., 29 pp.

Klemp, J. B., R. B. Wilhelmson, and P. S. Ray, 1981: Observed and numerically simulated structure of a mature supercell thunderstorm. *J. Atmos. Sci.* 38, 1558-1580.

Kohn, N. M., A. L. Johnston, and C. Mohr, 1978: MUDRAS: Multiple Doppler radar analysis system. NOAA Tech. Memo No. ERL WPL-35, National Oceanic and Atmospheric Administration, Wave Propagation Laboratory, Boulder, Colo.

Kropfli, R. A., and L. J. Miller, 1976: Kinematic structure and flux quantities in a convective storm from dual-Doppler radar observations. *J. Atmos. Sci.* 33, 520-529.

Kyle, T. G., W. R. Sand, and D. J. Musil, 1976: Fitting measurements of thunderstorm updraft profiles to model profiles. . *Weather Rev.* 104, 611-617.

Lamb, H., 1945: *Hydrodynamics.* Dover Publications, New York, N.Y., 738 pp.

Lemon, L. R., 1976: Wake vortex structure and aerodynamic origin in seven thunderstorms. *J. Atmos. Sci.* 33, 678-685.

-----, and C. A. Doswell III, 1979: Severe thunderstorm evolution and mesocyclone structure as related to tornadogenesis. *Mon. Weather Rev.* 107, 1184-1197.

Lhermitte, R. M., and M. Gilet, 1975: Dual-Doppler radar observation and study of sea breeze convective storm development. *J. Appl. Meteorol.* 14, 1346-1361.

Ludlam, F. H., and R. S. Scorer, 1973: Convection in the atmosphere. *Q. J. R. Meteorol. Soc.* 79, 317-341.

Marwitz, J. D., 1972a: The structure and motion of severe hailstorms: Part I. Supercell storms. *J. Appl. Meteorol.* 11, 166-179.

-----, 1972b: The structure and motion of severe hailstorms: Part II. Multi-cell storms. *J. Appl. Meteorol.* 11, 180-188.

-----, 1973: Trajectories within the weak echo regions of hailstorms. *J. Appl. Meteorol.* 12, 1174-1182.

Middleton, G. V., 1966: Experiments on density and turbidity currents: I. Motion of the head. *Can. J. Earth Sci.* 3, 523-546.

Miller, L. J., 1975: Internal airflow of a convective storm from dual-Doppler radar measurements. *Pageoph* 113, 765-785.

Moncrieff, M. W., and M. J. Miller, 1976: The dynamics and simulation of tropical cumulonimbus squall lines. *Q. J. R. Meteorol. Soc.* 102, 373-394.

Nelson, S. P. 1977: Rear-flank downdraft: a hailstorm intensification mechanism. Prep. 10th Conf. on Severe Local Storms, Omaha, Nebr., 1977, Am. Meteorol. Soc., Boston, Mass., 521-525.

Paluch, I. R., 1979: The entrainment mechanism in Colorado cumuli. *J. Atmos. Sci.* 36, 2467-2478.

Ray, P. S., 1976: Vorticity and divergence fields within tornadic storms from dual-Doppler radar observations. *J. Appl. Meteorol.* 15, 879-890.

Rotunno, R., 1981: On the evolution of thunderstorm rotation. *Mon. Weather Rev.* 109, 577-586.

Schlesinger, R. E., 1978: A three-dimensional numerical model of an isolated thunderstorm: Part I. Comparative experiments for variable ambient wind shear. *J. Atmos. Sci.* 35, 690-713.

-----, 1980: A three-dimensional numerical model of an isolated thunderstorm: Part II. Dynamics of updraft splitting and mesovortex evolution. *J. Atmos. Sci.* 37, 395-420.

Schlicting, H., 1960: *Boundary Layer Theory*, Chapter 23. McGraw-Hill, New York, N. Y.

Simpson, J. E., 1969: A comparison between laboratory and atmospheric density currents. *Q. J. R. Meteorol. Soc.* 95, 758-765.

Sinclair, R. W., R. A. Anthes, and H. A. Panofsky, 1973: Variation of the low level winds during the passage of a thunderstorm gust front. NASA Contractor Rep. No. CR-2289, National Aeronautics and Space Administration, Washington, D. C., 65 pp.

Squires, P., 1958: Penetrative downdraughts in cumuli. *Tellus* 10, 381-389.

Turner, J. S., 1962: The 'starting plume' in neutral surroundings. *J. Fluid Mech.* 13, 356-358.

-----, 1963: The flow into an expanding spherical vortex. *J. Fluid Mech.* 18, 195-208.

-----, 1964: The dynamics of spheroidal masses of buoyant fluid. *J. Fluid Mech.* 19, 481-490.

Wilhelmson, R. B., and J. B. Klemp, 1981: A three-dimensional numerical simulation of splitting severe storms on 3 April 1964. *J. Atmos. Sci.* 38, 1581-1600.

Woodward, B., 1959: The motion in and around isolated thermals. *Q. J. R. Meteorol. Soc.* 85, 144-151.

CHAPTER 19

The 22 July 1976 Case Study: Storm Structure Deduced from a Penetrating Aircraft

Andrew J. Heymsfield and Dennis J. Musil

19.1 AIRCRAFT SAMPLING PATTERNS AND THE GENERAL FEATURES OF THE STORM AT THE MIDDLE LEVELS

The temporal and spatial changes in the character of the storm at the middle levels were investigated primarily from penetrations with the South Dakota School of Mines and Technology (SDSM&T) T-28 armored aircraft, with supporting conventional and Doppler radar measurements. Six penetrations were made through the updraft region over a 50-min period, from 6.0 to 7.3 km MSL, $-10\,°C$ to $-19\,°C$. The track of the T-28 aircraft was obtained from an M-33 tracking radar. The "skin paint" locations of the aircaft in the CP-2 radar data indicated that they were accurate to better than 0.5 km. The intrumentation and accuracy of the measurement systems on board the T-28 are considered in Appendix A; the calculations of vertical velocity have been discussed by Heymsfield and Parrish (1979).

Data from twelve penetrations through the storm inflow region by NCAR Queen Air N306D (see Chapter 17) complemented the T-28 data at the midstorm levels. These measurements indicated that the cloud base altitude varied from approximately 3.7 to 3.9 km and the corresponding cloud base temperatures varied from $+7.6\,°C$ to $+5.7\,°C$. The altitude of the $0\,°C$ level within the updraft core was approximately 4.7 km, as estimated from a pseudoadiabat and the measured cloud base conditions.

Prior to describing the T-28 measurements below, it will be useful to define the characteristics of the various storm regions at the middle levels during the measurement period. The measurements in Chapters 16-18, and those described here, indicate that the storm at these levels was composed of several identifiable regions or "zones," which were found throughout the measurement period. The approximate locations of these zones are indicated on a photograph of the storm taken at 1653 MDT and on a schematic plan view diagram (Figs. 19.1a and b).

Zones I and II composed the region of the most intense updrafts, and were distinguished from each other according to the T-28 measurements described later by a difference of about $0.5\ g\ m^{-3}$ in the measured cloud liquid water content and about 3 K in the equivalent potential temperature. Zone III was associated with downdrafts originating at about the 8-km level, as inferred from the Doppler data (see Chapter 18). These downdrafts did not extend to the surface directly below but were carried to the north into Zone IIIa, where they supplied one source of downdraft air for the gust front. This zone was not directly penetrated by the T-28 because

the radar echo typically reached 55 dBZ_e, and the region was considered unsafe for penetration. Zone IV contained the forward overhang region of the storm. New cumulus congestus clouds and new major convective impulses formed in this region, but it had weaker updrafts than the main updraft core (see Chapters 16, 17, and 18). These cells frequently developed through a region characterized by radar echo from 5 to 20 dBZ_e. Zone IV extended as a visual cloud band around the forward side of the storm, as seen in Fig. 19.1a, between 5 and 10 km. This cloud layer, which we call the precipitation debris region, apparently formed from the decay of convective elements in the cumulus congestus region of Zone IV and from debris from the main updraft core (Fig. 19.1b). Zone V had the weakest vertical motion measured with the T-28, and was associated with precipitation falling from the anvil.

The horizontal wind fields in all zones were relatively constant over the observational period, as noted in the discussion of Doppler radar data in Chapter 18 and by Heymsfield et al. (1980). Inflow into the updraft core was from the southeast and the storm moved toward the southeast (Chapter 16 and Fig. 19.1b). At the midstorm levels, the storm-relative windflow was from the west (Chapters 16 and 17, and Fig. 19.1b).

This chapter concentrates on interpreting measurements that were obtained during the T-28 penetrations into the various regions of the storm. The locations of these penetrations relative to the zones of the storm are indicated in Fig. 19.1b and in Table 19.1. T-28 penetrations 2, 4, and 6 went through the main updraft/downdraft core from west to east; penetration 3 intersected the forward overhang, the anvil. The measurements obtained during penetration 1 are not discussed in detail, since that penetration was not directly associated with or along the periphery of the main updraft region. In the discussion of the reflectivity data below, the labeling scheme used to classify the "impulses" or "cells" follows that for the major impulses given in Chapter 16, but smaller-scale impulses are labeled differently.

19.2 CHARACTERISTICS OF THE MAIN UPDRAFT REGIONS AND THE DOWNDRAFT CORES (ZONES I, II, AND III)

The T-28 encountered four updraft regions since, during penetration 6, the updraft region was split into two cells. Reflectivity data for these regions, or "cells," at the aircraft

163

A: COMPOSITE (WRAP-AROUND) SABRELINER PHOTOGRAPH

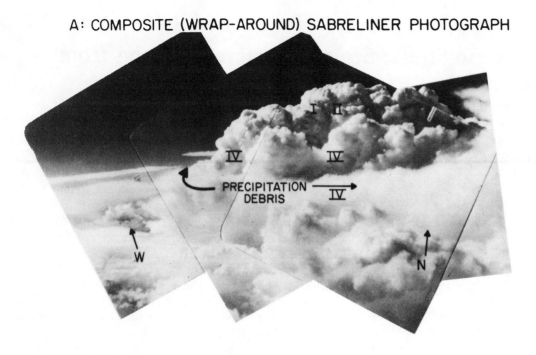

B: SCHEMATIC PLAN VIEW OF STORM

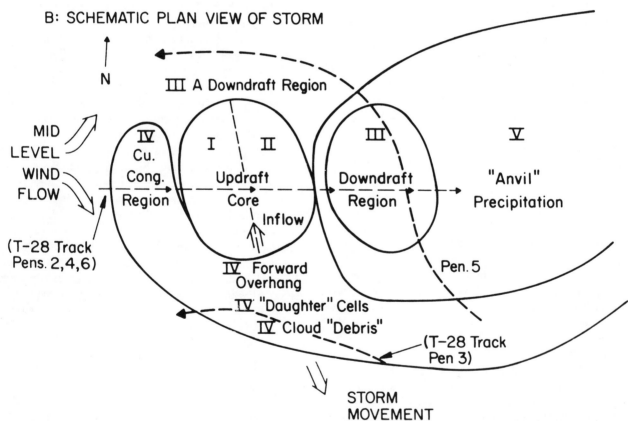

Figure 19.1 Depiction of the zones sampled on 22 July 1976 at the midstorm levels: (a) "Wraparound" photograph taken from the NCAR Sabreliner at 1653 MDT, with zones indicated, (b) schematic depiction of zones in (a) in a plan view. Approximate positions of T-28 penetrations relative to main updraft core are indicated.

altitude are given in Fig. 19.2 to show their staged development at the time of penetration. As indicated by the data obtained in the "weak echo" regions (Z_{min}) associated with each of these updrafts, the patterns of reflectivity development differed between cells. Cells I (penetration 2) and K1 (penetration 4) developed in regions of pre-existing, relatively high reflectivity. As these cells developed, each weak echo region (WER) initially became more distinguishable, and then

TABLE 19.1
Zones penetrated by T-28 aircraft

Zone description	Penetration number
I – West portion updraft core	2, 4, 6
II – East portion updraft core	2, 4, 6
III, IIIA – Downdraft region	2, 4, 6
IV – New cumulus congestus	1, 2, 3, 5, 6
New major convective impulses	
Precipitation debris from convective	
towers and main updraft cores	
V – Precipitation falling from anvil	1, 2, 3, 4, 5, 6

Penetration number	1	2	3	4	5	6
Times (MDT)	1630–1634	1640–1642	1650–1654	1656–1658	1705–1710	1714–1716

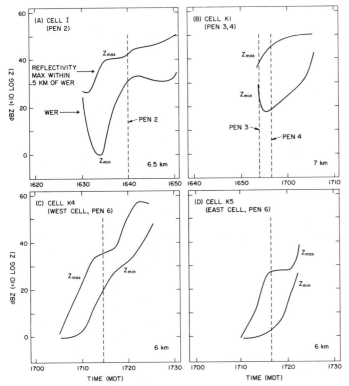

Figure 19.2 Temporal evolution of the radar reflectivity in the cells penetrated by the T-28 at the flight altitude. Data showing the minimum reflectivity (Z_{min}) and the maximum reflectivity (Z_{max}) within 5 km of Z_{min} are indicated. Times of the T-28 penetrations 2, 3, 4, and 6 are indicated with dashed vertical lines.

"filled in" within a few minutes. Cells K3-K4 (penetration 6) and K5 (penetration 6) were first detected in regions essentially devoid of pre-existing radar echo at the aircraft level. The WERs associated with these cells persisted for periods of nearly 10 min. The T-28 penetrations into each of the cells typically occurred several minutes after a minimum in reflectivity had occurred within the WER.

The maximum reflectivity (Z_{max}, Fig. 19.2) obtained within 5 km of each of the WERs discussed above indicates the stage of development of precipitation at the time that each cell was penetrated. Each penetration occurred at a time when the maximum reflectivity at the aircraft level was increasing; precipitation generated aloft apparently descended to the aircraft level following the time of each T-28 penetration. Peak reflectivities associated with the sampled cells ranged from 40 to 60 dBZ$_e$.

Vertical and horizontal sections through the penetrated cells, with radar echo contours and velocity vectors, are given in Figs. 19.3a and 19.4a. There were reflectivity maxima about 2 km above the aircraft level in all the cell penetrations (Fig. 19.3a). Local reflectivity minima (WER's) extended downward from the maxima. The T-28 (Fig. 19.3a) penetrated all four WERs. The attitude of the T-28 penetrations relative to the forward (southern) edge of the storm, as determined from the reflectivity measurements (Fig. 19.4a), ranged from 3 km for penetration 6 to 6 km for penetrations 2 and 4. Peak reflectivities were 45, 33, 53, and 43 dBZ$_e$ in the four cells sampled. The reflectivity minima coincided with the regions of strong updraft velocities as derived from Doppler radar analysis. The areas of updraft extended to the south of the positions of the T-28 penetrations (Fig. 19.4). Regions of downdraft tended to occur along the east side (Zone III), extending to the north (Zone IIIa) of each updraft region. Within the updraft cores, the horizontal wind was from the southeast during penetration 2 and from the west during penetration 6. A westerly flow extended in the low-reflectivity regions around the forward edge of the storm (Zone IV) during all penetrations.

The following discussion presents the measurements of the updraft structures, the thermodynamic characteristics, and the microphysical features of the storm during the three penetrations that encountered updraft cores and associated downdraft regions.

19.2.1 Vertical Air Motion Measurements

The vertical wind velocities derived from the T-28 measurements (see Appendix A) appear in Fig. 19.3b. The approximate bounds of the main regions of updraft, as determined from these measurements, are indicated as dashed vertical lines in Figs. 19.3b-f and the cells corresponding to each of these updrafts are identified in Fig. 19.3b. Peak updrafts of 19, 26, and 18 m s^{-1} were found during penetrations 2, 4, and 6, respectively. The locations of the strongest updrafts for each cell coincided with the WERs. The magnitudes and positions of the updrafts for penetrations 2 and 6 compare favorably with those found from the Doppler measurements (see the more complete discussion found in Chapter 18). Variations in the updraft strength were noted during each penetration, similar in scale to the fluctuations noted in the subcloud environment (Chapter 17). In the second updraft core sampled during penetration 6 a peak updraft velocity of 15 m s^{-1} was observed. As mentioned in Chapter 18, there is

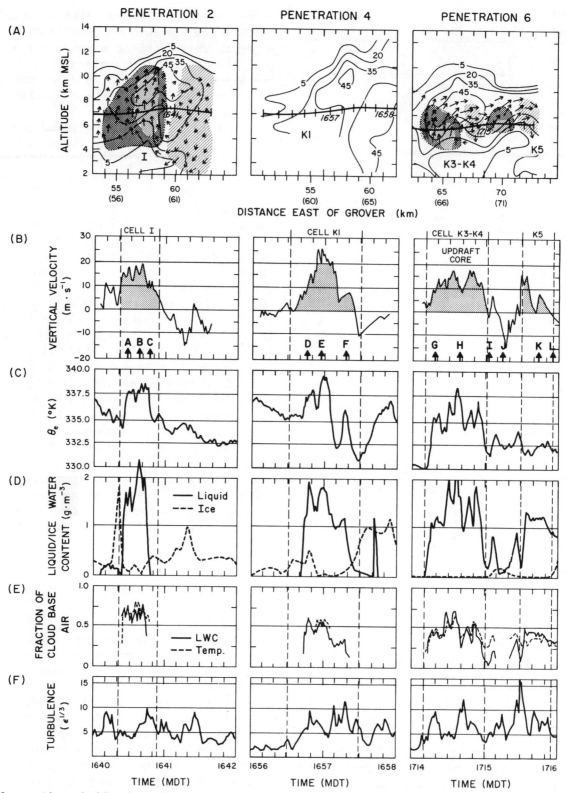

Figure 19.3 Summary of dynamical and thermodynamical data obtained from three T-18 penetrations into the updraft core and associated downdraft regions. Penetrations were from west to east. Regions between arrows associated with dashed vertical lines denote the approximate locations of the updraft region as determined from the T-28. (a) Vertical sections showing the measured radar reflectivity and the storm-relative wind fields derived from Doppler analysis at the times of the penetrations. Coordinates show the storm position at 1635 MDT for penetrations 2 and 4, and at 1700 MDT for penetration 6. Coordinates in parentheses are relative to ground coordinates. Times of penetrations and aircraft track are indicated. The regions of updraft > 10 m s^{-1} are surrounded with dotted lines, and regions > 20 m s^{-1} are stippled. Hatching represents regions of downdraft stronger than -5 m s^{-1}. Locations of cells sampled are designated. Wind vector lengths $= 15$ m s^{-1} km^{-1}. (b) Vertical air motion from T-28 measurements, updrafts shaded. Periods in which spectra were plotted in Fig. 19.5 (A-L) are indicated along lower axis. (c) θ_e. (d) Liquid and ice content (g m^{-3}). (e) Derived fraction of cloud base air at the aircraft level based on the measured liquid water and temperature. (f) Derived turbulence $\epsilon^{1/3}$, cm$^{2/3}$ s^{-1}.

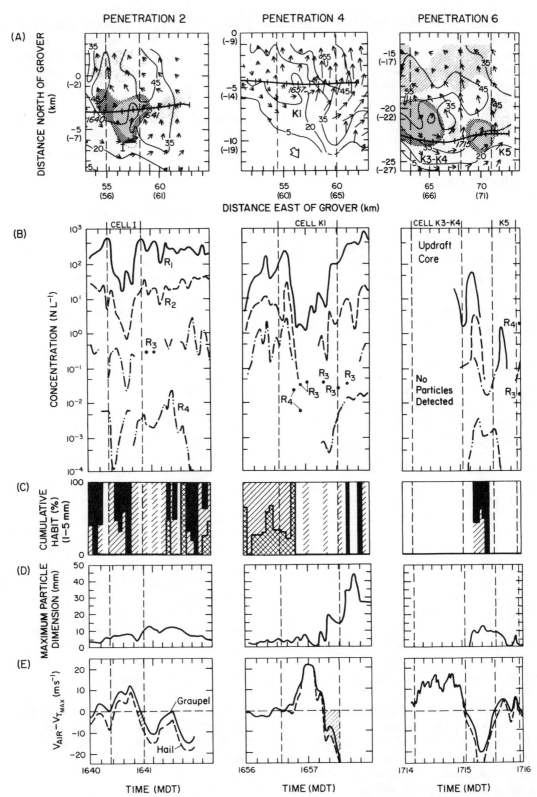

Figure 19.4 Summary of microphysical data obtained during the three T-28 penetrations into the updraft core. Dashed vertical lines show the bounds of the strong updraft regions. (a) Horizontal sections at the aircraft level showing the aircraft track, the radar reflectivity, and the storm-relative wind fields from the Doppler analysis. Symbols for Doppler-derived vertical velocities are the same as in Fig. 19.3a. Direction of storm motion is indicated with wide arrow. Locations of cells sampled are designated. (b) Measured ice particle concentrations based on 4-s averages over the size ranges designated as R_1 (25-250 μm), R_2 (250-1000 μm), R_3 (1000-5000 μm) and R_4 (> 5000 μm). Gaps in data indicated that no particles were sampled. (c) Percentage of each particle habit in size range R_3 for each 4-s spectrum in (b) above, expressed as a percentage of total. Solid bars indicate graupel particles; cross-hatching indicates unrimed aggregates of planar crystals; hatching indicates rimed aggregates of planar crystals. Gaps in data indicate regions where no particles of this size were sampled. (d) Maximum particle dimension. (e) Sedimentation velocity of the particle with the highest terminal velocity, V_{air}-V_T. The represented calculation assuming solid-line low-density graupel; the dashed-line solid-ice density, from each spectrum in (b) above. Shaded region indicates where some particles are falling through the updraft.

a discrepancy of about 1 km between the T-28 and Doppler determinations location.

The strongest observed downdrafts, 10-15 m s^{-1}, were located to the east of each updraft core; strong downdrafts were not observed to the west of any of the updraft regions. The triple-Doppler wind field analysis shows that extensive downdraft regions were also found to the north of the T-28 penetrations (Fig. 19.4a).

19.2.2 Equivalent Potential Temperature (θ_e) and Temperature

Considerable variability of θ_e was observed throughout the updraft regions (Fig. 19.3C).* Peak θ_e values coincided with the WER's for all cells except cell K3. There were higher values of θ_e on the western side of the updraft (Zone I) and lower values on the eastern side (Zone II) extending into the downdraft region during penetrations 2 and 4. This pattern may have been due to differing entrainment mechanisms because of differing wind fields along the western and eastern edges of the updraft (see Chapter 18). The values of 340-346 K for θ_e, computed for the inflow regions from measurements made by the subcloud aircraft, were consistently 2-5 K higher than the values of θ_e derived for the midstorm levels, indicating entrainment of environmental air throughout the updraft region.

The temperature "excess" within the updraft regions was estimated by comparing measured temperatures within the updraft regions to those found upwind and outside of cloud at about the same level. Peak positive temperature excesses of 2.5 °C, 2 °C, and 3 °C were found during penetrations 2, 4, and 6, respectively.

In order to estimate the degree of entrainment within the updraft region the trajectories of air parcels encountered along the T-28 track were computed backward in time to the point when they were located below cloud base. This trajectory computation, which used the Doppler-derived wind fields, was possible only for penetrations 2 and 6, since no vertical velocity data were available for penetration 4. The wind fields were assumed to be steady over the 2-3 min required for the air parcel to ascend from the cloud base to the T-28 level. The subcloud aircraft was within 1 km of the parcel's location and within 1-2 min of the time when the parcel was at that level. For penetration 2, source θ_e values were typically 341 K. This was considerably lower than the value of 346 K found below cloud base along the forward (southeast) portion of the inflow region, and apparently represented a mixture of inflow and outflow air from a different storm (see Chapter 17). At the origin of updraft air for

penetration 6, θ_e values showed wide variations—340-346 K and 340-344 K for the western and eastern updrafts. Comparisons between θ_e in the inflow region and at the midstorm levels suggest that some mixing occurred throughout the updraft region, but it must be kept in mind both that the single penetration of each updraft by the T-28 represents a limited sample and that the trajectory calculations rest upon possibly unrealistic assumptions about the storm steadiness.

The discussion of entrainment will be continued in Section 19.2.5.

19.2.3 Liquid Water and Ice Concentrations

The highest values of cloud liquid water content (LWC) obtained with the Johnson-Williams (J-W) probe were found in the strongest region of updraft (the WER) in each penetration (Fig. 19.3d). The peak LWC values of 2.5 g m^{-3} were 70% to 80% of the adiabatic values, which ranged from 3.1 g m^{-3} for penetration 2 to 3.2 to 3.6 g m^{-3} for penetrations 4 and 6. The uncertainty of ±20% in the J-W measurements suggests that regions of adiabatic LWC may have been present; however, the presence of adiabatic cores is unlikely in view of the evidence of mixing from the θ_e measurements.

Droplet concentrations (not indicated) averaged 750-800 cm^{-3} and the mean droplet diameter ranged between 15 μm and 18 μm within the updraft region, so that the cloud droplet spectra were "continental." Droplet diameters did not exceed 24 μm. The LWCs obtained with the Particle Measuring Systems (PMS) FSSP and the J-W probe were well-correlated through the updraft regions. The absolute accuracy of the FSSP liquid water content measurements was estimated to be ±50%.

The ice mass concentration, computed from the measured size spectra and derived particle habits over the entire size spectrum (see Appendix B), was typically less than 1.0 g m^{-3}. The ice content varied considerably between penetrations, from near 0 during penetration 6 to 1 g m^{-3} over an extensive area during penetration 4. The highest ice water content was noted on the eastern side of each updraft region except in penetration 6. The combined liquid and ice water contents were lower than the adiabatic water contents by at least 1.0 g m^{-3}. The low ice content within the peak updraft region (the WER) during each penetration suggests that depletion of the liquid water by the growing ice particles was negligible within the updraft core.

19.2.4 Turbulence

The derived values of turbulence, defined in terms of the eddy dissipation rate ($\epsilon^{1/3}$ in cm$^{2/3}$ s^{-1}), are presented in Fig. 19.3f. Turbulence values tended to be at a minimum in the strongest updraft regions (the WERs). This may reflect in part

*The θ_e values were computed from the measured temperature and pressure by assuming the environment was saturated with respect to liquid water. This assumption is valid only within the updraft regions. Outside of the updraft regions, in which high concentrations of ice particles are noted, the environment is likely to be close to ice saturation conditions. In these regions, θ_e is likely to be overestimated by at most 2 K.

the way the values were calculated, because gradients in airspeed can pass through the Fourier transformation and appear as turbulence (see Johnson et al., 1980). The highest turbulence values in each penetration were found in the eastern side of the main updraft region. These relatively high turbulence values were well correlated with marked reductions in the LWC from the values found within the strongest updraft region. The high $(\epsilon^{1/3})$ value noted along the western edge of the second updraft, sampled during penetration 6, coincided with the pilot's voice comments that the aircraft was barely controllable at that time.

19.2.5 Entrainment

Although entrainment must have taken place over a wide range of altitudes within the updraft region, the wind field data in Chapter 18 suggest that there were layers in which entrainment was especially active. To examine the possible origin and amount of the entrained air along the T-28 track within the updraft regions, the measured temperature and LWC were compared with those predicted from very simplified entrainment calculations. The calculations assumed that the entrained air originated at a single specified level and mixed with updraft air in varying proportions. Calculations were done for three levels: 4.7 and 5.6 km (below the aircraft level) and 7.6 km (above the aircraft level). For each specified level, the entrained air was assumed to mix either upward or downward to the aircraft level, and the LWC and temperature were calculated as a function of the assumed fraction of entrained air. For any calculated LWC at the aircraft level, mixing from above produced considerably warmer temperatures (1 °C-3 °C) than mixing from below (see Paluch, 1979). Matching the calculated and measured LWC and the calculated and measured temperatures provided a best estimate of (1) the fraction of cloud base air at the aircraft level, and (2) the altitude of origin of the entrained air.*

The resulting estimated fraction of cloud base air at the aircraft level appears in Fig. 19.3e. The calculations suggest that the eastern and western edges of these updraft regions were mixed more extensively than the main updraft core, but that the updraft regions were mixed throughout. Within and to the west of the region of the strongest updrafts (Zone I), the calculations suggest that the entrained air originated between the 4.7 and 5.6 km levels. Within the updraft core, the entrained air may have originated in a storm to the east and have been introduced into the updraft core below cloud base

(see Chapter 17). Along the eastern border of the updraft region (Zone II), the calculations suggest that the entrained air originated above the aircraft level. This level of entrained air is not surprising, in view of the strong downdrafts noted to be to the east of this region in Chapter 18.

19.2.6 Ice Particle Measurements

Ice particle size spectra measured with the PMS two-dimensional probe, the foil impactor, and the hail spectrometer were compared throughout the T-28 aircraft research period. These three instruments yielded consistent size spectra in size regions where measurements overlapped, giving a high degree of confidence in the data.

Examples of particle size spectra obtained within the updraft regions near the western and eastern peripheries and within the core of each updraft region appear in Figs. 19.5a-l. The times and locations at which the spectra were measured are indicated as A-L in Fig. 19.3b. The data are plotted in a log-log format, and are normalized according to the channel width. With the exception of the spectra measured in the western cell during penetration 6 (Figs. 19.5g-i), the western portion of each updraft core contained broad and continuous size spectra. Continuous, broad spectra were also noted along the eastern portion of the updraft core during penetrations 2 and 4 (Figs. 19.5c and f). Calculated radar reflectivities based on the measured spectra were within 5 dB of the actual radar reflectivities in these regions. With the exception of penetration 2, narrow spectra or no detectable particles were noted within the intense updraft cores. These particle-free regions were associated with radar WERs (Fig. 19.3a).

The particle habits that composed the measured spectra are indicated in Fig. 19.5, based primarily upon visual inspection of the 2-D data (see Appendix B). For particle sizes $< 200\ \mu m$ it was nearly impossible to determine the particle habits; these are referred to as "small particles" in the figures. Along the western periphery of the updraft core, the observed particles were primarily graupel and rimed particles (Figs. 19.5a, d, and j), some of which were aggregates.** Similar particles were observed within the eastern portion of the updraft region (Figs. 19.5c and f). Within the updraft core, the observed particles were primarily small in size and rounded (e.g., Figs. 19.5b and e), suggesting riming. Examples of 2-D images of particles obtained within the updraft regions appear in the figure.

The concentrations of ice particles measured during penetrations 2, 4, and 6 are indicated in Fig. 19.4b. The con-

---

*This technique does not account for the effects of depletion of liquid water by growing ice particles (which is negligible; see Section 19.2.8) and the consequent warming of the air due to the release of latent heat. Calculations indicated that if the liquid water concentration were reduced by 10% (approximately 0.3 g m⁻³) below the adiabatic value due to depletion, the predicted temperature would be underestimated by 0.1 °C to 0.3 °C. The ice content values were generally lower than 0.3 g m⁻³ within the strong updraft regions (Fig. 19.3c), with the exception of the east side of the updraft in penetration 2 and the west side of the updraft in penetration 4.

**The following three criteria were used in the identification of the particle habit and the degree of riming: graupel was distinguished by a relatively smooth surface with no identifiable structures or single crystals; branches of unrimed crystals or lightly rimed (referred to as unrimed) have the ability to transmit light because they are thin; and unrimed or lightly rimed aggregates have large open areas between branches. It was impossible to distinguish the habit of particles smaller than about 150 μm. However, the lack of any "small" particles measured with the 2-D probe in the updraft core during penetrations 4 and 6 suggests that droplets > 25 μm were not present, a conclusion that is consistent with the cloud droplet probe data.

UPDRAFT REGIONS

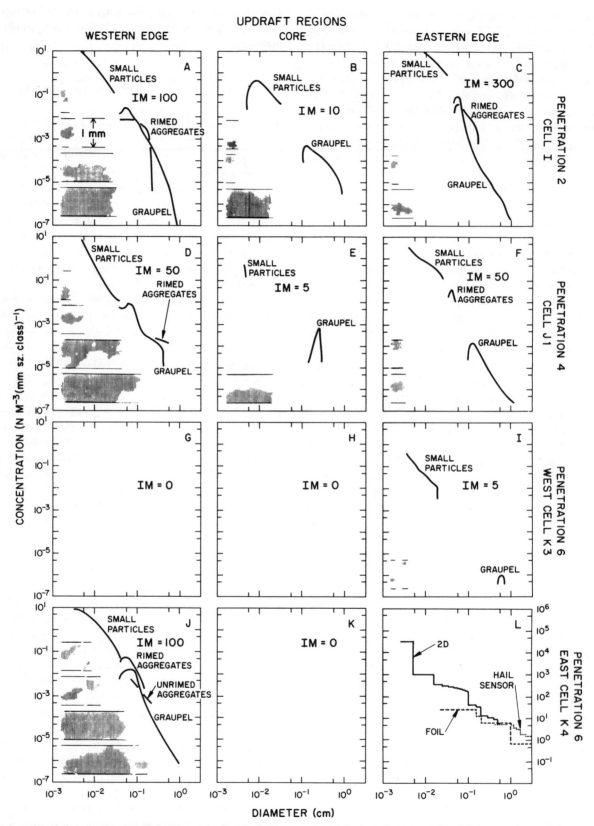

Figure 19.5 Examples of hydrometeor data obtained along the western edge (left column), in the core (central column), and along the eastern edge (right column) of the updraft region at times indicated in Fig. 19.3b. The first row is from Cell I, the second from Cell K1, the third from Cell K4, and the fourth from Cell K5. Examples are shown of two-dimensional particle images obtained from the indicated spectra. The scale is shown in A. Particle habits composing the observed spectra are indicated. IM refers to the ratio of the measured concentration of crystal sizes R_1 and R_2 to the "expected" ice crystal concentration based on primary ice nucleation. In (L), the bounds of the size categories used for each probe in deriving the spectra are indicated. The concentration (m^{-3}) that would be measured in each size category if one particle was sampled per 4-s interval is indicated.

centrations varied considerably during each penetration and from one penetration to the next. For "small" ice particles, represented by particles in the ranges 25-250 μm (referred to as R_1) and 250-1000 μm (R_2), the lowest ice particle concentrations were found within the updraft cores (the WERs). Virtually no ice particles of sizes R_1 and R_2 were found within the updraft cores sampled during penetrations 4 and 6. Comparatively high concentrations of particles of these sizes were found along the western portion of the updraft region during penetrations 2 and 4, and consistently high concentrations of particles were found along the eastern portion of each updraft region and within the downdrafts.

The spatial variability of the concentration of ice particles of "moderate" sizes, R_3, 1 to 5 mm, was similar to that of the smaller ice crystals, with the highest concentration in the downdraft region and the lowest concentration in the updraft core. Comparatively high concentrations of ice particles of size R_3 were found along the western portion of the updraft region during penetrations 2 and 4.

As noted previously, virtually no particles larger than 5 mm in diameter, R_4, were found within the updraft core (the WER) during any of the penetrations. Comparatively high concentrations of large particles were noted along the eastern portion of each updraft, in the region where the reflectivity cores were apparently descending (Fig. 19.3a). High concentrations of particles of size R_4 were commonly found in the downdraft regions and in the reflectivity cores (Fig. 19.3b).

The predominant particle types observed within and along the western and eastern peripheries of the updraft regions depended upon the cell sampled and the sampling location. Fig. 19.4c shows the percentage of each observed particle habit over the size range 1-5 mm, R_3. No single planar crystals were observed in size range R_3 at these times. Graupel and rimed aggregates of planar crystals were the predominant particle types observed on the western side of the updraft (Zone I) during penetration 2, while unrimed and rimed aggregates of planar crystals were the predominant forms observed in this region during penetration 4. Note that the reflectivity in this portion of the updraft region was considerably higher during penetration 2 than during the other penetrations (Fig. 19.3a). Regions of graupel, hail, and rimed aggregates of planar crystals were found along the eastern side of the main updraft regions (Zone II) and in the downdraft cores (Zone III) during each penetration, in association with the reflectivity maxima in these regions (Fig. 19.3a). When "large" particles were noted in the intense updraft core during penetration 2, they were graupel.

The maximum measured particle size from each 4-s distribution, plotted in Fig. 19.4d, showed that the largest particles were on the eastern side of the updraft (Zone II) and extended into the downdraft (Zone III). Some of the large ice particles noted in the downdraft during penetration 4 were a low-density form of hail or large graupel (Jameson and Heymsfield, 1980). All particles noted along the western side

of each updraft were smaller than 1 cm. Particles in the updraft core (the WER) were limited to a few millimeters in diameter.

19.2.7 Ice Particle Sedimentation in the Updraft

The terminal velocities of all measured particles were calculated to estimate density for each 4-s measurement, using the observed particle habits. The highest terminal velocity within each spectrum was combined with the measured updraft or downdraft to yield a "sedimentation velocity," (Fig. 19.4e). When the particle with highest terminal velocity was a graupel, two densities were used: solid ice, or an empirical value appropriate to northeastern Colorado (Heymsfield, 1978).

In most places within the strong updraft cores (the WERs), all measured particles were being rapidly lifted by the updraft; in the downdraft regions, particles were descending at velocities up to 3 m s^{-1}. Along the western portion of the updraft core during penetration 2 there was a small region (shaded in Fig. 19.4e) in which particles were descending in the updraft. The large particles noted during this penetration were able to enter the updraft core, as inferred from the reflectivity pattern in Fig. 19.3a and the wind field data. However, particles were not able to descend appreciably and enter the updraft during the other penetrations. Along the eastern side of the updraft during penetrations 2 and 4, a region 1 km wide was found in which some particles were falling through the updraft. Relatively high reflectivities were noted within these regions (Fig. 19.3a), and it is likely that particles in these regions were advecting around the forward side of the storm while descending and growing.

19.2.8 Depletion of Liquid Water

The depletion time constant, defined as the time required to reduce the LWC to 1/e of the initial value through particle collection, was calculated (Appendix B) from the size spectra measured throughout penetrations 2, 4, and 6. Computed values of τ_L in regions containing liquid water provide rough estimates of subsequent depletion of liquid water; τ_L was computed over the size ranges R_1 (25-250 μm), R_2 (250-1000 μm), R_3 (1-5 mm), and R_4 (>5 mm). The conversion of liquid water to precipitation mass at the ground is likely to be more effective for R_3 and R_4 particles than for R_1 and R_2, since most of the latter might well evaporate below cloud base. Calculations of τ_L are given in Fig. 19.6. The fraction of the total depletion by R_4 particles is also shown.

The computed depletion time constants within the updraft regions (Fig. 19.6) were mostly found to be > 10^3 s, and thus the conversion of liquid water to precipitation mass that would have subsequently taken place in these regions would have been relatively inefficient. Within the intense updraft

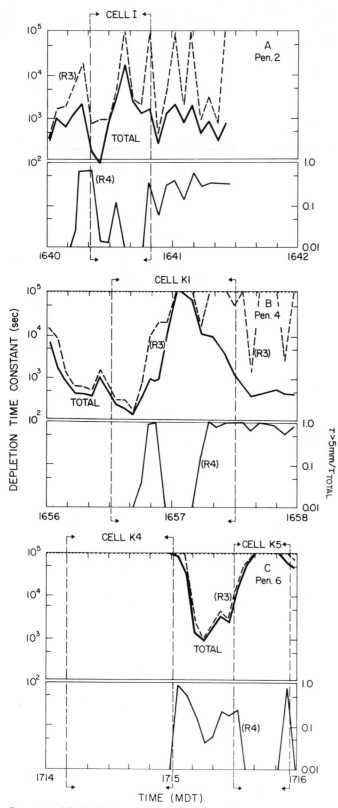

Figure 19.6 Calculated liquid water depletion time constants during three penetrations into the main updraft cores and associated downdraft regions. Dashed vertical lines show approximate boundaries of main updraft regions (same as Figs. 19.3 and 19.4). The three top panels (scales at left) give τ_L for the total particle ensemble as well as all in size range R_3. The lower panels (scales at right) give the fraction of total depletion by particles in size range R_4 (> 5 mm).

cores (the WERs), τ_L was found to be exceedingly long. Particles growing above the WER were likely to have been growing in LWC's that were largely undepleted by prior ice particle growth. The depletion time constants were lowest along the edges of the updrafts. Along the western portions of the updraft regions, most depletion was due to particles in the size ranges R_2 and R_3. Along the eastern portions of the updraft regions R_4 particles accounted for most of the depletion of liquid water; in these regions, the depletion times within the downdraft regions were $\approx 10^3$ s, suggesting that the conversion of liquid water to precipitation mass had been relatively effective. Most of the depletion of liquid water in these regions was due to particles of hail size, indicating that hail was relatively effective in converting liquid water to ice either around the forward edge of the storm or over the WER. Hail was relatively efficient in converting liquid water to ice mass. This is an exception to the rule that particles smaller than hail size do most of the depleting of liquid water, as was concluded in Chapter 7. It is just in the location where one might anticipate such an exception: near the top of a "collapsing" vault. Browning and Foote (1976) first pointed out this inflection of the radar vault structure.

19.2.9 Particle Temperatures

Particle temperatures were computed on the basis of the assumed particle densities and collection efficiencies, as discussed in Appendix B. Temperatures were lower than 0 °C for all particles sampled throughout the three penetrations into the strong updraft cores. This suggests that dry growth predominated at and above the aircraft level. The 2-D images of graupel particles with rough surfaces (Fig. 19.5) also suggest that dry growth predominated.

19.3 CHARACTERISTICS OF THE PERIPHERAL REGION OF THE STORM AT MIDDLE LEVELS

19.3.1 Forward Portion of the Storm (Zone IV)

The forward portion of the storm (Zone IV, the "forward overhang") of 22 July 1976 usually had the visual appearance of a glaciated cloud layer, but new convective towers were occasionally observed within this region (Fig. 19.1a). Radar reflectivity and air motion data indicated that the particles in the forward overhang region originated on the western side of the storm from decaying cloud turrets, from decaying regions of the feeder cells, or from the main updraft cores (see Chapters 16 and 18), and were advected around the storm by the environmental winds, which were diverging around the updraft regions. Occasionally, new convective towers developed through this region; the reflectivity data in Figs. 19.2a-b suggest that cells I and J1 developed through this region.

19.3.2 Cumulus Congestus Towers in Zone IV

Several weak updraft regions ("turrets") were penetrated along the western side of the main updraft region. Radar echo patterns associated with these turrets moved around the forward (southern) edge of the storm following their decay and contributed to the precipitation debris region (see Chapter 16). Typically, peak reflectivities did not exceed 20 dBZ$_e$, and the turrets appeared to have lifetimes of 5-10 min as determined from the reflectivity data. One turret was sampled by the T-28 prior to penetration 2. This turret, sampled at an altitude of 6.8 km, was characterized by weak vertical air motion at that stage of its lifetime. The maximum value of θ_e in this turret, 337 K, was comparable to that in the western portion of the main updraft region. The LWC was less than 0.1 g m^{-3}. The ice particles were small (< 1 mm) unrimed and lightly rimed planar crystals and aggregates, at a total concentration of about 10 ℓ^{-1}.

19.3.3 Characteristics of the Quiescent Regions of Zone IV

Measured reflectivities in these quiescent regions were typically 5-25 dBZ$_e$. Weak updrafts and downdrafts (1-3 m s^{-1}), low LWC (0.0-0.1 g m^{-3}), and low ice particle concentrations (< 1 ℓ^{-1}) were typical.

The particle size and habit data indicated that planar crystals were the predominant ice crystal types within these regions, as, for example, can be seen from the 2-D images in Fig. 19.7. Planar crystals would be expected to form at the sampling temperatures of −12 °C to −16 °C. A typical size spectrum measured in this region appears in Fig. 19.8. Generally, particles smaller than 0.5 mm were platelike forms and dendrites or stellars. For sizes between 0.5 and 1.0 mm, dendritic and stellar forms composed the spectra. The maximum sizes of planar crystals and dendrites believed to have developed within the storm were calculated based on the times available for particles to advect around the forward side of the storm (approximately 20 min) and assuming water saturation. Calculated sizes were somewhat larger than those found from the measurements. Particles larger than about 1 mm were generally aggregates, some of which were larger than 5 mm. Particle spectra were frequently bimodal (see Fig. 19.8). The sampling volume of the 2-D probe is limited, and it is likely that aggregates considerably larger than those indicated in Fig. 19.8 were present.

19.3.4 New Updraft Impulses in Zone IV

The most complete measurements within new convective impulses associated with the propagation of the main updraft region took place during penetration 3, the location of which

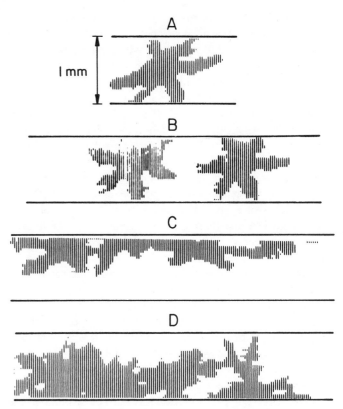

Figure 19.7 Examples of two-dimensional particle images obtained in the precipitation debris region: dendrites (stellars) and aggregates.

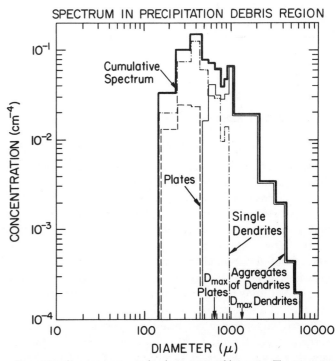

Figure 19.8 Size spectrum measured in the precipitation debris region. The size spectrum for each of the observed particle habits (plates, dendrites, and aggregates of dendrites) and the cumulative spectrum are indicated.

appears in Fig. 19.1. For the purpose of consistency with the data presented from penetrations 2, 4, and 6, data from penetration 3 (Fig. 19.9) are plotted with time increasing from right to left as though the aircraft penetrated from west to east, even though the sampling was from east to west. A region of updraft approximately 5 km wide with velocities of 8-14 m s^{-1} was encountered (Fig. 19.9a). This updraft developed through the precipitation debris region, and was associated primarily with the development of cell J1 (Fig. 19.2b), which was sampled again during penetration 4. There were comparatively high values of θ_e (Fig. 19.9b) and of LWC (Fig. 19.9d) along the eastern portion of the updraft and lower values of θ_e and LWC along its western portion. The ice content was low throughout the penetration (Fig. 19.9d). Turbulence values were highest toward the eastern boundary of the cell (Fig. 19.9c).

Particle concentrations, habits, and sizes indicated considerable variability throughout the cell. The concentration of ice particles (Fig. 19.9e) decreased from west to east. A well-defined region of rimed aggregates and graupel particles (Fig. 19.9f) was observed in the eastern part, while unrimed and rimed aggregates were observed within the western part. 2-D images of unrimed or lightly rimed dendrites and a rimed aggregate (western portion) and a heavily rimed aggregate (eastern portion) appear in Figs. 19.10a-c. The largest particle sizes observed at any location in the cell ranged from 1 to 5 mm (Fig. 19.9g). It is likely that particles larger than 5 mm were present, but in concentration below that detectable with the 2-D probe.

Most particles were rising in the updraft (Fig. 19.9h). The high concentration of particles along the western portion of the penetration (Fig. 19.9e) was responsible for the low values of τ_L (Fig. 19.9i) observed in this region.

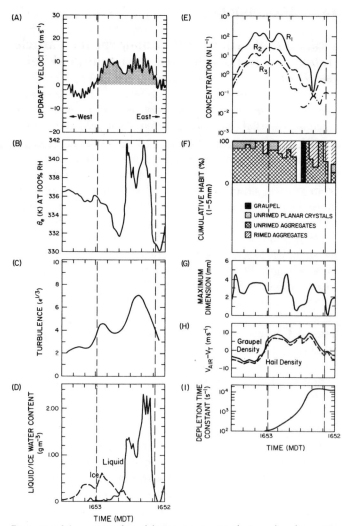

Figure 19.9 Measurements obtained during penetration into the forward overhang region of cells. Data are plotted backward in time so that penetration is similar to that presented in Figs. 19.3 and 19.4 (from west to east). The parameters are the same as those described for Figs. 19.3, 19.4, and 19.6.

19.3.5 Anvil Region (Zone V)

Information on the characteristics of precipitation falling from the anvil region was obtained to the east of the main updraft region. The most extensive measurements were obtained during penetration 5, which commenced on the southeast side of the main updraft/downdraft core, and, circling around the core, concluded on the northwest side (Fig. 19.1b). The vertical motions were generally quite weak throughout this region. Several areas of positive vertical velocities of 4-6 m s^{-1} were noted, in addition to several small regions with downdraft velocities of 5 m s^{-1}. LWC between 0.1 and 0.2 g m^{-3} was occasionally noted.

Rimed aggregates of dendrites and aggregates of irregular particles were the predominant particle habits. Ice particle concentrations frequently exceeded 100 ℓ^{-1} and maximum particle sizes frequently exceeded 5 mm throughout the anvil. The ice content ranged between 0.2 and 1.0 g m^{-3}.

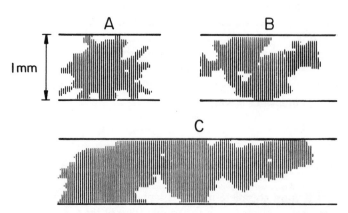

Figure 19.10 Examples of two-dimensional particle images obtained in the updraft region in the forward overhang (penetration 3).

19.4 DISCUSSION

19.4.1 Relationship of Predominant Ice Particle Habits and Maximum Sizes to Position Within the Storm

Contrary to previous research indicating that most precipitation-sized particles within thunderstorms were heavily rimed crystals or graupel (e.g., Musil et al., 1976), the T-28 observations here show that a spectrum of particle types existed. To understand more fully the reason for the variability in particle types, the observed particle habits and the measured sizes of particles were compared to the corresponding radar reflectivities. For all 4-s spectra obtained throughout the flight, corresponding to distances of 400 m, the percentage of particles of specified habits and sizes observed* within each 5-dB interval from 0 to 55 dBZ_e was determined (Fig. 19.11). Considerable insight into the relationship between the measured particle forms and sizes and the reflectivity has been gained from these data, hereafter referred to as frequency-reflectivity data. At radar reflectivities below 20 dBZ_e, large unrimed aggregates (> 0.5 mm) were the most frequently observed particles (Fig. 19.11c), far outnumbering unrimed and rimed dendrites > 0.5 mm (Fig. 19.11a,b). Small rimed aggregates and small graupel particles ($\approx$ 0.5 mm) were also measured in these regions, although less frequently. Between 20 and 40 dBZ_e, the observed particles tended toward larger sizes and were more heavily rimed than at lower reflectivities. In these regions, rimed aggregates and graupel particles (> 0.5 mm) were the most commonly observed particles (Fig. 19.11d-e). The dip in the observed frequency of rimed aggregates in the 30-40 dBZ_e range coincided with sampling in the updraft core (WER); in this reflectivity range graupel particles were predominant. As the radar reflectivity increased to the 40-55 dBZ_e range, the observed habits tended toward large, rimed particles including large graupel and hail (Fig. 19.11e) and large, rimed aggregates; few or no unrimed particles were observed. The frequency curve for particles of 12 mm size (Fig. 19.11e) observed in the 40-55 dBZ_e range was similar in form to that inferred for 12-mm particles from dual-wavelength-radar hail signals (i.e., Y' > 3 dB) on 22 July 1976 (Jameson and Heymsfield, 1980).

Assuming that this description of radar-particle habit tendencies is valid everywhere in the middle levels of the storm, it was possible to construct an approximate composite figure comparing habits to the position of the main updraft

---

*The following factors should be considered, when applying this figure toward estimating the types and sizes of particles at a location for radar reflectivity data. One aspect is that large differences existed between the sampling volumes of the probes (see Fig. 19.6l) used to derive the probability curves for planar crystals and aggregates (2-D probe) and those of the probes used for graupel particles and hail (foil impactor and hail spectrometer). The second aspect is that large aggregates (> 1.0 cm) would have been in concentrations below the sampling volume threshold of the 2-D probe, and therefore would not have been detected. The third aspect is that there are also very large differences ($\approx 10^7$) between the radar and the aircraft probes.

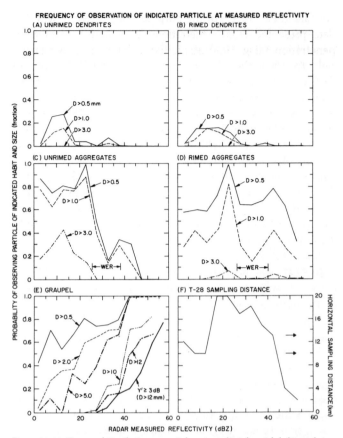

Figure 19.11 Curves relating the frequency of observing indicated particle habits and sizes at the indicated radar reflectivities based on the in situ aircraft measurements and radar data. Horizontal distance over which the T-28 sampled in each 5-dBZ_e reflectivity range is indicated (f). Data obtained in the weak echo regions are in the reflectivity range indicated by WER in C and D.

core at the midstorm level (Fig. 19.12) during the observational period, assuming a degree of steadiness in the reflectivity and wind field patterns, the conditions for which were nearly met during the observational period (see Chapters 16 and 18).

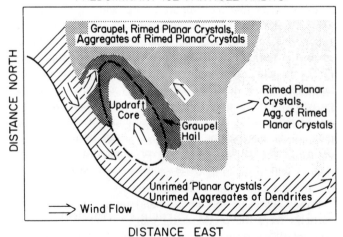

Figure 19.12 Summary of observed ice particle habits in the middle level of the storm; based upon the correlation between radar echo intensity and habit. Predominant windflow in each region is indicated.

A region containing primarily unrimed single dendrites and aggregates of dendrites was situated upwind of and extending around the forward portion of the storm (Zone IV). This was the region characterized by low radar reflectivities which extended around the forward portion of the storm (e.g., Fig. 19.4a). A region containing graupel particles and hail extended around the western (Zone I), northern, and eastern (Zones II and III) peripheries of the main updraft core. A forklike pattern of graupel and hail extended around the updraft core, apparently indicating that particles were growing along the edges of the updraft or were falling from above the WER. Within the intense updraft core (WER) few particles were observed (see Fig. 19.4b, penetrations 4 and 6). Graupel particles, rimed dendrites, and aggregates of rimed dendrites composed a broad region that extended generally to the north of the main updraft core. Rimed dendrites and aggregates of rimed dendrites were the primary forms of particles falling from the anvil region sampled to the east of the main updraft (Zone V).

19.4.2 Ice Particle Aggregation

The presence of large ice particle aggregates within the precipitation debris region and the updraft cores suggests that under some conditions these particles are in a favorable location for subsequent growth into graupel and hail. It is valuable to trace the probable growth modes and trajectories of the ice particle aggregates within the precipitation debris region more fully. A schematic representation of the growth of hydrometeors in this region appears in Fig. 19.13, which describes the development of aggregates as they advect around the forward side of the storm. The hydrometeor growth processes operating in this region were inferred from the T-28 measurements, and the locations of the initiation regions and the trajectories followed by the ice particles were estimated from the radar and wind field measurements. Initially, a cumulus congestus tower decays, or liquid water is detrained from the main updraft core, and ice crystals are nucleated (A, Fig. 19.13). Subsequently, particles grow as planar crystal forms, and the relative humidity drops to ice saturation through depletion of the excess vapor supply by the ice particle growth and through mixing with environmental air (B). When the excess vapor supply is totally exhausted, growth continues only through aggregation (C). Since this region is primarily an area of subsidence, the ice particles may also be evaporating during their transit. Some of the ice particles can either fall into or be entrained into the main updraft region or into newly developing feeder cells (D).

19.4.3 Secondary Ice Particle Production

The data obtained within the updraft region suggested that a secondary ice particle production process was operating. As

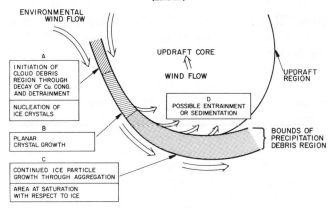

Figure 19.13 Schematic showing the growth modes of particles in the precipitation debris region (shaded), and their subsequent introduction into the updraft core. Arrows show predominant windflow.

indicated from the size spectra in Fig. 19.5 and from the sum of ice particle concentrations of sizes R_1 and R_2 in Fig. 19.4b, relatively high concentrations of ice particles with $D < 0.1$ cm were observed within these regions. The discussion below presents arguments that these particles probably originated through a secondary production process within the updraft regions.

The concentrations and types of ice particles observed in the updraft regions cannot be accounted for on the basis of their formation on primary ice nuclei. The primary ice nucleus spectrum thought to have been representative for the case would have produced a concentration of primary ice crystals (c_P) of at most 0.5 ℓ^{-1} at $-15\,°C$, of which only about one-tenth would be large enough to be detected (see Heymsfield et. al., 1979). The concentration is of such primary ice crystals, usually appreciably lower than that which was measured in the updraft regions (see Fig. 19.4b). Most ice particles produced during this growth period would have been lightly rimed hexagonal crystals and not the rounded, heavily rimed ice particles that were observed.

These particles were not likely to have been introduced into the updraft core from the environment surrounding the updraft region. The nearly vertical structure of the updraft core at the midstorm levels (Chapter 18), and the rapid ascent of the ice particles (Fig. 19.4e), would not have enabled them to have advected inward appreciably from the edges of the updraft regions to their observed locations within the updraft cores, or to have fallen into the updraft core from above. Turbulence could indeed have also led to the diffusion of these small ice particles (and ice nuclei) inward from the edges of the updraft into the core. If it is assumed that this diffusion took place in the same way as would a plume with the same turbulence values, particles could have penetrated inward by a distance of 0.5 km. However, even this process would not have enabled these particles to have advected to the observed locations within the updraft core. If turbulence or entrainment had introduced these particles into the updraft cores, the measured ice particle spectra would have been more similar in shape both inside and outside of the updraft cores. The discontinuously distributed spectra that were frequently observed within the updraft cores were very unlike the continuously distributed spectra (not shown) that were typically observed along the peripheries (see Fig. 19.5) of the updraft. If additional ice nuclei had been introduced into the updraft regions through turbulence, the crystals formed on these nuclei would not have been of detectable size because the mean updraft velocities are so large (see Heymsfield et al., 1979).

The enhancement in the ice particle concentration within the updraft core, IM ("ice multiplication" factor), due to this secondary ice-crystal-production mechanism can be estimated from the data in Fig. 19.5, using the ratio of the measured concentration of ice particles smaller than 0.1 cm to the value of C_p determined for the measurement temperature. Calculated values of IM (Figs. 19.5a-l) ranged up to 10 within the intense updraft cores and up to 300 in the weaker updraft regions. The actual values of IM would have been higher than those indicated, since the sizes attained by most of the ice particles nucleated on "primary" ice nuclei would not have been detectable with the 2-D probe.

Several mechanisms have been proposed to explain secondary ice particle production, but most of these are not applicable to the present study. Secondary ice particle production through the Hallett-Mossop mechanism (Hallett and Mossop, 1974) was unlikely because of the absence of droplets $\geq 24\ \mu$m. Shedding of water from large ice particles and the subsequent nucleation of these particles was also an unlikely source of additional ice particles. The large particles observed had rough edges, not rounded edges, which would be expected for particles capable of shedding liquid water. Computed particle temperatures (Section 19.2.9) also suggested that the particles were not shedding liquid water. Additionally, the IM factor was highest in the lower-liquid-water regions of the updraft, where shedding is not as likely as within the core.

Three independent sets of data, discussed below, indicated that the secondary ice particles were produced within the updraft regions through a collisional breakup process; the process envisaged is similar to that described by Vardiman (1978). While it is conceivable that the fragmentation of weak portions of rimed ice particles by aerodynamic processes alone can explain the observations, there is no evidence that rime can be weak enough to fragment in that way.

Correlation between the concentrations of small and large ice particles within the updraft regions

The correlation between the concentrations of small and large ice particles is consistent with a mechanism for the generation of small particles through collisions between large particles. In the discussion below, we show this correlation, by comparing IM to the concentrations of large particles.*

The most convincing evidence of the correlation between IM and the measurements of large particles was noted within the intense updraft cores, where the growth times of particles were the shortest and the introduction of the small particles from the environment would have been least likely. IM was 0 when no large particles were observed in these regions (Figs. 19.5h and k). The lack of any small particles in these regions indicated that any ice particles that were generated from primary ice nuclei were not yet up to detectable sizes. IM was relatively low when few large ice particles were observed (Fig. 19.5e), but was relatively large when high concentrations of large particles were observed (Fig. 19.5b). A distinct break in the size spectra between about 0.02 and 0.1 cm can be noted in Figs. 19.5b and e. This gap suggests that few secondary ice particles larger than about 0.02 cm were produced in the framentation process.

IM was generally higher in the weaker updraft regions along the western and eastern boundaries of the updraft cores, apparently because longer times were available in these weaker updrafts for small, secondary ice particles to grow to detectable size, and for more collision between the large particles. However, the correlation between IM and large particles was also observed in these regions. IM was either 0 or very low when few large particles were observed (Figs. 19.5g, i and L). IM was relatively high along the boundaries where fairly strong updrafts and broad spectra of large particles were observed (Fig. 19.5a, c, and f). The spectra observed in these regions were discontinuous, and again indicated that the fragments generally had diameters of 0.02 cm. IM was also relatively high along the boundaries in regions with fairly weak updrafts wherein large particles were observed (Figs. 19.5d and j), but continuous spectra were noted in these cases. The similarity of the spectra and particle habits observed in the downdraft regions to those reported above suggests that secondary ice-particle production was also occurring in the downdraft regions.

Riming of ice particles at unexpectedly small sizes

If the small particles were generated through collisions between large rimed particles, the newly generated particles would be irregularly shaped and rimed. Particles of sizes from 100-200 μm, for which the shapes could be identified, consistently had irregular edges (e.g., see Fig. 19.5). If these particles had formed through primary nucleation, they would have been hexagonal, since particles that form on primary nuclei subsequently develop as expected and retain their hexagonal shape to sizes exceeding 0.2 mm. We therefore conclude that these particles were indeed produced through fragmentation.

---

*Detailed calculations of the frequency of particle collisions can be used to show this correlation more directly. However, these calculations require information on the trajectories followed by particles at times prior to their observation.

Observations of "fragmented" ice particles

Branches missing from dendritic crystals and branches of dendrite crystals, were frequently observed along the peripheral regions of the storm. The nonspherical, irregular shapes of the ice particles, which were thought to have been produced from the collisional-breakup process (see Fig. 19.4), also provide evidence of their mode of formation through the breakup of rime during particle collisions. Calculations of the frequency of particle collisions that are under way appear to indicate that collisions between large particles may be frequent enough for the preface mechanism to work. These calculations provide additional insight into the particular characteristics that are most conducive to the process.

19.4.4 Efficiency of Conversion of Liquid Eater to Precipitation Mass Within the Storm

Data obtained within and around the main updraft cores suggest that the conversion of liquid water to precipitation mass was inefficient at the aircraft altitude. We will discuss here the particle sizes that were responsible for most of the depletion and the role of secondary particles in depleting the liquid water within the updraft regions.

Most of the depletion of liquid water at the T-28 altitude and at the time of the penetration occurred along the peripheral regions of the updraft. As indicated by detailed trajectory calculations (in progress), particles found in these regions had developed either along the western side of the updraft core, or around the forward overhang and then along the eastern side of the updraft core prior to being observed. Along the eastern boundary of the updraft region, most of the depletion was due to particles of hail size, while along the western boundary, the depletion was due to particles of submillimeter sizes. Within the intense updraft cores, the depletion times were exceedingly long. As air ascended within these cores from the aircraft level to the top of the radar-detected WER, insignificant depletion of the liquid water would have occurred. Therefore, particles located directly above the WER grew in a region where liquid water had not been depleted by ice particle growth. In these regions, only entrainment would have reduced the LWC's below the 70%-80% of the adiabatic values found at the midstorm levels.

Secondary ice particles are thought to have had an important role in converting liquid water to ice within the updraft regions. The effect of the additional depletion by secondary ice particles was estimated by assuming that these particles had a bulk density of 0.3 g cm^{-3} and were spherical; if it was assumed that these particles were hexagonal and had typical ice crystal densities, their collection efficiencies would have been close to 0. The depletion times calculated for spectra obtained along the upwind and downwind edges of the updraft cores were reduced by 10%-50% and by 10%-60%, respec-

tively, assuming that these particles were not collecting water drops. We therefore conclude that fragments are contributing significantly to the depletion of liquid water within the updraft regions.

19.5 CONCLUSIONS

In this multicellular hailstorm in northeastern Colorado, a primary region of strong updraft (updraft "core"), which moved with approximately the velocity of the storm, was found to exist continuously over the observational period. This region was maintained through a succession of "cells," each of which was characterized by transient radar weak-echo regions (WER). The aircraft penetrations provided the first definitive measurements of the structure and composition of the WER, which coincided with the regions of the strongest updrafts, from 15 to greater than 25 m s^{-1}. The highest liquid water contents were found within the WER, with highest values up to 70%-80% of the adiabatic liquid water content. The absence of adiabatic "cores" may be due to mixing and turbulence, which began within the updraft cores below the cloud base level. As indicated in Chapter 17, the subcloud-base environment was likely to contain surface air mixed with outflow air from another storm; relatively strong turbulence was also noted in these regions. Within the WER, all hydrometeors were rising rapidly, and few ice particles larger than 1 mm were found. The observed ice particles were heavily rimed irregular and graupel forms. Secondary ice-particle production appears to have been a minor factor in this region. Because of the relatively low ice particle concentrations and the small sizes, the conversion of liquid water into precipitation mass was extremely inefficient. Therefore, particles located directly above the WER grew in an environment in which the liquid water was reduced below the adiabatic value only by entrainment, if at all.

The processes operating within the weaker updraft regions adjacent to the WER differed appreciably from those within the WER. Peak vertical velocities of 10-15 m s^{-1} were measured; strong downdrafts were consistently found near the downwind (eastern) portions of the updraft regions. The liquid water contents were only 25%-50% of the adiabatic values in these regions. Within the upwind regions, the entrained air apparently originated at low to midstorm levels (4.5-6.0 km), while that in the downwind portions originated at upper storm levels (>7 km). The largest observed particles, which were smaller than 1.0 cm in the upwind region and up to 4 cm in the downwind region, were descending through the updrafts. The observed particle types were graupel, unrimed and rimed planar forms, and aggregates of planar forms. A secondary ice particle production mechanism operated vigorously in these regions, generating concentrations of ice particles up to 300 times more than that expected on the basis of primary ice nuclei. The secondary ice particles were apparently generated through a collisional-breakup mechanism, and required broad spectra of relatively high ice particle concentration for the process to operate. Fragment sizes were apparently generally less than 200 μm. The depletion of the liquid water in these regions was appreciable; in the upwind regions, depletion was primarily by relatively small ($<$ 5 mm) particles, while in the downwind regions, depletion was primarily by particles of hail size. The secondary ice particles were found to be a significant factor in the depletion of the liquid water.

The southernmost extent of the forward portion of the storm provided an environment in which particles grew primarily through aggregation and subsequently were in a favorable location to serve as graupel hail embryos. Particles found in these regions initiated their growth in short-lived turrets, out of debris from the main updraft core or from feeder cells. Initially, particles grew as planar crystals, and then grew through aggregation to sizes $>$ 5 mm. Some of these particles were probably ingested by the main updraft cores, and by feeder cells, where growth continued through accretion.

Appendix A

T-28 Instrumentation

The T-28 was equipped with instrumentation that permitted detailed measurements of the thermodynamic structure and hydrometeor distributions during the cloud penetrations. The aircraft and measurement capabilities have been described by Sand and Schleusener (1974), and more recently by Johnson and Smith (1980). The measurements include pressure, temperature, vertical air motion, liquid water concentration, and hydrometeor size distribution. The instruments for measuring hydrometeor size distributions include a PMS (Particle Measuring Systems, Boulder, Colorado) FSSP cloud droplet probe sizing in the range 3-45 μm, a PMS 2-D probe capable of determining particle images as well as size for particle sizes from 25 μm to more than 1 mm, a foil impactor sizing from less than 1 mm to several centimeters, and the SDSM&T "hail" spectrometer which is similar in design to the PMS 1-D optical array device (Knollenberg, 1970) and was capable of sizing from 5 mm to 10 cm. The sampling volumes of these four probes are presented in Heymsfield and Parrish (1978). All data except those from the foil impactor were recorded on magnetic tape during the sampling period.

The accuracy of the measurements discussed in this paper has been difficult to evaluate. The consistency of similar or complementary measurements provided the most direct evidence that individual instruments performed satisfactorily. The accuracy of the Johnson-Williams (J-W) liquid water device is $\pm$ 20 % according to the manufacturer, and we have no reason to doubt this from the present study. Because the

sampling volume of the FSSP probe was artificially limited by an electronic thresholding circuit, it was necessary to correct data from the probe for droplet coincidence errors by a technique described by Heymsfield and Parrish (1978). Loose optics in the FSSP probe also made calculation of the sampling volume subject to error, but the sizing of droplets was not affected by this problem. A detailed comparison of measurements obtained from the reverse flow temperature probe with the expected temperatures based on one-dimensional entrainment calculations (Heymsfield et al., 1979) and the measured liquid water (Section 19.2.3) suggests that the temperature measurements may not have been affected by wetting or icing of the element. The temperature measurements are thought to be accurate to within $\pm$ 10 °C. Vertical velocity measurements are among the most important to the analyses described in this study and yet are the most subject to error. Several approaches were used in calculating the vertical velocities, and no single method consistently provided the best correlation with the liquid water and temperature trends. The accuracy of the vertical velocity measurements is discussed in Heymsfield and Parrish (1979). Measurements obtained with the 2-D probe and the hail spectrometer were consistent in the regions of sizing overlap, except for cases in which the predominant particle types were aggregates, which are not well detected on the hail spectrometer.

Appendix B

Calculations Based on Particle Size Spectra

Techniques employed in the processing of hydrometeor data from the T-28 aircraft are described by Heymsfield and Parrish (1978, 1979). Particle sizes and habits were determined from the 2-D data though visual inspection.[*] Since habit information for particles sampled with the foil impactor and hail sensor was not directly available, it was necessary to make several assumptions regarding particle habit when processing the data from these probes. For size ranges that overlapped with the 2-D probe, habit information was based on simultaneous 2-D data. At larger sizes, particles were assumed to be graupel or hail; this assumption was usually confirmed from the pilot reports of particle types and data from a "hail" microphone affixed to the aircraft windscreen.

Measured particle size spectra and derived particle habits averaged primarily over 4-s periods (or approximately 400 m of flight path) were used to compute several parameters of the size distribution throughout the flight. These parameters

---

[*] Shedding of liquid water from the 2-D probe tips in high-liquid-water-content regions resulted in some artificially generated particles, which were sampled as though they were real particles. It is believed that most of these artifacts were recognizable, and were eliminated from the data set through careful visual inspection.

included the ice mass concentration, the precipitation rate, the particle sedimentation velocities within the updraft, the depletion time constant, and the derived particle growth temperatures. For these calculations, it was necessary to compute the particle mass and terminal velocity. Particle mass (m) and terminal velocity (V_T) were calculated according to the measurements described in Heymsfield and Parrish (1979); the calculations of V_T were based on the pressure at the sampling altitude. The calculations also assumed a solid ice density and spherical shape for calculating the m and V_T values for graupel and hailstones.

Particle fall velocities (V_f) were computed relative to the ground as the difference between the air velocity (V_{air}) and the terminal velocity. The speed at which the largest particles were falling within updraft and downdraft regions was expressed as the difference between the air velocity and the highest terminal velocity computed from the measured particle size spectra ($V_{T_{max}}$).

The depletion time constant* (τ_L), defined as the time required to reduce the liquid water content (LWC) to a value equal to 1/e of the initial value through collection of droplets by the growing ice particles (see Paluch, 1978), was computed from each 4-s average size spectrum. The depletion time constant can be computed from the equation

$$\tau = \text{LWC}(gm^{-3}) \sum_{I=1}^{n} \sum_{J=1}^{J_{max}} N_{IJ} \frac{dm_{IJ}}{dt}\Bigg|_{\text{accretion}} (gm^{-3}s^{-1}), \quad (A19.1)$$

where N_{IJ} is the number of particles (m^{-3}) of size I and habit J and where $dm_{IJ}/dt|_{\text{accretion}}$ is the accretional mass growth rate of particles of size I and habit J. The technique used in calculating the accretional mass growth rate is described in Heymsfield and Parrish (1979).

Computation of the particle temperatures was based on the measured ice particle sizes and habits over specified averaging periods to determine whether particle growth was primarily wet or dry in different regions of the updraft. The techniques used are discussed in Heymsfield and Parrish (1979).

References

Browning, K. A., and G. B. Foote, 1976: Airflow and hail growth in supercell storms and some implications for hail suppression. *Q. J. R. Meteorol. Soc.* 102, 499-533.

Hallett, J., and S. C. Mossop, 1974: Production of secondary ice crystals during the riming process. *Nature* 249, 26-28.

Heymsfield, A. J., 1978: The characteristics of graupel particles in northeast Colorado cumulus congestus clouds. *J. Atmos. Sci.* 35, 284-295.

-----, and J. L. Parrish, 1978: A computational technique for increasing the effective sampling volume of the PMS two-dimensional particle size spectrometer. *J. Appl. Meteorol.* 17, 1566-1572.

-----, and J. L. Parrish, 1979: Techniques employed in the processing of particle size spectra and state parameter data obtained with the T-28 aircraft platform. NCAR Technical Note (NCAR/TN-137+IA), NCAR, Boulder, Colo., 78 pp.

-----, C. A. Knight, and J. E. Dye, 1979: Ice initiation in unmixed updraft cores in northeast Colorado cumulus congestus clouds. *J. Atmos. Sci.* 11, 2216-2229.

-----, A. R. Jameson, and H. W. Frank, 1980: Hail growth mechanisms in a Colorado storm: Part II: Hail formation processes. *J. Atmos. Sci.* 37, 1779-1807.

Knollenberg, R. G., 1970: The optical array: An alternative to scattering and extinction for airborne particle size determination. *J. Appl. Meteorol.* 9, 86-103.

Musil, D. J., E. L. May, P. L. Smith, Jr., and W. R. Sand, 1976: Structure of an evolving hailstorm: Part IV. Internal studies from a penetrating aircraft. *Mon. Weather Rev.* 104, 596-602.

Paluch, I. R., 1978: Size sorting of hail in a three-dimensional updraft and implications for hail suppression. *J. Appl. Meteorol.* 17, 763-777.

-----, 1979: The entrainment mechanism in Colorado cumuli. *J. Atmos. Sci.* 36, 2467-2478.

Sand, W. R., and R. A. Schleusener, 1974: Development of an armored T-28 aircraft for probing hailstorms. *Bull. Am. Meteorol. Soc.* 55, 1115-1122.

Vardiman, L., 1978: The generation of secondary ice particles in clouds by crystal-crystal collision. *J. Atmos. Sci.* 35, 2168-2180.

---

*In calculating the depletion time constant, it was assumed that ice particles remain at a constant size and concentration over the 1/e time period and that the liquid water content is not reduced except by depletion. Over short time periods up to 10³ s, τ_L is thought to be accurate to approximately a factor of 2. At time periods of 10⁴ or longer, τ_L will only be accurate to a factor of 5. The depletion rate was estimated in regions that did not contain liquid water; to provide an estimate in these regions, average values of the ratio of the effective to the measured liquid water content within updraft regions were used in conjunction with equation (A19.1).

CHAPTER 20
The 22 July 1976 Case Study: Hail Growth

G. Brant Foote, Harold W. Frank, Andrew J. Heymsfield, and Charles G. Wade

The previous four chapters have treated various aspects of Storm III, a storm that occurred within the NHRE network on 22 July 1976. The storm passed over a small town called Westplains, and for brevity we henceforth refer to it as the Westplains storm. In the current chapter we review the pattern of airflow deduced for the storm. Then, using a simple hail growth model utilizing the Doppler wind fields, calculations of the possible hail growth trajectories are presented. Finally, the topic of hail suppression is considered.

20.1 AIRFLOW MODEL

The circulation of air in and around the Westplains storm has been discussed extensively in the previous four chapters. Since the field of air motion forms the basis for the hail growth calculations to be presented here, it is appropriate to review the basic structure. Figure 20.1 shows in perspective view the main branches of the circulation. The drawing is based primarily on the Doppler measurement of winds relative to the storm for the 1626-1640 period reported in Chapter 18, both supported and augmented by the surface mesonet, Queen Air, and T-28 data reported in Chapters 17 and 19. The dominant feature of the circulation is the large updraft, labeled A in the figure, that originates in the low levels from the south-southeast, rises abruptly as it is lifted over a surface gust front and passes through cloud base, and finally turns back in the direction of the storm-relative winds in the upper troposphere to form the anvil streaming away toward the northeast. The cyclonic twist depicted in the updraft ribbon in Fig. 20.1 is meant to illustrate qualitatively the cyclonic vorticity in the updraft. While the updraft itself possesses cyclonic rotation, individual streamlines (for example, the edges of the updraft ribbon shown) actually turn anticyclonically, as discussed in Chapter 18. Klemp et al. (1981) have illustrated the same phenomenon in cloud model and multiple-Doppler results.

On the shoulders of the updraft the horizontal momentum

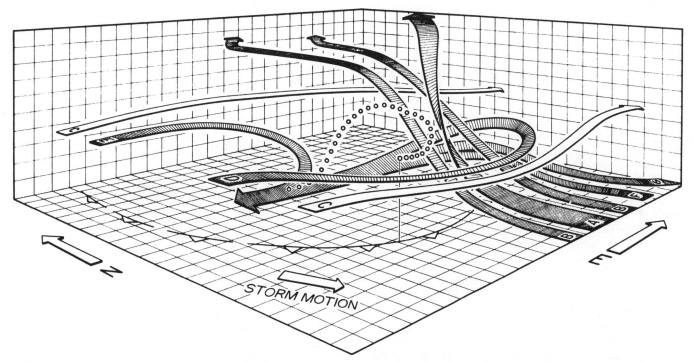

Figure 20.1 Major components of the airflow in the Westplains storm. The strong updraft is depicted by the ribbon labeled A, which starts in the low levels to the south-southeast of the storm, rises sharply in the storm interior, and leaves the storm toward the northeast to form the anvil outflow. On the flanks of the strong updraft the air rises more slowly and penetrates farther to the rear of the storm before also turning to the northeast. In the middle levels there is a tendency for the westerly environmental flow to be diverted around the sides of the storm (streamlines labeled C) but some air also enters the storm (streamlines D and E) and contributes to the downdraft. A contribution to the downdraft flux is also made by air originally in the low levels to the southeast and east of the storm (streamlines F and G), which then rises several kilometers before turning downward in the vicinity of the echo core. The small circles indicate the computed trajectory of a hailstone, discussed later. The various streamlines are depicted relative to the storm, which is moving toward the south-southeast as shown, rather than relative to the ground.

is similar to that in the updraft core, while of course the updraft strength is reduced. The streamlines shown alongside the broad updraft, labeled B, are tilted more toward the back of the storm and eventually enter the lower part of the anvil.

In the middle levels the environmental flow is from the west, and is to an extent forced to flow around the storm as if it were a solid barrier (streamlines labeled C). As a barrier, the storm is, in fact, rather porous, and streamlines such as D and E represent air entering the storm on its upwind flank much as packets of chaff have been observed to enter a storm's radar echo (Fankhauser, 1971). Flow along the cyclonic streamline D is substantially stronger than that along E, a result that is general for right-moving storms. As D passes around the east side of the updraft it turns sharply downward to form a principal branch of the storm downdraft and surface outflow, similar to Browning's (1964) depiction. During the period 1626-1640, streamline E was also observed to contribute to a downdraft in the vicinity of the echo core, just north of the updraft. However, in the 1707-1718 Doppler analyses this branch of the flow was not present, and the downdraft in that region was fed partly by air that had traveled along a path initially like D, but extending around the storm to its north side. Part of the northern downdraft at both times also originated with air initially rising on the northern side of the updraft. This air can be traced backward along streamline F, shown in Fig. 20.1, to a source in the low-level environment to the southeast of the Westplains storm. The surface measurements reported in Chapter 17 indicate that this air was too dry to be convectively unstable. Some of the downdraft air may even have originated in the low-level outflow from an adjacent storm just to the east, as depicted by streamline G. Though the downdraft air near the surface is shown in the figure as moving primarily toward the west, there is in fact some spreading of the cold downdraft in all directions as it encounters the earth's surface (see Chapter 17). The position of the cold air boundary is noted in the figure.

From the point of view of understanding hail growth, the processes occurring along paths A and D in Fig. 20.1 are the most important. The liquid water contained in the supercooled droplets carried by the updraft along path A provides suitable conditions for hail growth, while path D provides a mechanism for transporting potential hail embryos to the west, south, and east sides of the updraft periphery. In the hail model proposed by Browning and Foote (1976) an emphasis was placed on the ability of embryos to sediment into a sloping updraft while being carried along in the cyclonic flow, such as path D, that traverses it. However, a slope in the updraft core in the lower part of the cloud, though it may be common, is not a necessary condition for the transfer of particles from path D to path A, and the mechanism detailed by Browning and Foote is probably not the fundamental one. Rather, the important fact is that the lower half of the updraft is horizontally convergent. For the Westplains storm the detailed discussion in Chapter 18 shows, for example, that

the region of convergence includes all altitudes below 6 to 7 km. Thus, the implication is that even for those cases in which the updraft core is vertical, in the low to middle part of the cloud the updraft will have convergent streamlines around its periphery that will tend to transport into the updraft those hydrometeors with terminal velocities comparable to the updraft speed (see the sketch in Fig. 20.2). This convergent flow can be seen in the velocity data shown in east-west cross section in Fig. 18.14. The data in the north-south section, Fig. 18.13, do not appear to extend far enough south to reveal it (in Fig. 18.9 the simple extrapolation of the north-south component of velocity in the data void on the south of the updraft may be incorrect in not showing this convergence). It then appears that a natural mechanism exists for introducing particles into the lower regions of updrafts where the vertical acceleration of air and the associated horizontal convergence will be commonplace. While the mechanism amounts to a form of entrainment, it does not rely on small-scale turbulent mixing as one often visualizes entrainment through cloud boundaries.

The lower region of the cloud that is favored for the ingestion of particles is, of course, ideal from the standpoint of producing hail, since embryos introduced near the foot of the updraft may be expected to enjoy the longest growth times. The role of the cyclonic streamline D, then, is primarily to transport potential growth embryos from their source region to the periphery of the large updraft, where in the zone of strong convergence the hydrometeors will be urged into the updraft core.

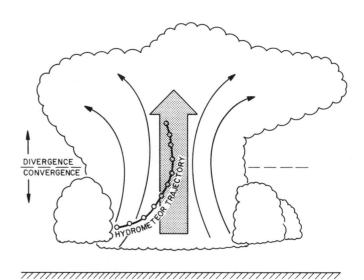

Figure 20.2 Sketch illustrating the tendency for hydrometeors to move into the updraft core in the lower convergent region of updrafts. If the particle terminal velocity is close to the updraft speed, then the particle will tend to sediment across streamlines and move toward the center of the updraft. This mechanism is thought to be fundamentally important in supplying the updraft with collection centers.

20.2 HAIL GROWTH CALCULATIONS

An extensive literature exists on the subject of modeling hail growth in clouds. The summaries of Orville (1977), Browning (1977), and Macklin (1977) are still reasonably up to date, though the laboratory work of Lesins et al. (1980) has raised further questions regarding the interpretation of hailstone structure. In studies that have treated hail trajectories explicitly, the work has tended to be either qualitative (e.g., Browning et al., 1976; Browning and Foote, 1976) or involve the use of idealized flow fields (Browning, 1963; Musil, 1970; English, 1973). Paluch (1978) calculated hail growth trajectories in a storm for which the motion field had been obtained by dual-Doppler radar techniques (Kropfli and Miller, 1976). Though she made the unrealistic assumption that the motion field was in a steady state, the results were revealing with regard to the complexities of the trajectories and the size sorting that might occur.

The hail trajectory computations reported here are similar to those of Paluch with several important exceptions discussed below, including the treatment of the liquid water content and the particle drag law, and the use of time-varying flow fields. The growth equations for a single hailstone in a known environment are solved to give the particle size, temperature, and mass as a function of time, with specified initial conditions of particle size, density, and position. In addition, the motion of the particle is calculated from its terminal velocity and from the flow field as specified by the Doppler results from Chapter 18 so as to give its path through the storm. The growth equations are in principle those of Mason (1971). Using the sounding from Chapter 16, the particle temperatures computed for the Westplains storm are less than 0 °C during growth so that one need not be concerned about spongy growth (except in a small region of negligible importance near the 0 °C isotherm).

The flow fields employed here are from the 1626, 1635, and 1640 Doppler analyses of Chapter 18. Each flow field is used for a representative period of time, and no attempt is made to interpolate between them.

20.2.1 Particle Density

The density of the rime accreted by the growing hailstone is taken from Macklin's (1961) empirical expression involving the hailstone temperature and the droplet size and impact velocity. Based on an extension of Pflaum's (1978) data to higher Reynolds numbers, it is assumed that there is negligible deceleration of the droplet just before impact, and that the impact speed is thus the same as the terminal fall speed of the hailstone. The droplet size is computed by assuming that the liquid water content (discussed below) is spread equally over a specified concentration of droplets, in the present case 600 cm⁻³.

The initial density of the embryo is arbitrarily specified to be 0.4 g cm^{-3} in all the calculations reported here. Typical particle densities that result are shown as a function of size in Fig. 20.3. The solid curve represents the mean density for all particles of a given size. The standard deviation about the mean is generally about 0.1 g cm^{-3}, but the total range of values, as indicated by the area shaded, is much larger than this. If one fits a power law to the solid curve, the density is found to depend on diameter to the power 0.44. This is very close to the power 0.46 that one can deduce for a similar dependence from Fig. 3 of Heymsfield's (1978) study of natural graupel collected inside clouds in NHRE. The wide range of densities produced by the use of Macklin's (1961) accretion formula is also similar to the range exhibited in Heymsfield's data. Apparently the model calculations reported here lead to reasonable ice densities.

20.2.2 Drag Law

A number of expriments have been conducted to measure the terminal velocity of ice particles and simulated ice particles (crystals and conical and ellipsoidal figures representing graupel and hail). A wide range in values has been obtained, as summarized, for example, by Heymsfield (1978). In an attempt to consider the effect of different terminal velocities on hail growth, calculations have been made here using the three different drag laws shown in Fig. 20.4 The curve labeled "low drag" uses a constant drag coefficient of 0.6 for high Reynolds number (Re), as is often assumed for hailstones. For $10^2 < \text{Re} < 10^3$ it is in roughly the middle of the range of values obtained by List and Schemenauer (1971) from laboratory experiments on conical models. For Re between

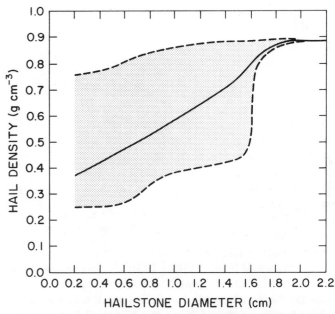

Figure 20.3 Range of final hail densities predicted by the growth calculations as a function of maximum hailstone diameter. The mean hail density is shown by the solid curve.

10 and 100 it is close to unpublished values obtained by John Pflaum (University of Oklahoma, Norman, Okla., private communication) for graupel grown in the laboratory.

The curve labeled "high drag" corresponds to the range of values for natural graupel and small hail in free fall as measured by Zikmunda and Vali (1977) and Auer et al. (1971) and summarized by Heymsfield (1978). The "medium drag" curve is in roughly the middle of the field measurements reported by Matson and Huggins (1980), though these authors found a wide range of drag coefficients encompassing both the low drag and high drag curves of Fig. 20.4.

Fig. 20.5 shows an example of the terminal velocities predicted by the three drag laws under the conditions noted

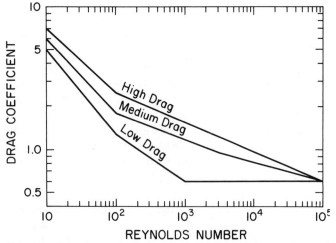

Figure 20.4 *Curves showing three different drag laws used in the hail growth calculations.*

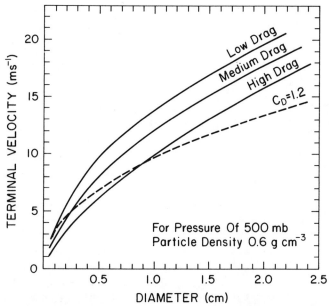

Figure 20.5 *Examples of hydrometeor terminal velocities produced by the three drag laws assuming an atmospheric pressure to 500 mb and a particle density of 0.6 g cm⁻³. The dashed curve results from assuming a drag coefficient of 1.2 independent of Reynolds number.*

there. An additional curve corresponding to a constant drag of 1.2, the value employed by Paluch (1978), is included for comparison.

The doubling of the drag coefficient between the low and high drag curves is perhaps surprising, and such high drag coefficients appear to be unsupported by available laboratory measurements on smooth objects of roughly conical and ellipsoidal shape (List and Schemenauer, 1971; Stringham et al., 1969). If the measurements are accurate then it appears that their explanation must be in terms of the surface texture of the particles. For example, the small feathery protrusions extending from the surface of the graupel reported by Heymsfield (1978) and seen also in the in situ photograph in Pflaum (1980) might have an important effect on the skin friction and the size of wake. On the other hand, the experiments of Pflaum (1978) with low-density graupel did not reveal such high drag coefficients.

If the surface characteristics do affect the drag then one might expect to see a difference in drag for particles grown at different temperatures, since presumably the character of the deposit, and in particular its density, will be a function of particle temperature. However, an examination of Pflaum's data for particles grown at different temperatures in the range −3°C to −11°C fails to show any dependence of drag coefficient on temperature.

There is thus some uncertainty in the drag law appropriate to natural graupel and small hail in the atmosphere. Some of this is surely a result of real differences in free-fall behavior resulting from roughness and shape effects. Calculations will be discussed in the next section for the different drag laws of Fig. 20.4, and the results compared. The effect on the computed trajectories is not negligible, but is within the range of the uncertainty produced by other unknown quantities, such as the particle density, and by uncertainties and errors in the Doppler data (for example, errors in updraft speed, and the effect of small, unresolved scales of motion). Further work should be conducted on the terminal velocity of graupel, with more attention paid to the surface roughness.

20.2.3 Liquid Water Content

In the case studied by Paluch (1978) there were no direct measurements within the storm, and she made the assumption that it was adequate to specify the liquid water content as a function only of height, that is, as some fraction of the adiabatic value. In the present case it is clear from the T-28 measurements presented in Chapter 19 that there is considerable variation in the liquid water content in the horizontal, with appreciable values reported only in or near strong updrafts.

In the present calculations a formulation has been adopted that gives the spatial distribution of liquid water as follows. If w is the updraft velocity at some point in the storm, and if w_c is the maximum vertical velocity in the core of the updraft at the same altitude, then we define $w' = 1.5\, w/w_c$. The liquid water content can then be specified in terms of the fraction, f, of the adiabatic value by the expression:

$$f = \begin{cases} a + b\sqrt{w'}, & o < w' \leq 1 \\ f_{max}, & w' > 1 \end{cases} \qquad (20.1)$$

where f_{max} is the adiabatic fraction in the center of the updraft. In most of the calculations here, f_{max} is taken as 0.6, based on the T-28 measurements, the constant a is taken as 0.03, and $b = f_{max} - a$. According to this formulation, the liquid water content varies in the vertical as some fraction of the adiabatic value, and in the horizontal as the square root of w for w less than 2/3 of its maximum value. For regions where w is greater than 2/3 of w_c, f is set equal to f_{max}.

The postulated dependence of liquid water on updraft speed is illustrated by the curve in Fig. 20.6, which is superimposed on a scatter diagram of T-28 measurements collected in horizontal passes (the T-28 data were obtained between 6.3 and 7.3 km MSL where the adiabatic liquid water content was approximately 3.0 g m⁻³; only points where $w > 5$ m s⁻¹ are shown). Wide scatter is seen in the original data, but the curve probably represents mean structure well enough. In any case, on the basis of the results in Chapter 19, the point of the present formulation is more to restrict the liquid water to the updraft region than to specify accurately its variation within the updraft.

The variation of the maximum updraft, w_c, with height is needed to use equation (20.1) and is taken as $w_c = 14.12(z - 3) - 1.66(z - 3)^2$, where z is in kilometers MSL and w_c is in meters per second. This relationship gives a reasonable fit to the profiles of the large updrafts described in

Chapter 18, with a maximum value of 30 m s⁻¹ occurring at an altitude of 7.25 km.

20.2.4 Relative Humidity in the Downdraft

In an attempt to model downdraft thermodynamics the environmental sounding has been modified below 550 mb (5.1 km MSL) to produce temperatures and dew points consistent with pseudo-adiabatic descent starting at 550 mb and maintaining the relative humidity at 75 %. The reason for doing this is that the warm temperatures in the low levels on the Sterling 1605 sounding used in the calculations would have led to more melting of hail than would occur if the hail fell through the cool downdrafts of the storm, as presumably happened in fact.

The choice of 75 % as the downdraft humidity is based on the mesonet data of Chapter 17, though, as pointed out by Paluch (1978), the value assumed is not too critical. The reason is that if the humidity is assumed to be maintained at a lower value during descent, a reduced evaporation rate is implied, so that the computed air temperature will be higher. As a result, the increased heat loss due to evaporation from a melting hailstone is largely compensated by the added heat gain through conduction.

In fact, the terms don't exactly balance, and there is a tendency for less melting at the lower humidities, though the effect tends to be small unless one changes the humidity drastically. For example, sample calculations have been made with humidities of 75 %, 50 %, and 25 %. For hail with a final diameter at the ground of 6 mm or larger, there is no more than a 2-mm increase in size as the humidity is decreased. For the smaller stones that normally melt a greater fraction of their mass, the effect can be somewhat larger, with some stones that would have been 2 mm or smaller if falling through air at 75 % humidity instead maintaining against melting a size as large as 6-8 mm at 25 % humidity.

20.3 RESULTS OF HAIL GROWTH CALCULATIONS

Numerous authors have noted that a hailstone experiences a variety of growth conditions during its lifetime (e.g., Ludlam, 1958; List, 1960; Browning et al., 1963; Carte and Kidder, 1966; Knight and Knight, 1970). The need for different environmental conditions basically stems from the fact that the terminal velocity of an ice particle can change by a factor of 10 to 100 as it grows from small to large size. The requirement that a hailstone must remain in the mixed-phase region of the cloud if it is to grow by accreting supercooled water then means that its terminal velocity must be comparable to the updraft speed. The result then follows that a region of cloud with a given updraft speed cannot provide a suitable environment for significant growth of both large and

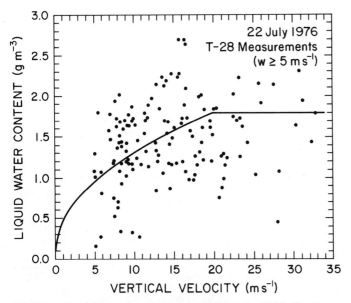

Figure 20.6 Scatter diagram of measurements from the T-28 aircraft of liquid water content (from the Johnson-Williams instrument) and updraft speed. The measurements were collected during several quasi-horizontal penetrations through the storm updraft at altitudes of 6 to 7 km MSL (see Chapter 19). Only data for which the updraft speed exceeded 5 m s⁻¹ are included.

small particles, since one or the other will have too short a residence time. It is customary to call the early phase of hail growth the embryo stage, with embryos thought of as having sizes on the order of millimeters. Examination of the structure of individual hailstones gives some direct justification for emphasizing that different stages of growth take place, since a distinct early growth unit is commonly identifiable in the growth center of large hailstones.

It is possible to use the Doppler data from Chapter 18 to study the growth of hail after the embryo stage, but the motion fields are poorly suited for considering the embryo growth itself in a deterministic way. Many embryo sources are probably linked to smaller scales of motion that are not resolved either temporally or spatially in the present Doppler analyses.

In Chapter 19 it is concluded that the riming of aggregates of crystals probably led to most of the graupel that served as hail embryos in the Westplains storm. These crystals are thought of as originating in cumulus towers that flank the mature storm. Mature cells located upwind also seem to have contributed, and some embryos also may have originated along the upwind, western edges of the main updraft of the Westplains storm.

The approach in the present study is to bypass a quantitative consideration of the early stage of growth. Instead, for the purpose of the calculations, the assumption is made that graupel of density $0.4\ \mathrm{g\ cm^{-3}}$ are widely distributed throughout the storm, and the analysis proceeds by examining which of these potential embryos are predicted to grow, and what their growth trajectories are. In a parallel study Heymsfield et al. (1980) sought to carry this approach a step further by using dual-wavelength radar measurements to determine whether graupel actually existed in the storm at positions that would be favorable for hail growth. However, Rinehart and Tuttle (1981) have recently presented evidence that their technique for detecting graupel was erroneous, and the results relating to graupel should probably be viewed with some skepticism.

Hail growth calculations have been made here for four sizes of embryos (0.2, 0.4, 0.6, and 0.8 cm in diameter) positioned 1 km apart throughout the area shaded in Fig. 20.7, and with starting altitudes of 5, 6, 7, and 8-km MSL. (In the present discussion we disregard any size convention in the definition of hail.) The horizontal wind field and three contours of updraft speed at 7 km are included in the figure for context. Detailed wind fields at other heights are displayed in Chapter 18. Hail greater than a centimeter or so in diameter tends to originate in the hatched region to the southwest of the updraft, where embryos are in the best position to be advected into the updraft core. With the right combination of particle size, starting altitude, and drag law, however, it is found that hail can originate from virtually any position in the region shaded except near the extreme southern corners.

A representative sample of the computed hail trajectories is shown in plan view in Fig. 20.8, with details of the

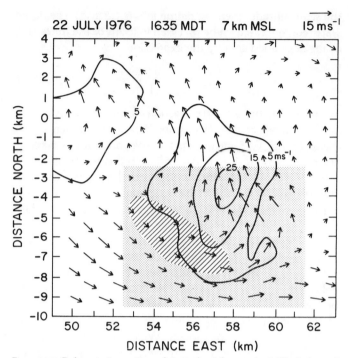

Figure 20.7 Embryo starting positions relative to the airflow at 7 km MSL. Embryos of various sizes are inserted into the flow over a 1-km grid defined within the shaded region. Horizontal wind vectors relative to the storm are shown (scale in upper right), and updraft contours are indicated. The largest hail originated within the hatched region, where embryos were in the best position to be transported into the updraft core.

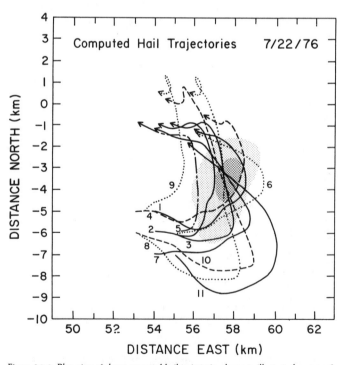

Figure 20.8 Plan view of eleven computed hail trajectories chosen to illustrate the range of behavior. The trajectories are numbered for ease of discussion, and the 15 and 25 m s⁻¹ updraft contours are repeated from Fig. 20.7 to provide a spatial reference. The trajectories are shown relative to the storm.

hailstone growth histories listed in Table 20.1. The medium drag law from Fig. 20.4 is assumed. The trajectories are numbered such that the final hailstone size decreases monotonically with increasing number. As seen in the figure, hailstones can enter the updraft from practically any position on its west, south, and east sides. The tracks of the largest hailstones, 1.9 cm in diameter, are repeated for clarity in Fig. 20.9. They grow from the two larger embryo sizes by entering the strong updraft on its southwest and southeast flanks and then turning northward and traveling along the long dimension of the updraft with the southerly flow there (trajectories 1, 2, and 3). For these trajectories, which are shown in north-south vertical section in Fig. 20.10 relative to the observed radar echo, the updraft speed and particle terminal velocity are matched well enough that the hailstones are lifted less than 2.0 km above their initial position. After passing to the north through the updraft core, the hailstones fall to the ground in the region of high reflectivity. Trajectory 2 is also sketched in perspective in Fig. 20.1. The maximum updraft speed experienced along these three trajectories, 29 m s^{-1} (see Table 20.1), is approximately 1/3 larger than the hailstone terminal velocities at any time. It is of little significance in determining the final hailstone size because the hail moves too quickly through the region of peak updraft.

To illustrate the temporal growth history, trajectory 2 is shown in Fig. 20.11 in time-height section. The size of the hailstone is indicated at various times along the curve, and

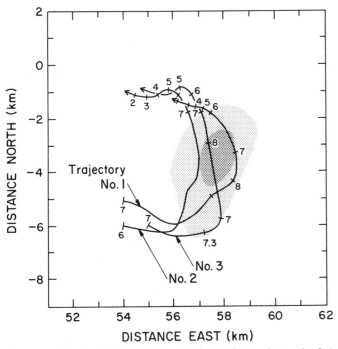

Figure 20.9 Plan view of trajectories 1, 2, and 3 relative to the 15 and 25 m s^{-1} updraft contours, as in Fig. 20.8. These are examples of trajectories that produced the largest hail. Heights are indicated alongside the tracks in units of kilometers above mean sea level.

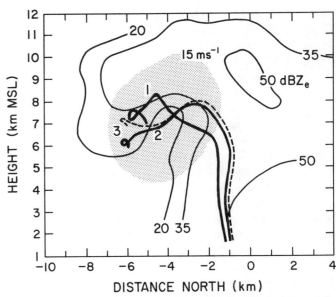

Figure 20.10 Trajectories 1, 2, and 3 projected onto a north-south vertical section. Reflectivities are shown at three levels of intensity for a north-south plane passing through the updraft maximum. The region of updraft exceeding 15 m s^{-1} in this same plane is also indicated. After entering the updraft from the west, the hailstones grow while making a simple traverse to the north, falling out in the vicinity of the radar echo core.

TABLE 20.1
Hailstone growth histories of 11 modeled trajectories for 22 July 1976

Trajectory	Embryo diameter	Final diameter	Maximum diameter	Melted fraction	Starting height	Maximum height	Maximum updraft	Elapsed time
	(cm)	(cm)	(cm)		(km)	(km)	(m s^{-1})	(min)
1	0.6	1.9	2.0	0.11	7.0	8.4	23.7	22.3
2	0.8	1.9	2.0	0.13	6.0	7.9	28.8	23.0
3	0.8	1.9	2.0	0.15	7.0	8.0	29.2	20.7
4	0.2	1.7	1.7	0.15	6.0	8.4	21.2	29.0
5	0.8	1.2	1.3	0.26	6.0	6.6	16.4	18.3
6	0.4	1.1	1.3	0.26	6.0	8.9	20.9	21.8
7	0.8	0.9	1.1	0.45	5.0	6.6	21.8	25.9
8	0.2	0.8	1.0	0.51	6.0	9.1	29.9	32.0
9	0.2	0.8	1.0	0.51	5.0	8.1	12.0	23.9
10	0.2	0.6	0.8	0.65	5.0	8.1	28.1	32.0
11	0.4	0.5	0.8	0.68	7.0	7.6	11.9	24.0

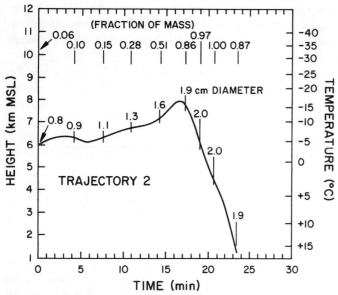

Figure 20.11 *Trajectory 2 in time-height section. The hailstone size is indicated at various times along the curve. In the upper part of the figure the mass of the hailstone is represented in terms of its ratio to the maximum mass attained before any melting occurs. In this case 50% of the hailstone mass is accreted in less than 7 min. About 13% of the mass is lost through melting during fallout.*

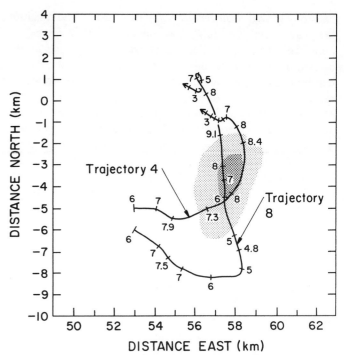

Figure 20.12 *Trajectories 4 and 8 in plan view, similar to Fig. 20.9. Heights (km MSL) are indicated along the tracks. Both trajectories are initiated with 2-mm embryos at 6 km MSL, and the only difference is the initial 1-km separation in the horizontal.*

the numbers in the upper part of the figure represent the ratio of the mass of the stone at the indicated time to its maximum value. Though the whole trajectory occupies 23 min, most of the mass is accreted during a much shorter time. For example, 80% of the mass is acquired in 11 min, 50% in less than 7 min, and the final 20% in approximately 4 min, numbers which are typical for almost all eleven trajectories in Fig. 20.8 and Table 20.1. After reaching maximum size, the hailstone falls to the ground in only 2.3 min, aided by a local downdraft. During this time melting decreases the size from 2.0 to 1.9 cm, amounting to a 13% reduction in the mass of the stone. Most of the hail growth for trajectory 2 takes place between temperatures of −10°C and −20°C (see scale on right of Fig. 20.11). The same is true for trajectories 1 and 3.

In contrast to the trajectories of stones that quickly enter the strong updraft, like trajectories 1 through 6, many hailstones grow while traversing the updraft periphery. For example, number 9 grows to a final diameter of 0.8 cm during a simple up-and-down traverse along the west side of the updraft (see Fig. 20.8). Number 7, which grows to a final size of 0.9 cm, acquires most of its mass (about 80%) during its east-northeast pass through the southern edge of the updraft, and then, after turning to the northeast, acquires the additional 20% while passing, at an altitude of 5-6 km, underneath the updraft maximum.

Trajectory 8, which is shown with height indicated along the track in Fig. 20.12 (trajectory 4 is also included there for later comparison), starts out as a 2-mm embryo at an altitude of 6 km, imbedded in the cyclonically streaming flow. After sedimenting to 5 km, it encounters the inflow and is then swept through the core of the updraft in a manner identical to

the hail growth model proposed by Browning and Foote (1976). Fig. 20.13 shows its trajectory in north-south vertical section. Unlike trajectories 1-3, it experiences a large excursion in the vertical. Indeed, if all hail trajectories were similar to number 8, one would expect the storm's radar echo to display a large vault. The vault in the present storm is not very large and is quite transitory (Chapter 16). In spite of passing through the updraft maximum, this trajectory does not lead to very large hail: only 0.8 cm in diameter, as opposed to 1.9 cm for trajectories 1-3. It is of interest to examine the reason for this.

Fig. 20.14 shows trajectory 8 in time-height section. Trajectory 4, included for comparison, is initiated at the same height and has the same embryo size of 2 mm, but is located 1 km further north (see Fig. 20.12) and leads to hail 1.7 cm in diameter. By examining Figs. 20.12 and 20.14 together the reason for the difference in growth histories is clear, and in this case is connected to the early growth before entering the strong updraft. Trajectory 4 remains in regions of moderate updraft and liquid water content for the first 10 min or so of its history, so that by the time the hailstone enters the strong updraft it has already grown from 0.2 cm to about 0.8 cm, and is then large enough to have a substantial terminal velocity and growth rate. On the other hand, the hailstone following trajectory 8 is carried around the updraft by the cyclonic flow and grows substantially less in the first 10-15 min. By the time it encounters the strong updraft on turning northward it is only about 0.5 cm in diameter. The combination of its smaller size and lower terminal velocity then produce a

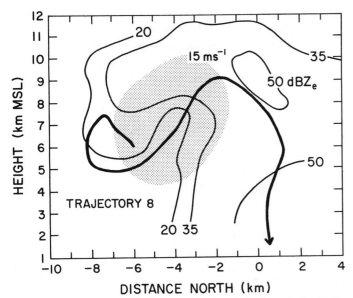

Figure 20.13 *Trajectory 8 projected onto a north-south vertical section, as in Fig. 20.10. Unlike trajectories 1 to 3, trajectory 8 experiences a large excursion in the vertical because of the smaller hailstone size upon entering the strong updraft.*

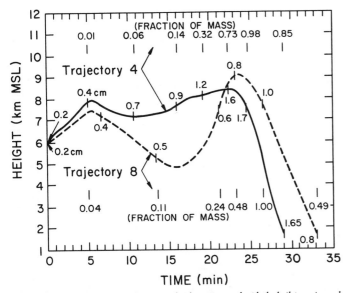

Figure 20.14 *Trajectories 4 and 8 in time-height section, and with the hailstone size and mass indicated as in Fig. 20.11.*

significantly smaller growth rate than for trajectories 1-4, and in the 8 min or so that it takes to pass through the region of strong updraft ($w > 15$ m s^{-1}) and high liquid water content it can grow to a maximum size of only 1.0 cm. It subsequently loses half of its mass in melting to a size of 0.8 cm during fallout, in contrast to the larger hailstones, which lose only 15% or less of their mass.

The divergence in trajectories 4 and 8 associated with a shift in the initial horizontal position is also found in association with initial height differences and differences in embryo size. For example, trajectories 4 and 9 start at the same

horizontal position (53 km, −5 km) and with the same embryo size (2 mm), but at altitudes of 6 and 5 km respectively. With reference to Fig. 20.8 and Table 20.1, trajectory 4 at the higher altitude moves eastward into the updraft core and leads to hail 1.7 cm in diameter. Trajectory 9 at the lower altitude experiences at all times somewhat smaller updraft speeds and stronger southerly velocities. Though it also rises to over 8 km, it stays well to the west of the updraft core, and grows to only 0.8 cm.

Trajectories 5 and 6 are started at the same horizontal position and the same altitude, but with embryo sizes of 0.8 and 0.4 cm, respectively. They grow to about the same final size of 1.1 to 1.2 cm, but trajectory 5 accomplishes this by moving at relatively constant altitude into the southerly flow of the updraft, and remaining on the west side of the updraft core. In contrast, the smaller embryo of trajectory 6 rises sharply to an altitude of almost 9 km in passing over the southern part of the updraft, where it accumulates most of its growth. Having diverged by 3-4 km from its companion particle on trajectory 5, it then sinks on turning to the northwest and passes through the lower region of the updraft before falling to the surface.

The calculated positions (based on the starting conditions previously discussed) at which hailstones reach the ground are summarized for embryo sizes of 0.4 and 0.8 cm in Figs. 20.15 and 20.16. For ease of plotting, the final hailstone radii are shown in units of millimeters. For both embryo sizes, the notable result is the strong tendency for the hail to fall within the strongest part of the echo (> 50 dBZ$_e$). Such a relationship between reflectivity and hailfall has, in fact, been observed for the Westplains storm, as described in Chapter 16. The degree of coincidence between the observed high reflectivity and hailfall predicted here is surprisingly good, and may be regarded as supporting the consistency of the derived wind field, the hail growth calculations, and the independent reflectivity measurements.

While hail predicted to be co-located with the low-level echo core can arrive from a variety of starting locations because of the strong convergence into the downdraft (see Chapter 18), the particles seen in the southeast extension of the echo in Figs. 20.15 and 20.16 grow predominantly along trajectories like numbers 8, 10, and 11 of Fig. 20.8, though they fall to the ground somewhat earlier.

Little if any size sorting is evident in the hailfall pattern, with the exception that the largest hail, of 8-, 9-, and 10-mm radius, is observed in Fig. 20.16 to fall only in the very southern part of the echo core. Given the amount of mixing and interweaving of trajectories associated with the large velocity gradients in the storm, as illustrated earlier, this result is not surprising. The largest hailstones predicted by the calculations are about the same size, 2 cm, as the largest stones actually sampled from the storm (see Chapter 16).

The main differences between Figs. 20.15 and 20.16 are that the predicted modal hail size is larger with the 0.8-cm embryos, and a much larger proportion of the hailstones

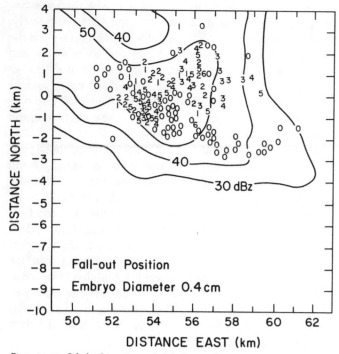

Figure 20.15 *Calculated positions at which hailstones fall to the ground. The positions are shown by a number indicating the final hailstone radius in millimeters. The calculations represented here were all initiated with embryos 0.4 cm in diameter. The 1640 MDT radar echo at 2 km MSL, about 500-600 m above the surface, is also included. Most of the hail is predicted to fall within the 50-dBZ contour. The extension of the 30- and 40-dBZ$_e$ contours to the southeast corresponds to a region where the hailstones tended to melt completely.*

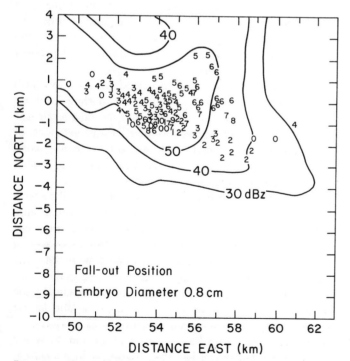

Figure 20.16 *Computed hailstone fall-out positions, as in Fig. 20.15, except for initial embryos of 0.8-cm diameter.*

grown on 0.4-cm embryos melts completely before reaching the ground. This trend continues to smaller embryo sizes also, with an even larger fraction of the 0.2-cm embryos than the 0.4-cm embryos leading to rain rather than hail because of their smaller average size before falling into the warm region.

The degree of melting as a function of hail size is illustrated in Fig. 20.17, which shows the average results for trajectories computed over the whole area indicated in Fig. 20.7. The curves starting in the upper left indicate the fraction of the hailstone mass that is lost on the average for hail of the indicated maximum size. Thus, using the medium drag law discussed in the previous section, hailstones of 1.0 cm diameter just above the melting level lose about 73% of their mass during descent through the warm air. The ascending curve interprets this in terms of final diameter rather than mass loss: on the average a 1.0-cm hailstone melts to become 0.65-cm hail. The melting becomes more extreme for hail sizes less than 1 cm, but even the largest hail considered here melts by at least 0.1 cm.

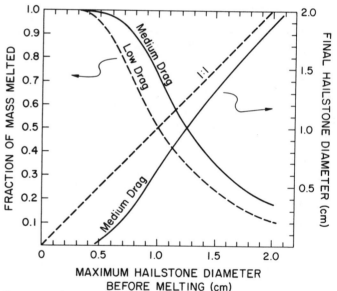

Figure 20.17 *Average amount of melting during fall to the ground for hailstones of different sizes. The curves starting in the upper left indicate the fraction of the stone mass that is lost. Curves resulting from using two different drag laws are shown. The curve starting in the lower left gives the final hailstone size as a function of the maximum size before melting.*

The dashed curve in Fig. 20.17 shows the mass loss due to melting when the low drag formulation is used. The higher fall speeds and consequently the shorter residence times in the warm air lead to significantly less melting under the low-drag conditions: a 1-cm stone loses only 50% of its mass, rather than 73%.

In addition to drag-law effects, variations in the amount of melting from one trajectory to another, and hence deviations from the curves in Fig. 20.17, are induced by differences in the particle density and in the vertical air velocity experienced. Hail falling in strong downdraft has a significant advantage over hail that does not, in that it can spend considerably less time in the warm air before reaching the ground.

In the previous discussion, variations in trajectories associated with changes in embryo size and position have been considered. Variations also result from changes in the particle drag law. For example, Fig. 20.18 shows trajectory 3 recomputed with the other two laws. With the lower terminal velocity, the trajectory for the high drag case rises more sharply and moves farther to the south in response to the relative flow at the 8-km level. For the low drag case the embryo sediments more quickly into the southerly flow and consequently stays somewhat further west. In this case the final hailstone sizes do not vary dramatically, being 1.5, 1.9, and 1.8 cm, respectively, for the cases of high, medium, and low drag.

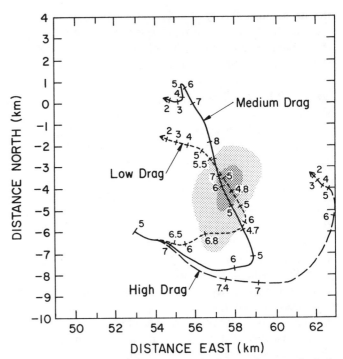

Figure 20.19 Sensitivity of trajectory to changes in drag law, as in Fig. 20.18, but for the initial conditions of trajectory 10.

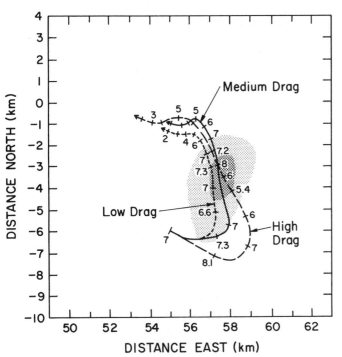

Figure 20.18 The sensitivity of the computed trajectories to changes in the drag formulation is illustrated using the initial conditions for trajectory 3 but with three different drag laws. The trajectories are shown in plan view as in Fig. 20.8.

Fig. 20.19 shows the response of trajectory 10 to changes in the drag law. In this case the trajectory calculated with the high drag law diverges markedly from the other two, and the hailstone melts almost completely during fall (final diameter 0.1 cm). In the low drag case, the embryo grows more rapidly initially than in the medium drag case. It enters the strong updraft at a similar position but grows slightly faster because of its size advantage, though this is largely offset by its staying somewhat lower in the updraft. It leads to hail of 0.7 cm, only slightly larger than the 0.6 cm for the medium drag case.

In general, calculations using the low-drag formulation predict larger hailstones, and, as a consequence, fewer starting positions that lead to complete melting. Smaller stones and more melting tend to result from using the high drag law in this flow field.

20.4 ASPECTS OF THE HAIL SUPPRESSION PROBLEM

With the realization that available artificial ice nucleants like AgI are not sufficiently active to freeze the majority of cloud droplets at temperatures as cold as $-15\,°C$ or $-20\,°C$, the idea of a competition mechanism for hail suppression became popular. According to the competition hypothesis (e.g., Dennis, 1977), if more hail embryos are produced by cloud seeding, these embryos will compete with one another and with their naturally formed counterparts for the available supercooled water. The reduction of the liquid water that would result is then supposed to lead to lower growth rates and smaller final sizes for all the hailstones. If the stones are small enough, they will melt before reaching the ground and hail will be suppressed entirely. One notes from this description that the competition hypothesis can be stated so simply that it is possible to avoid any explicit discussion whatsoever of the storm circulation, in spite of the pivotal role it plays in determining such things as the length of time the hailstone can spend in the growth region, and the nature of the transition in growth environments that is necessary as the particle goes from the embryo stage to the hail stage. It would appear that an adequate theory of hail suppression ought to consider more of the details of storm processes, including the embryo stage, the hailstone growth trajectories, and the overall efficiency of the precipitation process.

20.4.1 The Graupel Complication

Perhaps the most widely published theory of hail suppression is that of the Soviets, involving an accumulation zone of large, supercooled drops, in which most of the hail growth is theorized to take place (Sulakvelidze et al., 1974). If their hail growth theory is correct, then hail could be suppressed by seeding the accumulation zone and freezing the large drops. If, on the other hand, the supercooled water is dispersed in the form of many small cloud droplets, instead of a few raindrops, then seeding would likely have little effect. Insufficient time would be available in a strong updraft to freeze more than a small fraction of the total number of droplets.

Studies in Colorado carried out in NHRE do not support the idea of an accumulation zone, but rather have given strong evidence that precipitation forms via an ice process, that supercooled water is almost entirely in the cloud droplet sizes, and that the large hydrometeors in the cloud are graupel and hail, not supercooled raindrops (Dye et al., 1974). The calculations in the preceding section show that there is no problem in accounting for hail growth in subadiabatic water contents.

The "cold rain" process poses special problems for hail suppression, as discussed in some detail by Knight et al. (1979). Seeding in the strong updraft is then likely to do little good. In order to produce embryos that can compete within the strong updraft, the seeding must be done elsewhere, in a region where (a) sufficient time is available for graupel to form after the nucleation of ice crystals by seeding, and (b) the air motion will carry the artificially-formed graupel into the hail growth region. This has often been interpreted as meaning that seeding should occur in the same region as that in which the natural nucleation events leading to hail are taking place. However, natural embryo sources are still not well understood and may be highly variable, so that general seeding strategies have little, if any, empirical basis. It should also be pointed out that, since there is a 10 to 15 min delay between seeding and the formation of graupel, one must anticipate, a considerable time in advance, just where the collection centers that will provide the competition will be needed and where the seeding should be done. Such a time delay does not arise if raindrops are present in the updraft, for once frozen they are large enough and in the right position to start competing immediately.

20.4.2 Hail/Rain Efficiency and the Depletion Theory of Hail Suppression

Precipitation efficiency can be usefully defined as the ratio of the total precipitation (dominated by the rainfall) to the water vapor entering the cloud, or similarly a ratio in terms of the instantaneous fluxes of these quantities. Decreases in efficiency below 100% result from (a) the entrainment of dry air into the cloud, with the accompanying evaporation of some water, (b) evaporation in downdrafts, and (c) an inefficiency in the conversion of cloud droplets to particles of precipitation size. In cold, continental clouds it is thought that most of this latter conversion takes place by the collection of cloud droplets on growing ice particles—rimed ice crystals and graupel. We refer to this as depletion. If too few collection centers exist, then many of the cloud droplets will be carried into the upper part of the cloud where they will form tiny ice crystals and be lost to the precipitation process (Foote and Fankhauser, 1973).

Precipitation efficiencies have been determined by moisture budget studies, and can be very low, $\approx 10\%$, particularly for hailstorms (e.g., Marwitz, 1972a). Lack of a sufficient number of collection centers can also be deduced by computing a depletion time scale $\tau = LWC/\Sigma\, dm/dt$, where the numerator is the liquid water content and the denominator is the sum of the hydrometeor mass growth rates (see Chapter 19). Using the growth equations, τ can be expressed as $\tau = 4/(\Sigma \pi\, D^2 N V_t E)$, where D is the particle diameter, N is the particle concentration, V_t the terminal velocity, and E the collection efficiency. To first approximation, τ is the time to reduce a given liquid water content by the factor $1/e$ (actually one can show that a closer approximation to the $1/e$ time can be obtained by multiplying the previous expression by $ln(2 + \alpha)/[(1 + \alpha)ln2]$, where α is the liquid water content (assumed to be in cloud droplets) divided by the precipitation content; the fractional error is within 30% for $\alpha < 2$ or so). Calculations in Chapter 19 show that τ can be very large, $> 10^3$ s. Since the time required for hail growth can easily be as short as 500 s, it is clear that hailstones can grow to their final size during a period in which the liquid water content is not substantially depleted by their grow. In other words, the hailstones are not competing effectively, a result that should be general for storms that have a low precipitation efficiency and produce a large cirrus canopy.

The significance of this situation in the context of hail suppression is that if one increases only modestly the number of hail embryos in the updraft, then more hail rather than less hail will be the result, since an abundance of supercooled water will be available. Not until enough embryos are added that the depletion time scale is drastically reduced, to perhaps 10^2 s or so, will final particle sizes be greatly affected. The possibility of increasing hail amounts by seeding is entirely consistent with the competition theory of hail suppression. In fact, if one's intention were to increase the amount of hail from a storm, then, based on first principles, one might go about it in the same way, that is, by attempting to increase the number of hail embryos. Of course, in the competition theory one strives for such a high concentration of collectors that the precipitation efficiency is markedly increased and the size of hydrometeors reduced, but there is no assurance that this can be achieved in practice.

In considering these ideas further it is instructive to ask which part of the hydrometeor size spectrum contributes most to the depletion. As discussed in Chapter 15, the fact that in general much more rain than hail reaches the ground is evidence that up in the cloud the depletion is dominated by those collectors that are destined to fall as rain, rather than by growing hailstones, that is, the more numerous smaller particles must be accounting for most of the depletion. Calculations in Chapter 19 based on measured ice particle spectra in the Westplains storm also show that the depletion tends to be dominated by particles a few millimeters in diameter or smaller. This result comes about because the measured particle number density per unit diameter commonly varies as the diameter raised to a power between about −3 and −4, so that the depletion time scale due to particles in a given diameter range varies as the diameter to the power 0.5 to 1.5. Thus, the greater number of smaller particles generally gives the shorter depletion times.

On the basis of these results a revised suppression concept seems in order. The goal of hail suppression should not be to duplicate the natural hail cycle by introducing hail embryos and asking them to compete for the supercooled water. Rather it is the natural rain cycle that one should try to duplicate. In the first place, assuming that the spectral slope following seeding is not greatly altered, these "rain embryos" (typically ice particles in cold, continental clouds) are what accomplish most of the depletion anyway. In this sense the hail competition idea seems to have been incorrect from the start. Second, by adding smaller "depletors" rather than "competitors" one reduces the risk of increasing hail, because, according to the calculations in the previous section, the smaller embryos have a weaker tendency to grow to hail size anyway, even in the presence of high water contents. The third advantage that may be inherent in this concept is that the targeting of seeding material into potential embryo source regions may be less difficult. While there appears to be a selection mechanism whereby only a subset of the potential embryos becomes hail, based on fortuitous initial sizes, positions, and growth trajectories, the restraints are reduced for rain. The trajectories in the previous section do not indicate that there is any dynamical problem in introducing rain embryos into the updraft from a variety of positions. In the case of storms with large radar vaults, however, as discussed by Browning and Foote (1976), this presumably is still a major problem.

The depletion theory of hail suppression just discussed is, of course, not ready for application, and much more thought needs to be given to it. While embryos 1-2 mm in diameter or smaller might effectively deplete the supercooled water in the Westplains storm without themselves producing a significant hailfall, there is no assurance that this would be the case in a different storm possessing a different updraft structure.

The fundamental idea of attempting to suppress hail by duplicating the natural rain cycle, however, seems worthy of further consideration.

References

Auer, A. H., J. D. Marwitz, G. Vali, and D. L. Veal, 1971: Final Report to the National Science Foundation. Dept. of Atmospheric Resources, Univ. of Wyoming, Laramie, Wyo., 94 pp.

Browning, K. A., 1963: The growth of large hail within a steady updraught. Q. J. R. Meteorol. Soc. 89, 490-506.

-----, 1964: Airflow and precipitation trajectories within severe local storms which travel to the right of the winds. J. Atmos. Sci. 21, 634-639.

-----, 1977: The structure and mechanisms of hailstorms. Meteorol. Monogr. 16 (38), 1-43.

-----, and G. B. Foote, 1976: Airflow and hail growth in supercell storms and some implications for hail suppression. Q. J. R. Meteorol. Soc. 102, 499-533.

-----, and F. H. Ludlam, 1962: Airflow in convective storms. Q. J. R. Meteorol. Soc. 88, 117-135.

-----, J. C. Fankhauser, J-P. Chalon, P. J. Eccles, R. C. Strauch, F. H. Merrem, D. J. Musil, E. L. May, and W. R. Sand, 1976: Structure of an evolving hailstorm: Part V. Synthesis and implications for hail growth and hail suppression. Mon. Weather Rev. 104, 603-610.

-----, F. H. Ludlam, and W. C. Macklin, 1963: The density and structure of hailstones. Q. J. R. Meteorol. Soc. 89, 75-84.

Carte, E., and R. E. Kidder, 1966: Transvaal hailstones. Q. J. R. Meteorol. Soc. 92, 382-391.

Dennis, A. S., 1977: Hail suppression concepts and seeding methods. Meteorol. Monogr. 16(38), 181-191.

Dye, J. E., C. A. Knight, V. Toutenhoofd, and T. W. Cannon, 1974: The mechanism of precipitation formation in northeastern Colorado cumulus: III. Coordinated microphysical and radar observations and summary. J. Atmos. Sci. 8, 2152-1259.

English, M., 1973: Alberta hailstorms: Part II. Growth of large hail in the storm. Meteorol. Monogr. 36, 37-98.

Fankhauser, J. C., 1971: Thunderstorm-environment interactions determined from aircraft and radar observations. Mon. Weather Rev. 99, 171-192.

Foote, G. B., and J. C. Fankhauser, 1973: Airflow and moisture budget beneath a northeast Colorado hailstorm. J. Appl. Meteorol. 12, 1330-1353.

Heymsfield, A. J., 1978: The characteristics of graupel particles in northeastern Colorado cumulus congestus clouds. J. Atmos. Sci. 35, 284-295.

-----, A. R. Jameson, and H. W. Frank, 1980: Hail growth mechanisms in a Colorado storm: Part II. Hail formation processes. J. Atmos. Sci. 37, 1789-1807.

Klemp, J. B., R. B. Wilhelmson, and P. S. Ray, 1981: Observed and numerically simulated structure of a mature supercell thunderstorm. J. Atmos. Sci. 38, 1558-1580.

Knight, C. A., and N. C. Knight, 1970: Hailstone embryos. J. Atmos. Sci. 27, 659-666.

-----, G. B. Foote, and P. W. Summers, 1979: Results of a randomized hail suppression experiment in northeast Colorado: Part IX. Overall discussion and summary in the context of physical research. J. Appl. Meteorol. 18, 1629-1639.

-----, N. C. Knight, J. E. Dye, and V. Toutenhoofd, 1974: The mechanism of precipitation formation in northeastern Colorado cumulus: I. Observations of the precipitation itself. J. Atmos. Sci. 31, 2142-2147.

Kropfli, R. A., and L. J. Miller, 1976: Kinematic structure and flux quantities in a convective storm from dual-Doppler radar observations. *J. Atmos. Sci.* 33, 520-529.

Lesins, G. B., R. List, and P. I. Joe, 1980: Ice accretions: Part I. Testing of new atmospheric icing concepts. *J. Rech. Atmos.* 14, 347-356.

List, R., 1960: Growth and structure of graupels and hailstorms. *Physics of Precipitation, Geophys. Monogr.* 5, 317-324.

-----, and R. S. Schemenauer, 1971: Free-fall behavior of planar snow crystals, conical graupel and small hail. *J. Atmos. Sci.* 28, 110-115.

Ludlam, F. H., 1958: The hail problem. *Nubila* 1, 12-96.

Macklin, W. C., 1961: Accretion in mixed clouds. *Q. J. R. Meteorol. Soc.* 87, 413-424.

-----, 1977: The characteristics of natural hailstones and their interpretation. *Meteorol Monogr.* 16(38), 65-88.

Marwitz, J. D., 1972: Precipitation efficiency of thunderstorms on the High Plains. *J. Rech. Atmos.* 6, 367-370.

Mason, B. J., 1971: *The Physics of Clouds.* Clarendon Press, Oxford, England, 671 pp.

Matson, R. J., and A. W. Huggins, 1980: The direct measurement of the sizes, shapes, and kinematics of falling hailstones. *J. Atmos. Sci.* 37, 1107-1125.

Musil, D. J., 1970: Computer modeling of hailstone growth in feeder clouds *J. Atmos. Sci.* 27, 474-482.

Orville, H. D., 1977: A review of hailstone-hailstorm numerical simulations. *Meteorol. Monogr.* 16(38), 49-61.

Paluch, I. R., 1978: Size sorting of hail in a three-dimensional updraft and implications for hail suppression. *J. Appl. Meteorol.* 17, 763-777.

Pflaum, J. C., 1978: A wind tunnel study on the growth of graupel. PhD. dissertation, Univ. of California, Los Angeles, Calif., 107 pp.

-----, 1980: Hail formation via microphysical recycling. *J. Atmos. Sci.* 37, 160-173.

Rinehart, R. E., and J. D. Tuttle, 1981: The effects of mismatched antenna beam patterns on dual-wavelength processing. Prepr. 20th Conf. on Radar Meteorol., Boston, Mass., 1981, Am. Meteorol. Soc., Boston, Mass., 676-682.

Stringham, G. E., D. B. Simons, and H. P. Guy, 1969: The behavior of large particles falling in quiescent liquids. Geological Survey Professional Paper 562-C, U. S. Govt. Printing Office, Washington, D.C., 36 pp.

Sulakvelidze, G. K., B. I. Kiziriya, and V. V. Tsykunov, 1974: Progress of hail suppression work in the USSR. In *Weather and Climate Modification,* W. N. Hess (ed.), John Wiley and Sons, New York, N.Y., 410-431.

Zikmunda, J., and G. Vali, 1977: Corrigendum to "Fall patterns and fall velocities of rimed ice crystals." *J. Atmos. Sci.* 34, 1483.

The 25 July 1976 Case Study

CHAPTER 21
The 25 July 1976 Case Study: Environmental Conditions, Reflectivity Structure, and Evolution

James E. Dye, L. Jay Miller, Brooks E. Martner, and Zev Levin

21.1 INTRODUCTION

Little research using coordinated Doppler radars and aircraft has been done on the early stages of storm development or on weak thunderstorms that produce precipitation but little if any hail. The 25 July 1976 storm that is the subject of this and the following two chapters is substantially smaller than the other two cases in this volume. As such it represents a more populous class of convective phenomena.

The systems used to obtain measurements on the storm are listed in Table 21.1, and their locations with respect to the storm are shown in Fig. 21.1. Unfortunately, the 25 July 1976 storm occured 10 to 20 km west of the western boundary of the surface network, so that surface measurements were of limited use in this case study.

TABLE 21.1
Types of data, their sources, and times

Data type	Source	Time (MDT)
Environmental characteristics	(1) Rawinsondes released from Grover, Co, and Potter, NB	1730
	(2) Surface mesonetwork	continuous
	(3) National Weather Service	
	(a) Surface observations	1900
	(b) 500-mb chart	1300
Radar reflectivity structure	(1) Grover 10-cm radar	1845–2000
	(2) NOAA/WPL 3-cm radar	1855–2000
Cloud base properties	(1) NCAR Queen Air N304D, N306D	1900–1935
	(2) Univ. of Wyoming Queen Air N10UW	1940–1946
Microphysical, vertical motion, and thermodynamic properties	(1) NCAR/NOAA sailplane	1900–1933
	(2) Univ. of Wyoming Queen Air N10UW	1900–1933
Internal air motions	(1) NCAR/FOF 5-cm radar	1859–1940
	(2) Two NOAA/WPL 3-cm radars	

Equivalent reflectivity factor measured by the CP-2 radar provides the context within which the aircraft and other data are analyzed. The radar had a 1° pencil beam and a minimum detectable reflectivity of $-6\ dBZ_e$ at 30-km range. The 600-m range resolution produced by the DADS processor is used in this study, except where the finer-scale 150-m resolution of the MINA II processing is noted.* During most of its

*DADS: Data Acquisition and Display System. MINA II: Multiplexed Input NHRE Averager (Model II).

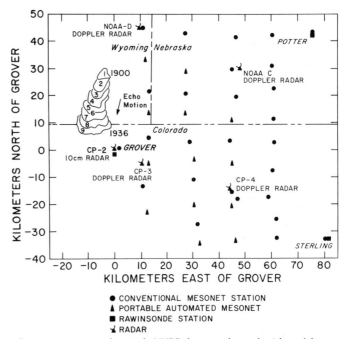

Figure 21.1 Location, relative to the NHRE observational network, of the 25 July 1976 storm used for the first-echo case study. The 5-dBZ_e reflectivity contours and the core of the reflectivity (designated by sequential numbers) at 4 km are shown at about 5-min intervals from 1900 to 1936.

life cycle the 25 July 1976 storm was within 35 km of the CP-2 radar, so that high spatial resolution was possible. The radar scanned in elevation steps such that the entire storm was viewed about every 2 min.

Tracks of the penetrating aircraft from X-band M33 tracking radars were compared with occasional radar return observed with CP-2 and adjusted to achieve an accuracy of about 500 m. Tracks of the cloud base aircraft were generated from the Inertial Navigation System (INS) on each aircraft and corrected for INS drift by comparison with radar returns observed with the CP-2. Further details on the measurement system used in this study are discussed in Dye et al. (1980).

21.2 WEATHER SYNOPSIS FOR 25 JULY 1976

A minor shortwave trough was located over the Montana-Dakota region on this day, with very weak height and temperature gradients aloft near the NHRE project area in

Colorado. The 500 mb map (Fig. 21.2) for 1800 (0000 GMT, 26 July 1976) shows a trough of cold air extending from the Dakotas into Colorado. The north-northeasterly flow over the NHRE area drew cooler air southward, helping to destabilize the middle and upper levels and also influencing storm motion.

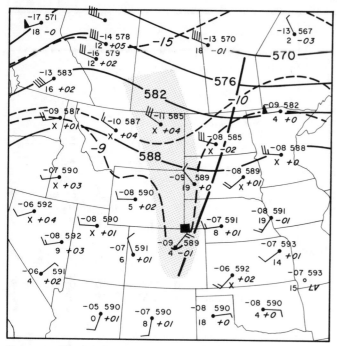

Figure 21.2 The 500-mb map for 1800 local time, 25 July 1976 (0000 GMT, 26 July 1976) showing a trough of cold air (shaded area), a weak short wave (north-south dashed line), and the NHRE operational area (heavily shaded area). The station model is standard, with temperature, dew point depression, winds (full barb = 5 m s⁻¹), height in decameters, and 12-h height change.

A weak cold front at the surface (Fig. 21.3) was associated with the shortwave aloft. The front extended from a surface low pressure center in Manitoba, Canada, into northeastern Colorado, where its location was discernible only by the wind shift across it. A trough extended southward from the front through eastern Colorado. The front entered the northwest corner of the NHRE mesonetwork at about 1000, slowed down, and then stagnated in the eastern part of the area by midafternoon. Surface mesonet stations on the western side of the network recorded wind shifts from southwesterly and westerly to north-northeasterly, with increases in wind speeds from 2-3 m s⁻¹ to 6-7 m s⁻¹ as the front passed. These northerly winds persisted across the NHRE mesonetwork during early afternoon. Dew points on both sides of the front were between 11°C and 14°C, thus providing ample moisture for convective activity. Weak wind convergence and cyclonic circulation seen across the NHRE area were factors in the development of the numerous thunderstorms in the NHRE area that began in the afternoon and persisted well into the night. Note the wind confluence between Scottsbluff, Nebraska (C), and Cheyenne, Wyoming (A), and Grover, Colorado (B). Many storms, including the

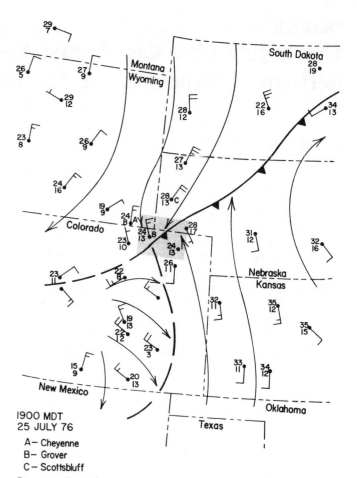

1900 MDT
25 JULY 76

A— Cheyenne
B— Grover
C— Scottsbluff

Figure 21.3 Regional surface map for 1900 MDT, 25 July 1976, with winds (full barb = 5 m s⁻¹), temperature (°C), dew point (°C), and airflow patterns as shown. The NHRE operational area is shaded. The position of the front is shown by a barbed line, and the trough across eastern Colorado is shown by a dashed line. The letters A, B, and C show the locations of Cheyenne, Wyoming, Grover, Colorado, and Scottsbluff, Nebraska, respectively.

case study storm, formed along this confluence line, which will be discussed in greater detail in Section 21.6.

Comparison of morning soundings for 25 and 26 July revealed little or no change in the temperature structure above 500 mb but did show 2°C to 3°C of cooling below 500 mb as a result of the frontal passage and associated thunderstorm activity.

21.3 THE STORM ENVIRONMENT

Because of its proximity in both time and space, the Grover 1729 sounding is considered to be that most representative of the storm environment and is presented in Fig. 21.4. Since the Grover sounding extended only to about 400 mb, the Potter 1730 temperature sounding has been used at levels above 400 mb. Winds from both Potter and Grover are plotted in Fig. 21.4 and shown as a hodograph in Fig. 21.5. The winds at all levels were light, with north-northeasterly winds in the lowest levels backing (on the Potter sounding) to

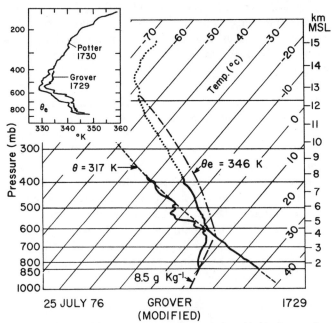

Figure 21.4 *The 25 July 1976 Grover sounding at 1729, modified by the 1730 Potter sounding above 400 mb. The equivalent potential temperature derived from the Grover and Potter soundings is shown in the inset in the upper left.*

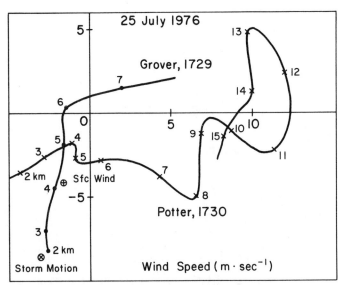

Figure 21.5 *The wind hodographs for 25 July 1976 from soundings taken at Grover (1729) and Potter (1730). The motion of the reflectivity core of the storm is shown by x and the surface wind at Grover at 1900 by +.*

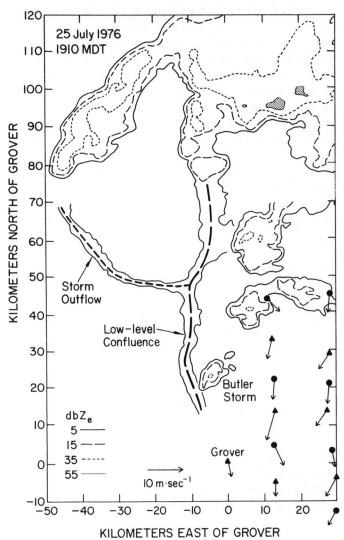

Figure 21.6 *A composite of PPIs at about 1910 near cloud base (≈ 4 km), showing the location and scale of the 25 July 1976 first-echo case study (Butler storm) relative to other convective activity in the area. Reflectivity contours are as indicated. Surface winds observed by the northwestern part of the NHRE mesonetwork (▲ designates PAM sites, ● conventional sites) at 1910 are as shown. The weak reflectivity feature associated with low-level confluence and the outflow from the large storm in the upper portion of the figure are shown by long- and short-dashed heavy lines, respectively (see text).*

predominantly westerly by 9 km and above. Although the Grover sounding shows average winds of about 10 m s⁻¹ in the lowest kilometer, the surface winds at Grover and at the mesonetwork stations closest to the storm decreased to 4 to 5 m s⁻¹ (Fig. 21.6) after 1800.

The sounding shows that a lifted parcel (dot-dash curve) will have 3 °C to 4 °C of instability from about 600 mb up to 250 mb. Out-of-cloud temperature observations made by the sailplane and N10UW at altitudes up to 400 mb agree well with the temperatures shown on the sounding. The cloud base of the storm was observed by N304D to be near 3.7 km (650 mb and 6.5 °C to 7.0 °C). This is consistent with a parcel of air at the surface, with a mixing ratio of 9.0 to 9.5 g kg⁻¹ and temperature of about 32 °C, being lifted unmixed up to cloud base. These values give a potential temperature and an equivalent potential temperature at cloud base of 317 K and 346 K, respectively. The sounding and N304D measurements show virtually no negative energy area at cloud base. Thus, relatively small impulses could initiate convection and, as might be expected, there were many clouds. The Potter and Sterling soundings show a 1 °C to 2 °C inversion at 400 and 370 mb (7.7 and 8.2 km) respectively, which would tend to suppress the early ascent of convective elements trying to rise through this level. Based on the sounding, cloud tops were expected to extend above 12 km, the top of the positive energy area.

21.4 RADAR HISTORY OF 25 JULY 1976

The first radar echoes of the day over the plains near the NHRE area appeared at about 1530 in the South Platte River Valley southwest of Sterling, Colorado, and along the North Platte River west of Scottsbluff, Nebraska. In the immediate NHRE area first echoes appeared about 1 h later.

There were two main regions of development. One began 40-70 km southeast of Grover about 1630. The other area of development was along a line extending from Scottsbluff, Nebraska, through Grover, with the closest cell being 35 km north of Grover. Activity continued along this line for several hours, with isolated cells appearing, growing, moving south-southwestward, and then dissipating. The cell that became the case study storm appeared along this line at about 1840, nearly over the Butler private airstrip from which the sailplane operated; consequently this storm will be called the Butler storm. A major feature in this northern region of development was a large intense storm near Scottsbluff. Radar coverage of this storm was poor because of both extreme range and commitment of the radar to cover storms southeast of Grover, but it moved toward the south and apparently persisted for several hours.

At about 1850 radar coverage was concentrated on the Butler storm, which first appeared as a relatively isolated, small cloud system along a line of weak radar echo shown in Fig. 21.6. Motion of the storm was toward 200° at 9-10 m s⁻¹ between 1855 and 1940. New growth and the most intense part of the storm were on the southwest side, as was true for the earlier storms. The storm persisted for about 2 h and by 2040 had almost dissipated.

Radar coverage showed that thunderstorms continued to develop in the NHRE area well into the night, probably due to the convergence associated with the stationary surface front. Moderate to heavy precipitation (up to 60 mm h⁻¹) and light hail fell in the eastern portion of the NHRE precipitation network between 2100 and 2300.

21.5 STORM ORIGINS

As early as 1840, the time when CP-2 radar began complete scans of the region of the Butler storm, weak radar return was observed from the surface up to nearly 4 km along a north-south line extending at least 100 km north of Grover. (A similar feature was seen on the western edge of the storms southeast of Grover, but continuity between the two areas could not be established because of the lack of continuity in the radar coverage.) The Butler storm moved along this feature with the line being underneath the westernmost part of the storm echo from 1840 to at least 1940. The axis of the reflectivity feature is shown in Fig. 21.6 by a heavy long-dashed line.

This low-level radar return might possibly have resulted from insects being swept upward from the surface by weak vertical motions associated with a confluence line. This contention is supported by the large-scale confluence observed at the surface (Fig. 21.3), the fact that the NOAA-D Doppler radar data show confluence across the line, the intermittent updrafts of 1-3 m s along this line observed and used by the sailplane on tow from 1820 to 1900 to climb to cloud base, and aircraft data obtained by N304D and N306D as they flew into the region to investigate the Butler storm. Their approach to the storm took them on an east-west path, approximately 20 km to the south, toward the southern extent of a flanking line building southward from the storm of interest. Although the CP-2 sector scan did not extend this far south, an extension of the line echo directly south for about 20 km is in the same region where horizontal wind measurements from both aircraft showed confluent flow near cloud base altitude (Fig. 21.7).

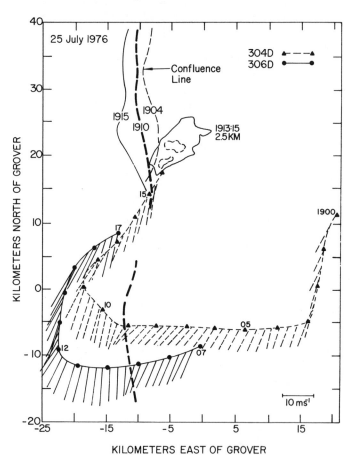

Figure 21.7 The Butler storm relative to the axis of the north-south oriented confluence feature shown at 1904, 1910, and 1915. The axis at 1910 is extended to the south, based on the wind shifts observed from N304D and N306D. The tracks of N304D and N306D are shown and winds are indicated with wind barbs originating on the track. Wind speed is given by the length of the barb.

Both aircraft were climbing at the time they crossed the confluence line, with N304D near cloud base at 4.5 km (650 mb) and N306D at 3.2 km (690 mb). Across the direction shift, the data from N306D showed a slight but definite increase in mixing ratio from 9.0 to 9.5 g kg⁻¹, corresponding to a change in equivalent potential temperature (θ_e) from 344.4 to 345.8 K and no change in the potential temperature. N306D measurements suggested weak upward motion (≈ 0.5 m s⁻¹) in this region. The measurements from N304D did not show a clear change in either mixing ratio or equivalent potential temperature across the wind shift line, but abrupt fluctuations of these parameters and of the vertical air motion suggested considerable vertical mixing. Observations from the cloud base aircraft consistently showed wind direction shifts of this nature on the western edge of the Butler storm from 1915 through 1935. Adjacent to the storm these wind shift regions were also the areas where updrafts were observed. Convergence values of 2×10^{-3} were indicated by NOAA-D single-Doppler data near the Butler storm.

At 1855, about 65 km north of Grover, the confluence line intersected another reflectivity feature—an arc extending from west to east (Fig. 21.6). This arc has the appearance of classic cold outflow from a severe storm and moved to the south at about 15 m s⁻¹. It was probably generated by the large storm shown in the northern part of Fig. 21.6. The eastern edge of this cold outflow reached the western mesonet stations, where the Grover PAM site recorded a

drop in temperature to 20°C, in θ to 308 K (from 317 K), and in θ_e to about 342 K. By about 2000 this cold outflow began to undercut the Butler storm, which was then 10-20 km due west of Grover. Thus, this outflow increased the stability in the lowest levels below the storm and may have caused the decay and dissipation of the storm by 2040.

The large-scale confluence at the surface, the Doppler radar indications of convergence, the continuous existence of the line echo feature for a period of more than 1 h, the fact that the updrafts of the storm were always found near or over this line, the occurrence of several storms along the line and the appearance of a new storm along it, all strongly suggest that isolated regions of convergence developing along a line of confluence were responsible for triggering the Butler storm as well as some others. The observed north-northeasterly surface winds imply some topographic lifting. Perhaps this contributed to the generation of the confluence line and isolated regions of convergence.

21.6 STORM STRUCTURE AND EVOLUTION

21.6.1 Reflectivity Measurements

The evolution of the maximum reflectivity with altitude and time for the entire storm for the period 1840 through 2000 is depicted in Fig. 21.8. During the period 1856 to 1930

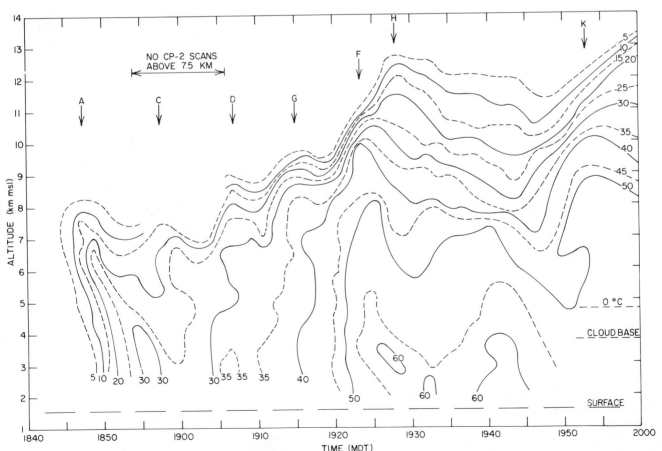

Figure 21.8 Time-altitude profile showing the maximum radar reflectivity history of the Butler storm as recorded by the MINA II processing system. Isopleths are graduated in dBZₑ. Letters at the top of the figure identify individual turrets responsible for reflectivity growth referred to in later figures.

201

areal extent, height of radar tops, and maximum reflectivity steadily increased as the storm grew (Fig. 21.9). Although low-altitude reflectivities did exceed 55 dBZ_e during the period 1925 to 1950, the areal extent of high reflectivities was rather small. Using a reflectivity-rainfall rate (Z-R) relationship ($Z_e = 615 R^{1.4}$) determined by Martner (1975) for northeastern Colorado thunderstorms gives rainfall rates of 7 and 38 mm hr^{-1} for reflectivities of 40 and 50 dBZ_e. Thus, moderate precipitation fell from this storm but only over a rather small region. Although a few small hailstones (8-9 mm) were observed at cloud base by N10UW, these stones probably melted before reaching the ground. The average rate of ascent of the radar top was about 1.5 m s^{-1} during the period 1900 to 1921, abruptly increasing to 5 m s^{-1} from 1921 to 1929. The rates of increase of the maximum reflectivity were 0.9 and 1.8 dB min^{-1} during these same periods of time.

multicell storm, and other cases that have appeared in the literature. In those cases, the maximum reflectivity in the storm tended to appear at a single altitude and spread vertically with time. As will be discussed later, the appearance of higher reflectivities almost simultaneously over a large altitude range also occurred in individual cells within the storm.

As seen in Figs. 21.1 and 21.10, the storm motion was at 9-10 m s^{-1} toward 200°. When compared with the surface winds of about 5 m s^{-1} toward 180°, it is evident that the storm was overtaking the low-level air. Thus, the low-level inflow to the storm was from the southwest (see the hodograph in Fig. 21.5). This storm is similar in this respect to the Raymer multicell storm described by Browning et al. (1976). In the present case the faster motion of the storm, compared with that of the low-level air, appears to be the result of both cell motion and a propagational component associated with the preferred region of subcloud convergence.

Although the mean motion of the high-reflectivity region of the storm was toward 200°-205°, the motion of individual

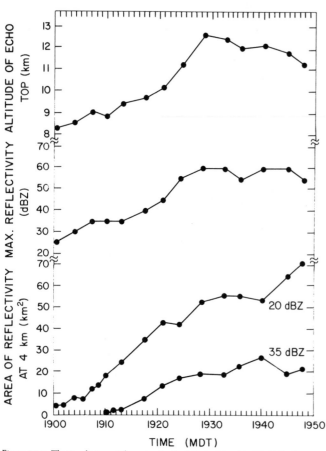

Figure 21.9 The time history of the areas of reflectivity ≥ 20 and > 35 dBZ_e, the maximum reflectivity, and the altitude of the echo top of the Butler storm.

The reflectivity history of the storm is interesting in that as higher reflectivities first appeared in the storm, they appeared almost simultaneously over a rather large altitude extent. For example, when the 40 dBZ_e contour first appeared at 1915, it extended from 2 km to almost 7 km within a period of 2 min. This is in contrast with similar observations, such as those recorded by Chalon et al. (1976) for the Raymer

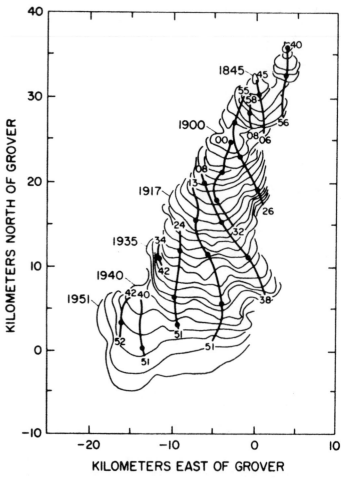

Figure 21.10 The evolution of the 5-dBZ_e contour of the Butler storm at 6 km for the period 1840 to 1952. Heavy lines show the motion of major reflectivity cores at 6 km. The beginning and ending times of the cores are shown in minutes after 1800 or 1900. Heavy dots show locations at multiples of ten minutes after 1800 or 1900.

reflectivity cells and cores* tended to be more eastward, with this effect increasing with altitude. Figure 21.10 shows the evolution of the 5-dBZ_e outer contour of the storm and the motion of major reflectivity cells and cores at 6 km for the period 1840 to 1952.

Reflectivity contours at a constant altitude of 6 km, generated by the technique described by Mohr and Vaughan (1979), are shown for the period 1856 to 1936 at approximately 4-min intervals in Fig. 21.11. Prior to 1925 the alignment of reflectivity cores tended to be along a

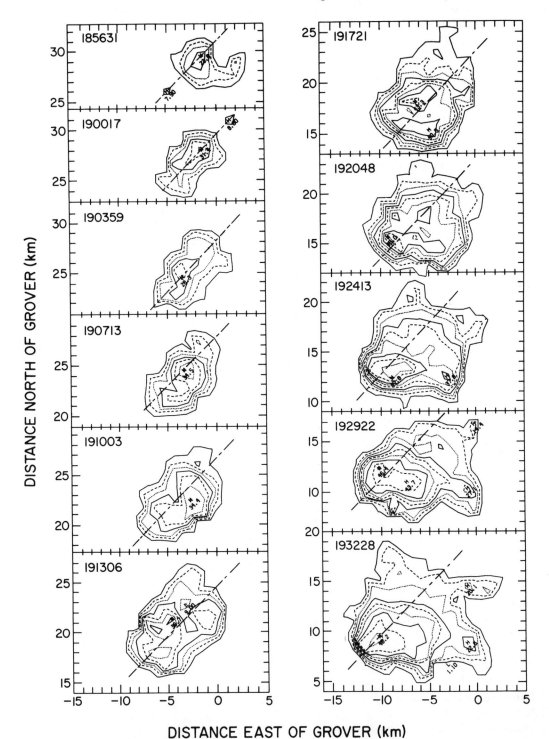

Figure 21.11 *Horizontal sections at 6 km of the reflectivity contours of the Butler storm at 3- to 4-min intervals from 1856 to 1932. The reflectivity contours start at 5 dBZ_e and the interval is 5 dB. The diagonal lines show the location of the vertical sections presented in Fig. 21.12.*

*The term "cell" is used for reflectivity features that form initially as separate areas of 5-dBZ_e radar echo. Features first detected within an existing echo and traceable for several minutes are called reflectivity "cores."

southwest-northeast direction, but around 1925 the reflectivity alignment shifted more to a west-east line. Apparently the alignment was a direct consequence of environmental wind shear through the cloud depth. As cloud tops got higher the precipitation fallout and the consequent reflectivity alignment were influenced by the westerly winds aloft. From the wind hodograph (Fig. 21.5) it can be seen that the lower-level (2-7 km, Grover sounding) wind shear vector was oriented

southwest to northeast, whereas the upper-level (6-10 km, Potter sounding) shear vector was oriented more west to east. After about 1920 a significant portion of the cloud was above 7 km.

The evolution of the storm, as depicted by vertical sections through the storm in a southwest to northeast direction, is shown in Fig. 21.12. The sections were selected to illustrate the evolution of the storm and the appearance of new reflec-

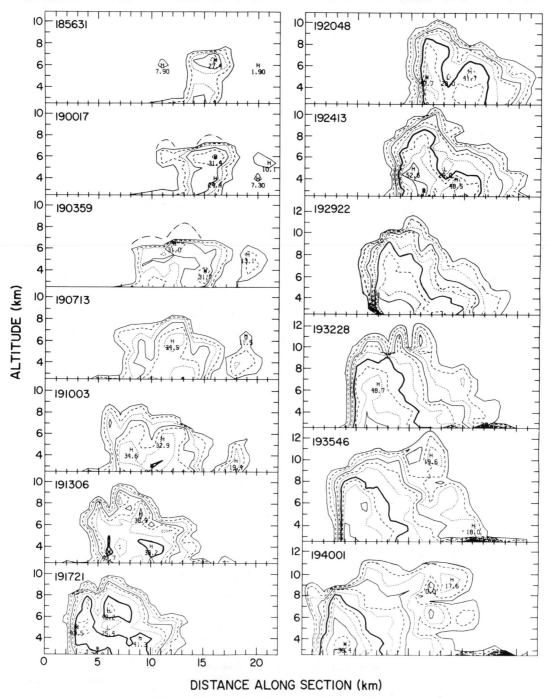

Figure 21.12 Vertical sections of reflectivity at 3- to 4-min intervals, showing the evolution of the storm from 1856 to 1940. Reflectivities are displayed as in Fig. 21.11. The locations of the sections from 1856 to 1932 through the storm are also shown in Fig.

21.11. Radar tops for the 1900 and 1904 sections were determined from the NOAA-C radar observations and are shown by dash-dotted lines.

tivity cores on the southwest side of the storm. The new growth all appeared at about 6 to 7 km on the southwestern side of the storm, intensified and moved to the east relative to the storm. Prior to 1924 individual reflectivity cores are readily distinguishable. From 1924 until about 1940 the storm had one main reflectivity maximum and a more steady state appearance, but individual reflectivity cores could be identified in the upper levels.

Note the low-level weak reflectivity on the far left side of the vertical sections prior to 1917. This tail is the upper portion of the reflectivity feature associated with surface confluence discussed above.

As mentioned previously, the average rate of ascent of the radar top from the period 1900 through 1921 was about 1.5 m s^{-1}. This growth was a result of individual turrets pushing up through the southwestern part of the storm and reaching to higher and higher altitudes. At any one time two or more turrets were emerging and growing out of the main body of the storm. This was also evident from the photographs taken from the towplane. Both the photographs and the radar observations (e.g., Fig. 21.12) show the horizontal dimension of the turrets to be typically 1.5 to 2.5 km. A photograph (Fig. 21.13) taken about 15 km due west of the storm at about 1935 shows the storm during its most intense

Figure 21.13 A photograph taken from the towplane at about 1935 looking east toward the Butler storm.

stage. Figure 21.14 shows a north-south vertical section of the reflectivity structure at approximately the same time as the photograph. Note the shelf of smaller clouds leading up to the storm from the south and the hard appearance of the southernmost turret in the photograph.

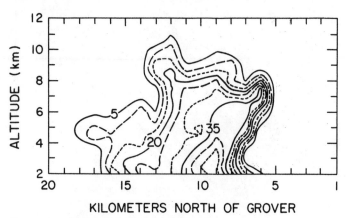

Figure 21.14 A north-south vertical section of reflectivity contours at 8 km west of Grover through the active region (western part) of the Butler storm. The vertical section was constructed from a CP-2 radar scan made between 1935:46 and 1937:42. This section corresponds approximately in time and view to the photograph in Fig. 21.13. The contours are labeled in dBZ$_e$.

The growth and decay of the individual, major radar echo turrets that were followed during the period 1850 through 1950 are shown in Fig. 21.15. The growth rate of individual turrets varied from 3-6 m s^{-1}. Note that successive turrets appeared at about 5- to 7-min intervals. Each turret rose for about 10 min and then began to decay, but following turrets pushed up higher than previous ones to give an overall growth of the radar top from 1900 to 1930. Prior to 1912

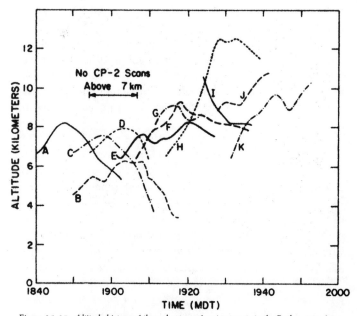

Figure 21.15 Altitude history of the radar tops of major turrets in the Butler storm from 1840 to 1950.

none of the turrets rose above 8 km (about 400 mb). The 1 °C to 2 °C inversion observed in the Potter and Sterling soundings at 400 mb apparently retarded the growth of the cloud top past this level.

The horizontal position and motion of the tops of turrets as they rose and decayed is shown in Fig. 21.16. For the most part, the turret tops moved to the southeast at about 8-12 m s^{-1}, which was consistent with the environmental winds at those altitudes. The early turrets tended to move slower and more southward than the later ones, primarily because they were lower and the winds at lower altitudes were less westerly and weaker than those at the upper levels. In the early stages of cloud development the turrets can be identified with the reflectivity cores at 6 km mentioned earlier and shown in Fig. 21.11. However, by about 1910 the association of turrets and cores is no longer unambiguous.

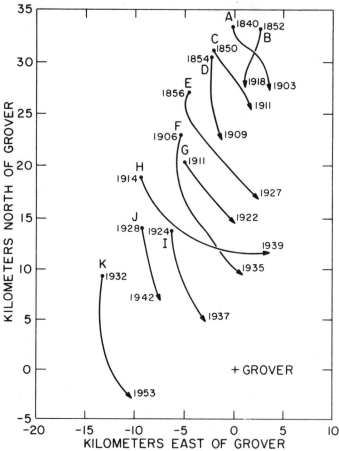

Figure 21.16 Horizontal location and motion of the radar tops of major turrets in the Butler storm.

21.6.2 Cloud Base Measurements

Thermodynamic and airflow measurements were made at cloud base from 1915 to 1938 by N304D and N306D and from 1940 to 1950 by N10UW. The passes of N304D and

N306D were made on northeasterly and southwesterly headings, which took them from the inflow region through precipitation into the outflow region. In general, the cloud base observations were characterized by weak vertical air motion. Observers on board the aircraft reported that, in contrast with the usual appearance of cloud base in northeastern Colorado, cloud bases in the inflow region were ragged and poorly defined, and precipitation was observed in some regions of updraft at cloud base.

Measurements of potential temperature, water vapor mixing ratio, and vertical air velocity from N306D are shown in Fig. 21.17 for the period 1923 to 1932. Measurements from

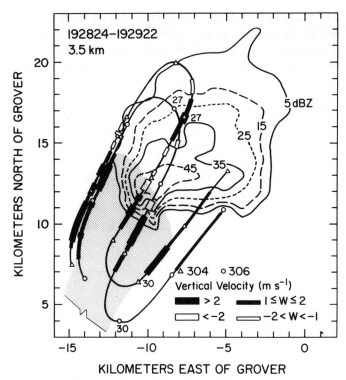

Figure 21.18 The tracks of N304D (△) and N306D (○) relative to the moving storm for the period 1923-1932. Locations of updrafts and downdrafts along the track are as indicated. The lightly shaded area shows the region, relative to the storm, where equivalent potential temperature was ≥ 345.5 K. The reflectivities are shown for a horizontal section at 3.5 km for the period 1928:24 - 1929:22.

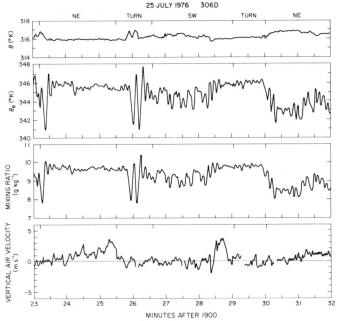

Figure 21.17 Measurements of potential temperature (θ), equivalent potential temperature (θₑ), mixing ratio, and vertical air velocity made by N306D slightly below cloud base (3.7 km) from 1923 to 1932 on 25 July 1976. The positions of the aircraft relative to the storm are shown in Fig. 21.18. The headings and turns are indicated at the top of the figure with corresponding tics at the top of each parameter block.

N304D were quite similar. Note that the values of θ and θ_e in upward-moving inflow air are relatively smooth, at about 316 and 346K. In the outflow, θ and θ_e are more variable, with values of roughly 316.5 K and 344.5 K. The thermodynamic parameters do show some contrast between inflow and outflow regions, but much less than in a more organized storm such as the Raymer storm (e.g., Fankhauser, 1976). The tracks of the aircraft relative to the storm are shown in Fig. 21.18, along with updrafts and downdrafts and the region in which $\theta_e \geq 345.5$ K.

Vertical motions were rather weak. Maximum updrafts were 3-4 m s⁻¹, with average values across the inflow of only 1-2 m s⁻¹. Maximum downdrafts were 1-2 m s⁻¹ and were present in rather small regions (≈ 1 km across, or less). Even near the region of maximum reflectivity at 3.5 km, strong downdrafts were not observed. The observations from N304D and N306D suggest that maximum updrafts at cloud

base increased from 1-2 m s⁻¹ around 1915 to 3-4 m s⁻¹ by 1925.

Measurements of vertical air motion at cloud base made from N10UW from 1940 to 1950 show maximum updrafts of 5-8 m s⁻¹ and downdrafts of 2-3 m s⁻¹. θ and θ_e values in the inflow were similar to those observed by N304D and N306D, i.e., 316 K and 346 K, but θ_e values in the outflow region dropped to 337 K. Thus, the measurements made by N10UW suggest that the large area of new growth (turret K), which was first detected by radar at about 1932, was more vigorous than earlier developments. This is consistent with the radar observations of this new development, which show tops in excess of 13 km by 2000 (Fig. 21.8) and an increase in areal extent.

The updrafts measured by N304D and N306D about 3 to 4 km west and south of the southwesternmost storm reflectivity contour seen in Fig. 21.18 at 1923-1924 are believed to be associated with the early stage of development of turret K. This conjecture is based on the speed and direction of motion of the storm and developing turrets on the western side of the storm and on the location of these updrafts relative to the subsequent appearance of the large area of reflectivity associated with turret K. Turret K first appears on radar at 1932 and is clearly distinct from storm growth associated with turrets D-J. Thus, there is at least a 9-min difference between the presence of updraft at cloud base and the appearance of the first echo at 6.5 km.

21.7 RADAR CHARACTERISTICS OF A DEVELOPING REFLECTIVITY CORE

The previous discussion has been devoted to a description of the reflectivity structure and evolution of the storm. We now confine our attention to the early stage of development of the reflectivity core in which the sailplane made its first ascent. This reflectivity core is the echo developing near 1900 at $(x,y) = (-3, 25)$ relative to Grover in Figs. 21.11 and 21.12, and labeled turrets D and E in Figs. 21.15 and 21.16. At about 1850, a region of new growth or first echo (at -5 dBZ_e) appeared 3 to 4 km southwest of the existing echo. Figure 21.19 shows the time-height profile of the maximum reflectivity of this developing echo from 1845 to 1915. Although the area covered by the first echo expanded after first detection, the reflectivity remained at -5 dBZ_e for 5 min. Then, during the period 1855 to 1910 it increased at the rate of 1.9 dB min^{-1} up to 25 dBZ_e.

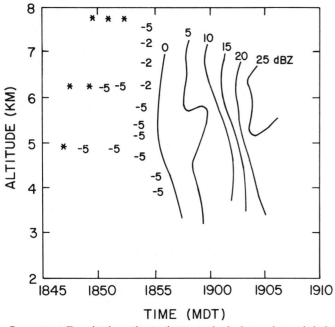

Figure 21.19 Time-altitude profile of reflectivity in the developing echo in which the sailplane made its first ascent. The data were recorded by the MINA II system and are shown in intervals of 5 dBZ_e. The value for each sweep above 4 km prior to 1855 is shown at the appropriate time and altitude. Asterisks (*) denote sweeps for which the return was below the minimum sensitivity of the radar (-6 dBZ_e).

Radar scans were widely separated prior to 1853, so that radar coverage during the first 3 or 4 min was marginal. However, detailed scans from 1853 to 1854 showed reflectivity extending from at least as low as 4.6 km to 7.5 km. Interpretation below 4.6 km is difficult because of ground clutter and the confluence reflectivity feature. As the reflectivity of this core increased, successive 5 dBZ_e contours tended to remain nearly vertical. This tendency for successive reflectivity contours to be almost vertical was an unusual feature of this storm.

The existence of a relative minimum (or hole) in reflectivity was noted in the early stages of development of the first echo. Vertical sections through this feature are shown in Fig. 21.20. This feature moved at about 9 m s^{-1} toward 200°, the same as the overall storm motion. Note that the maximum in reflectivity formed above a relative minimum in reflectivity and that, as the echo continued to develop, this maximum drifted further to the east while the reflectivity increased. The top of the weak-reflectivity region descended from 6.2 km at 1857:04 to 4.9 km at 1900:49. This minimum in reflectivity can be traced past 1905. Radar tracking of the sailplane showed that the sailplane made its first ascent in this region of minimum reflectivity. Indeed, the coordinated sailplane-radar measurements presented in Chapter 22 show this to be a region of updraft.

21.8 DISCUSSION

The Butler storm observed by the NHRE investigators on 25 July 1976 was one of many storms that formed on that day. Compared with a storm that was the focus of an NHRE study earlier that day and with a large storm near Scottsbluff, this storm was rather small, being only 15 km across and covering an area of only 200 km^2 during its most intense period. Investigations started very early in the life cycle of the storm, so that its growth and evolution are extremely well documented by conventional radar, Doppler radar, and aircraft observations. These observations provide a comprehensive picture of the life cycle of a small High Plains convective storm.

The conventional radar observations, which are the focus of this chapter, showed a line of weak radar return extending for tens of kilometers both north and south of the storm from the surface to near cloud base below the extreme southwestern part of the storm. It has been inferred that this feature was a result of localized regions of convergence along a confluence line. The Butler storm probably formed in a preferred region of convergence and moved along the confluence line during at least the first hour of its lifetime.

Perhaps as a result of both the low-level airflow and the westward motion of this confluence line, the storm moved toward the south-southwest at 9-10 m s^{-1}, overrunning the low-level flow. Consequently, inflow relative to the storm was from the southwest. This is consistent with the existence of vigorous convection above and extending southward along the confluence line and with the first appearance of new turrets on the southwest side of the storm.

During the period 1850 to about 1920 regions of new growth are readily distinguishable. They first appear at 2-4 km southwest of the main body of the storm at altitudes of 6-7 km. After 1920 individual turrets can be identified above 6 km but are no longer readily discernible in the low-level and midlevel reflectivity structure. It appears that during the entire lifetime of the storm there was a steady pro-

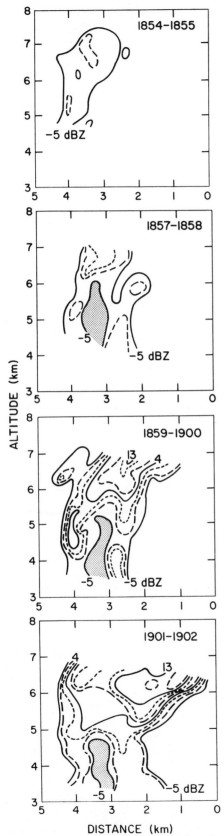

Figure 21.20 Vertical sections oriented northwest to southeast through the core of a developing echo showing a relative minimum (shaded) in reflectivity near the center of the echo with a maximum aloft. The sections were constructed from PPIs recorded by the MINA II system and are drawn with contour intervals of 3 dBZ$_e$.

gression of rising turrets separated by 5 to 8 min, but that as the storm grew in size and the reflectivity increased, individual precipitation cores (and resulting reflectivity) became less distinct due to the formation of new cells within existing cloud and the mixing within the storm. This will be discussed more in Chapter 23.

It appears that the storm was multicellular in nature, with new cells forming rather close (2-4 km) to the main radar echo, as contrasted with the Raymer storm where new turrets appeared at 5-10 km from the main radar echo (Chalon et al., 1976). A succession of turrets associated with each cell penetrated further and further into the drier air aloft. As the storm grew larger and individual turrets were rising partially through the remnants of previous growth, one area of convection became dominant. This is very much like Scorer and Ludlam's (1953) model of storms as successions of thermals. The description of the kinematic evolution of this storm based on the Doppler radar observations will be discussed in Chapter 23.

One of the more unusual observations made on this day was the appearance of successively greater reflectivity values almost simultaneously over a large altitude range rather than in a narrow range. Observations presented in Chapter 22 suggest that the particles responsible for the radar return in the developing first echo were already approaching a millimeter in size. Even in the earliest stages of detection incipient precipitation particles were detected over a large altitude extent of the cloud. This observation, combined with the direct microphysical observations and measured three-dimensional air motions, will be used in Chapter 23 to explore the importance of mixing and recirculation in precipitation development in this storm.

References

Browning, K. A., J. C. Fankhauser, J.-P. Chalon, P. J. Eccles, R. G. Strauch, F. H. Merrem, D. J. Musil, E. L. May, and W. R. Sand, 1976: Structure of an evolving hailstorm: Part V. Synthesis and implications for hail growth and hail suppression. *Mon. Weather Rev.* 104, 603-610.

Chalon, J.-P., J. C. Fankhauser, and P. J. Eccles, 1976: Structure of an evolving hailstorm: Part I. General characteristics and cellular structure. *Mon. Weather Rev.* 104, 564-575.

Dye, J. E., L. J. Miller, B. E. Martner, and Z. Levin, 1980: Dynamical-microphysical evolution of a convective storm. NCAR Tech. Note NCAR/TN-151+STR, National Center for Atmospheric Research, Boulder, Colo., 248 pp.

Fankhauser, J. C., 1976: Structure of an evolving hailstorm: Part II. Thermodynamic structure and airflow in the near environment. *Mon. Weather Rev.* 104, 576-587.

Martner, B. E., 1975: Z-R and Z-W$_0$ relations from drop size measurements in High Plains thunderstorms. Prepr. 9th Conf. on Severe Local Storms, Norman, Okla., Oct. 21-23, 1975, Am. Meteorol. Soc., Boston, Mass., 307-310.

Mohr, C. G., and R. L. Vaughan, 1979: An economical procedure for Cartesian interopolation and display of reflectivity factor data in three-dimensional space. *J. Appl. Meteorol.* 18, 661-670.

Scorer, R. S., and F. H. Ludlam, 1953: Bubble theory of penetrative convection. *Q. J. R. Meteorol. Soc.* 79, 94-103.

CHAPTER 22
The 25 July 1976 Case Study: Microphysical Observations
James E. Dye and Brooks E. Martner

22.1 INTRODUCTION

Previous studies in northeastern Colorado (Knight et al., 1974; Cannon et al., 1974; Dye et al., 1974) have shown that the precipitation there forms primarily by the ice process, with diffusional growth followed by riming. More recent studies in northeastern Colorado (e.g., Chapter 7) support this conclusion. By this process, it takes 10 to 25 min to form precipitation particles a few millimeters in diameter. Although the lifetimes of most clouds in northeastern Colorado are sufficiently long to allow growth for this period of time, little is known about the mechanisms that keep growing particles in the growth region of the clouds this long. With core updrafts of 5 to 15 m s^{-1} or more, which are commonly found in the cumulus congestus stage (Dye and Toutenhoofd, 1973; Kyle and Sand, 1973), there is not enough time for the particles to nucleate and grow by diffusion and accretion to millimetric sizes during one vertical transit through the main updraft region.

Deduction of the trajectories that allow individual particles to remain within cloud long enough to grow to the sizes that they do attain was one of the reasons for taking the aircraft observations described below, and will be the focus of the discussion to follow.

22.2 GROWTH TIME

To help in the interpretation of measurements presented in following sections, a general discussion of diffusional and accretional growth of precipitation particles is presented here. We will consider the length of time needed to grow graupel 5 mm in diameter. This size was arbitrarily chosen, but in the high cloud-base clouds of Colorado, precipitation particles must grow to about this size before much precipitation can reach the ground. Additionally, particles of this size can serve as effective hail embryos.

Relatively little is known about the initiation of the ice phase in the cumulus clouds of Colorado, but the dearth of large drops found in these clouds (Cannon et al., 1974) and examination of thin sections of hailstones and graupel (Knight et al., 1974) strongly suggest that diffusional growth of ice crystals to sizes large enough to start riming is usually a necessary step in forming precipitation in these clouds. There is considerable uncertainty concerning the temperatures at which significant concentrations (> 0.1 ℓ^{-1}) of ice crystals are nucleated and begin to appear. Cursory evidence presented by Dye et al. (1976) and ice nucleus measurements reported

in Chapter 3 suggest that temperatures of $-10\,°C$ or colder must be reached before nuclei become active in concentrations likely to lead to significant precipitation. However, to allow as much time as possible for growth in the updraft, we will assume that deposition nucleation occurs at $-5\,°C$ in an updraft.

Recently Ryan et al. (1976) measured crystal growth rates and water saturation in a laboratory near sea level, at temperatures from $-3\,°C$ to $-21\,°C$. Growth at the ambient pressure in the cloud would be 20%-40% faster (see Heymsfield et al., 1979), but, for the present discussion, the assumption of growth at 1000 mb is sufficiently accurate. The accelerated growth of broad-branch crystals near $-17\,°C$ reported by Ryan et al. (1976) was also neglected. Particle sizes were calculated as a function of time and altitude, assuming that the nucleated crystals were growing two-dimensionally and ascending in the updraft profiles listed in Table 22.1. Their terminal velocities were neglected. Transit times, average vertical velocities, and sizes reached during transit from the $-5\,°C$ level (6 km) to $-25\,°C$ (9 km) are also listed. Although some crystals reached diameters of 300 μm, which is approximately the riming threshold for plates (Ono, 1969), accretional growth was not included in these calculations, which are intended only to be estimates of the growth by diffusion. Even with these rudimentary calculations it is apparent that particles growing in updrafts of about 5 m s^{-1} or stronger do not reach the accretion threshold before they reach 9 km, near the cloud top. The time needed to grow to 500 μm (approximately the point where riming growth becomes efficient) is highly dependent upon the temperature,

TABLE 22.1
Growth of ice crystals in various updraft profiles

Updraft profile	Average vertical velocity (m s^{-1})	Transit time from $-5\,°C$ (6 km) to $-25\,°C$ (9 km) (s)	Diameter resulting from diffusional growth (μm)
Adiabatic (unresisted)	16.7	180	110
Lateral entrainment*	7.5	400	240
Doppler-observed (1905)**	1.25	2400	1600
Doppler-observed (1915)**	1.95	1540	940
Doppler-observed (1925)**	4.3	700	410

*A mixing coefficient of 0.2 was assumed.

**Taken from Fig. 23.4. The number in parentheses is the beginning time of radar volume scan.

211

with the fastest growth occurring at around $-16°C$. From Ryan et al. (1976) and the above examples we see that if a particle resided only in the $-15°C$ to $-17°C$ region, growth to 500 μm could take as little as 4 min. But for particles nucleated at $-5°C$ and ascending in the updraft, times of 10 min or more would be more typical.

After particles have reached a diameter of 500 μm, provided that supercooled water is present, accretional growth dominates. A continuous accretion model can then be used to estimate times required for further growth. In the continuous collection model for spheres, the rate of change of diameter with time is given by

$$\frac{dD}{dt} = \frac{V_T W_e}{2\rho_i} \qquad (22.1)$$

where V_T and ϱ_i are the terminal velocity and density of the ice particles and W_e is the effective liquid water content, which includes effects of collection efficiency of cloud droplets by the ice particles. On the basis of graupel collections made in northeastern Colorado by the sailplane, Heymsfield (1978) reported an empirical relationship between mass and diameter that was equivalent to $m = 0.210 D^{3.53}$, where m is in g and D in cm. This suggests that, on the average, particle density increases outward from the center, and it appears reasonable to interpret ϱ_i in the growth equation as superficial rather than average density, particularly since most graupel particle growth in this area is dry. If $\varrho_i(D)$ is the superficial density of a particle of diameter D, it follows that $\varrho_i (D) = 0.472 D^{0.53}$, where ϱ_i is in g cm^{-3} and D in cm.

Based on the results of Auer et al. (1971), Heymsfield deduced for these same graupel samples an empirical terminal velocity relationship—$V_T = 607 D^{0.77}$ (V_T in cm s^{-1} and D in cm). Using these expressions for ϱ_i and V_T, particle diameters were calculated as a function of time by integration of equation (22.1), with the above relationships of ϱ_i and V_T, values of 1, 1.5, and 2 g m^{-3} for W_e, and an initial particle diameter of 0.05 cm. The results of these calculations are presented in Fig. 22.1. For these conditions the growth to 5 mm takes from 10 to 20 minutes. Since these calculations are based on effective liquid water content (i.e., the product of collection efficiency and liquid water content), values of W_e between 1 and 2 g m^{-3} seem most appropriate for the Butler storm where, as will be shown later, measured liquid water contents were 1.5 to 3 g m^{-3} in the main updrafts over the height range in question.

Thus, if we use realistic conditions and combine the time required for diffusional growth with the time required for accretional growth, a period of 15 to 25 min is needed to grow 5-mm graupel. Under the most favorable conditions, i.e., diffusional growth near $-15°C$ and high liquid water contents, the growth time could be as short as 10 min. These times should be considered lower bounds, since it is unlikely that a particle could stay in regions of high liquid water content for periods of 10 to 20 minutes.

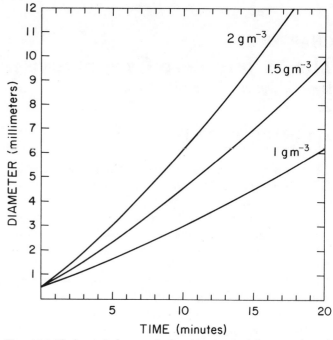

Figure 22.1 *The diameter is shown as a function of time for ice particles growing dry, with the specified shapes and liquid water contents, in a supercooled cloud. The particle densities were assumed from the mass-diameter relationship of Heymsfield (1978).*

The simplified calculations presented above are not intended to be accurate descriptions of the growth rate of graupel, but are estimates of the range of times that might reasonably be expected. The type of embryo, effects of collection efficiency, different growth environments, and other factors will give differing growth rates for individual particles. However, it is reassuring that more detailed calculations such as those of Harimaya (1981) yield comparable times of growth for graupel.

Since one focus of this study is the development of precipitation prior to and during the formation of the first radar echo in the cloud, it is of interest to examine the particle sizes and concentrations needed to produce a given reflectivity. By assuming the particles are spherical ice particles with the mass-diameter relationship reported by Heymsfield (1978), converting the ice particle diameter to melted equivalent water drop diameter, and correcting for the dielectric factor ($|K|^2$), for ice particles rather than water, one obtains the equation

$$Z_{ei} = 3.003 \times 10^{-3} \sum_{i=1}^{\infty} n_i D_i^{7.06} \qquad (22.2)$$

for the reflectivity factor for ice, Z_{e_i}, in terms of the concentration, m^{-3} (n_i), of ice particles of diameter D_i (mm). Calculations using this equation and assuming different concentrations of monodisperse particles are presented in Fig. 22.2

During 1976 the detection limit of the CP-2 radar was about -5 dBZ_e at a range of 30 km. From Fig. 22.2 it can be seen that, in order to detect a -5-dBZ_e return from 0.5 or

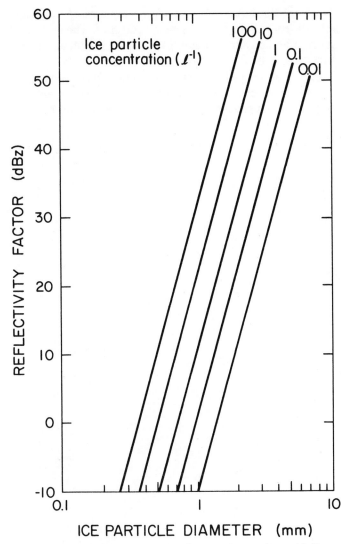

Figure 22.2 The equivalent reflectivity factor shown as a function of ice particle diameter for different concentrations. Reflectivities were calculated as discussed in the text.

22.3 AIRCRAFT OBSERVATIONS

Descriptions of the instrumentation and measurement capabilities of the four aircraft (NCAR/NOAA sailplane, University of Wyoming Queen Air N10UW, and NCAR Queen Airs N304D and N306D) used in this cloud study are presented by Dye et al. (1980). As shown in Chapters 21 and 23 the reflectivity and kinematic structure of the Butler storm evolved appreciably during the period of observations. The microphysical observations also show considerable evolution. For the sake of clarity, the microphysical observations obtained from the aircraft data will be divided into two parts: the early organizing stage before 1920 and the mature, more steady stage after 1920.

Cloud base measurements presented in Chapter 21 from N304D and N306D showed the base to be at 650 mb and 7 °C in the updrafts, resulting in a pseudo-equivalent potential temperature of 346.5 K. The peak updrafts were 4 m s^{-1}, with typical updrafts of only 1-2 m s^{-1}. The flight tracks were flown parallel to the direction of storm motion (200°), and in that direction the region of updraft was 2 to 4 km wide. Precipitation was noted in some regions of the updraft as well as in the downdraft by observers in the aircraft. Only weak downdrafts of 1-2 m s^{-1} were observed, even though the aircraft traversed most of the cloud base region at one time or another.

22.3.1 The Early Organizing Stage

On 25 July the sailplane released from tow at about 5.2 km and penetrated what appeared to be a vigorous turret, but during the first pass (1901:15 to 1902:30) observed only weak updrafts of 2-3 m s^{-1}. Equivalent potential temperature (θ_e) and liquid water content (q) measurements showed this region to be relatively inactive (Fig. 22.3). Although no particles were photographed by the camera, the CP-2 radar data indicated that the sailplane was in a region of reflectivity values from -5 to 0 dBZ$_e$.

Upon leaving cloud the pilot made a sharp turn, reentered cloud, and found sufficient lift to spiral just north of the first pass. The sailplane ascended within this updraft from 5.2 to 6.7 km on the northwest side of the highest reflectivities in the developing radar echo. The positions of the sailplane relative to the echo at four different times during the ascent are shown in Fig. 22.4.

It is apparent from Fig. 22.3 that the region in which the sailplane ascended was an active region of the cloud, with liquid water contents of 1-3 g m^{-3} and peak updraft speeds of 5-10 m s^{-1}. However, vertical airspeeds and liquid water contents were quite variable. Some narrow regions 400 to 800 m across had higher liquid water contents, stronger updrafts, and θ_e values approaching the adiabatic cloud base value of 346.5 K. A few precipitation-sized particles were observed during the period 1905 to 1911; all but one were less than

1 mm diameter ice particles, the concentrations must be 5 ℓ^{-1} or 0.05 ℓ^{-1}, respectively. The sample volume of the particle camera (Cannon, 1974) on the sailplane is approximately 0.5 liter per frame and operates at about two frames per second. Thus, the volume photographed in 1, 10, or 100 s is roughly 1, 10, or 100 ℓ, thereby giving a detection limit of 1, 0.1, or 0.01 particles ℓ^{-1} respectively for these time intervals. If the particle concentration averages 5 ℓ^{-1}, 1 s of data would show about five particles, but for a concentration of 0.05 ℓ^{-1}, photographs would have to be taken for 20 s to detect one particle. Thus, if particles creating the first radar echo are ≈ 0.5 mm in size there is a good chance of observing these particles with the camera. However, if the particles are 1 mm in diameter, or larger, the chances for camera detection are considerably less.

1 mm in diameter. The average concentration of such particles for this period was ≈ 20 m⁻³, although so few particles were seen that there is considerable statistical uncertainty in this value. Note that the few observed particles were associated with downdrafts or weak upward motion.

The droplet concentrations measured with the Particle Measuring Systems (PMS) Forward Scattering Spectrometer

Probe (FSSP) during the first several minutes of the ascent ranged from about 600 to almost 1200 cm⁻³. The regions of greatest liquid water content (LWC) were at altitudes from 5 to 5.5 km (1 to 1.5 km above cloud base) between 1902 and 1906. Mean droplet diameters in these regions were 12 to 16 μm. After about 1909 icing of the instrument began to affect the measurements seriously. Thereafter, the droplet con-

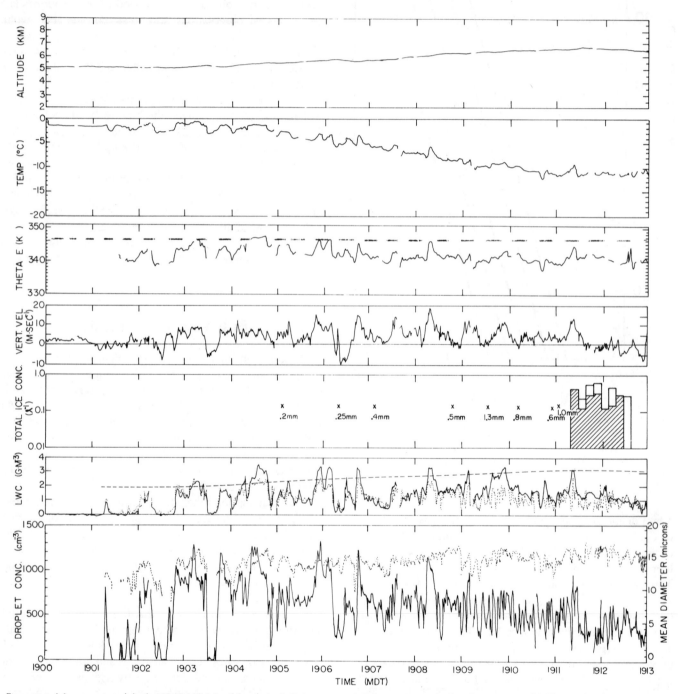

Figure 22.3 Measurements made by the NCAR/NOAA sailplane during its first ascent in the Butler storm. The total ice concentrations were determined from photographs from the Cannon camera. For the period 1905 to 1911 the times at which individual particles were observed are shown by an X, with the size shown below. After 1911 the 10-s average concentration of all particles and of all particles ≥ 1 mm in length are shown as unshaded and shaded areas, respectively. Liquid water contents (LWC) from the Johnson-Williams hot-wire (solid line) and the PMS FSSP (dotted line), and the expected adiabatic value (dashed line) are as shown. The total droplet concentrations and mean diameters observed by the PMS FSSP are shown as the solid line and the dotted line, respectively, in the bottom segment of the figure. Unless otherwise noted all data are 1-s averages.

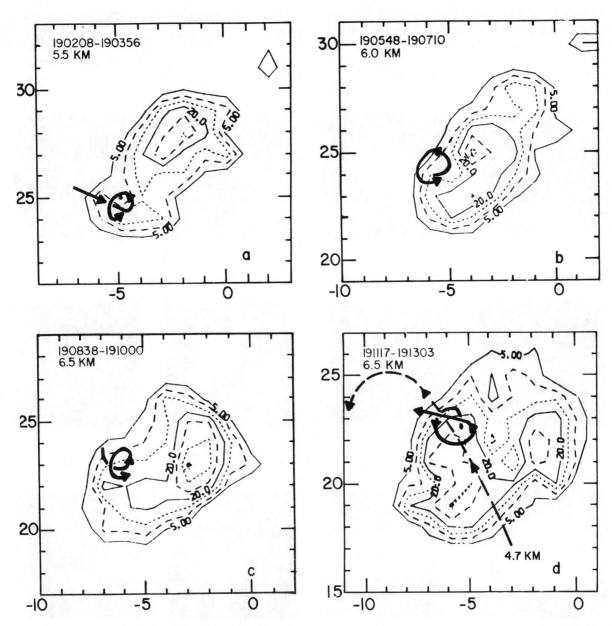

Figure 22.4 The flight track of the sailplane superimposed on horizontal sections of the reflectivity pattern closest to the altitude of the sailplane at each time interval. The arrows and beginning of each track show positions on the minute for the periods (a) 1901-1904, (b) 1905-1907, (c) 1908-1910, and (d) 1910-1913. The track of N10UW for its first pass at 4.7 km from 1910 to 1913 is shown in (d) by a heavy dashed line. The aircraft positions were determined from tracking radar and are accurate to 500 m, relative to the reflectivity contours.

centrations became more erratic and gradually began to decrease, so that LWC derived from the FSSP departed more and more from the values given by the Johnson-Williams (J-W) hot wire instrument.

Upon encountering downdraft at 1911:40 at an elevation of 6.7 km the pilot decided to leave the cloud on a westerly heading. Along this flight path the ice particle concentration increased to a few tenths per liter in the downdraft, and ice particles were observed in both updrafts and downdrafts, with the largest being 3 mm. A photograph of this particle taken by the sailplane's particle camera is shown in Fig. 22.5b. At the $-10\,°C$ level the dot pair signature of one 500-μm water drop was photographed (Fig. 22.5c). Ice particles $\leq 200\ \mu m$ were not observed, suggesting that negligible

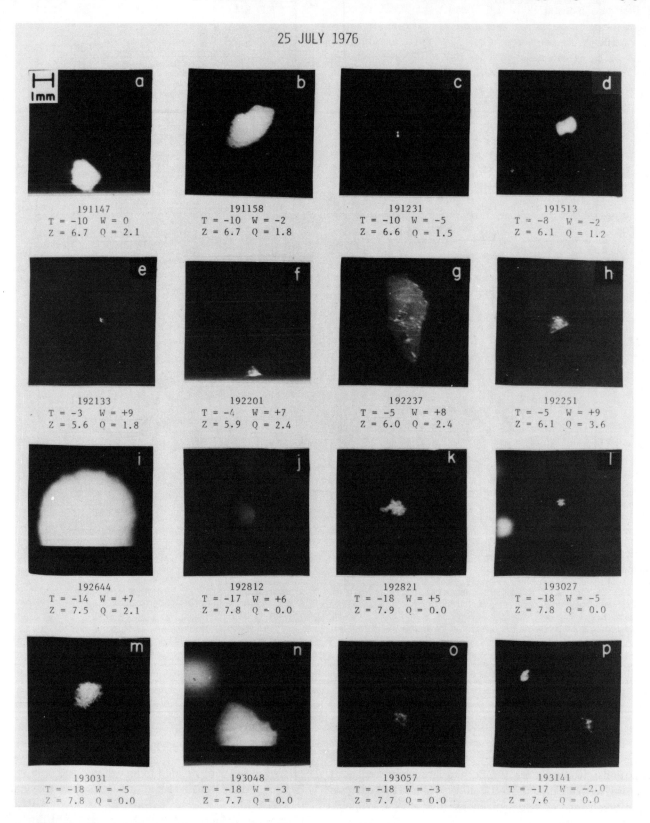

25 JULY 1976

a 191147 T = -10 W = 0 Z = 6.7 Q = 2.1	**b** 191158 T = -10 W = -2 Z = 6.7 Q = 1.8	**c** 191231 T = -10 W = -5 Z = 6.6 Q = 1.5	**d** 191513 T = -8 W = -2 Z = 6.1 Q = 1.2
e 192133 T = -3 W = +9 Z = 5.6 Q = 1.8	**f** 192201 T = -4 W = +7 Z = 5.9 Q = 2.4	**g** 192237 T = -5 W = +8 Z = 6.0 Q = 2.4	**h** 192251 T = -5 W = +9 Z = 6.1 Q = 3.6
i 192644 T = -14 W = +7 Z = 7.5 Q = 2.1	**j** 192812 T = -17 W = +6 Z = 7.8 Q = 0.0	**k** 192821 T = -18 W = +5 Z = 7.9 Q = 0.0	**l** 193027 T = -18 W = -5 Z = 7.8 Q = 0.0
m 193031 T = -18 W = -5 Z = 7.8 Q = 0.0	**n** 193048 T = -18 W = -3 Z = 7.7 Q = 0.0	**o** 193057 T = -18 W = -3 Z = 7.7 Q = 0.0	**p** 193141 T = -17 W = -2.0 Z = 7.6 Q = 0.0

Figure 22.5 Photographs taken with the Cannon particle camera on the sailplane. The time and measured temperature (T), altitude (Z), vertical speed of the air (W), and liquid water content (Q) in the location of each particle are given.

nucleation of ice had occurred in this region.

The University of Wyoming Queen Air, N10UW, made its first pass from 1911 to 1912 at 4.7 km (1 °C), directly below the sailplane, which was at 6.5 km. The observations from N10UW are presented in Fig. 22.6 and summarized in Table 22.2. As can be seen from Fig. 22.4d (see also Fig. 23.10), the penetration was made between relative maxima in reflectivity and between two active updraft regions of the storm; low liquid water contents and negligible vertical airspeeds were observed. Ice particles were observed with a PMS two-dimensional (2-D) optical imaging probe in concentrations of a few tenths per liter. One particle was as large as 2.4 mm, but others were primarily between 0.5 and 1 mm diameter. The smallest particle observed was 0.3 mm.

The second pass of N10UW (1913-1915, 5.3 km, −2 °C) was closer to the updraft core observed by Doppler radar, but still somewhat to the northeast of the broadest updraft (see Fig. 23.11). Measurements made from N10UW during this and other passes are summarized in Table 22.2 (see also Dye et al., 1980, for more detail). During the second pass one graupel particle of 2.2-mm diameter was observed, and the concentration of ice particles ≥ 1 mm was about 0.1 ℓ^{-1}. Some of the ice was observed in weak downdraft on the northwestern side of the cloud but the highest concentrations of ice were observed in regions of updraft, which covered ≈90 % of the cloud pass.

The second penetration by N10UW was vertically coordinated with and parallel to the sailplane's reentry into cloud on a southeast heading. During this penetration the sailplane pilot did not find much lift, and, upon entering an 8 m s^{-1} downdraft on the southeastern flank of the cloud, immediately made a 180° turn and left the cloud in the opposite direction. Most of this pass was in downdraft with only a few brief periods of weak updraft. Liquid water contents up to 1 g m^{-3} were observed during the first part of the pass. Ice particle concentrations were as high as 1 ℓ^{-1}, with ice particles as large as 3 mm in the broad downdraft in the middle of the cloud. However, it was surprising that few ice particles were observed in the stronger downdraft on the southeastern side. Apparently this downdraft originated in a region devoid of ice particles.

In summary, during this early stage of cloud growth, observations by the sailplane and N10UW show the start of precipitation development in the first echo region. With a lower limit of detection of about 10 m^{-3}, precipitation particles were not observed from the sailplane in the 5- to 6-km altitude region (0 °C to −5 °C level) of the first echo until about 1905. From 1905 to 1910, 0.2- to 1.0-mm particles were observed in concentrations of ≈20 m^{-3}, although a few larger particles must have been present to account for the measured reflectivity. By 1912, 2- to 3-mm particles were observed at the 5-km level by N10UW while 2- to 4 mm particles were observed at 6.5 km by the sailplane.

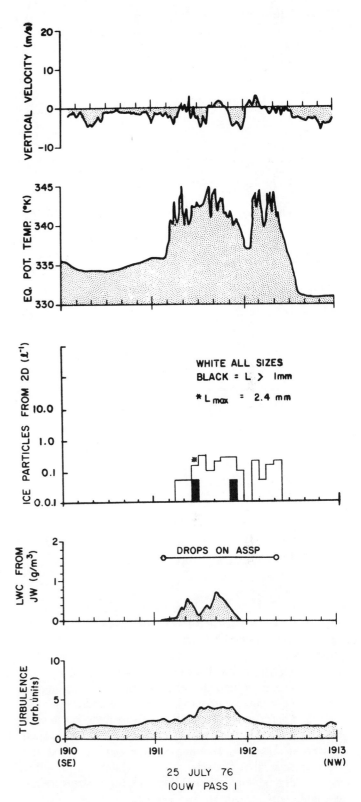

Figure 22.6 *Measurements made by N10UW during its first pass (1911-1913) through the Butler storm at approximately 4.6 km (−1 °C). Ice particle concentrations measured by the PMS 2-D probe are as shown for all particles (unshaded) and all particles ≥ 1 mm length (shaded). The line labeled "Drops on ASSP" shows where the ASSP detected cloud droplets. Turbulence measured by the MRI turbulence meter is shown in arbitrary units.*

TABLE 22.2
N10UW cloud passes, 25 July 1976

Pass no.	Time (MDT)	Alt. (km)	Ave. temp. (C)	Vert. air-speed (m s⁻¹) Ave.*	Max.**	LWC from J-W (g m⁻³) Ave.	Max.	θ_e(°K) Ave.	Max.	Ice conc. from 2-D images (l⁻¹) Ave.	Max.	Largest ice particle size (mm)	Largest water-drop diam. (mm)	Total no. waterdrops
1	1911:05–1912:05	4.6	0	−1	+1	0.3	0.6	341	343.5	<0.1	0.2	2.4	—	0
2	1913:50–1914:50	4.9	−2	+3	+11	0.5	1.1	340	343.1	0.2	0.4	2.2	—	0
3	1922:30–1923:45	5.1	−4	+7	+15	1.1	1.7	341	344.7	0.2	0.7	5.5	0.5	4
4	1927:00–1928:05	5.1	−4	+7	+16	0.7	1.6	339	344.1	0.3	1.8	5.5	2.0	2
5	1931:25–1932:45	5.1	−5	+3	+10	0.7	1.6	339	344.0	0.8	4.1	3.3	3.5	12
A†	≈1940:00–1945:00	3.8	+5	+1	+6	<0.1	0.1	343	346.2	0.8	4.7	≈8.0 -Partially Melted-	1.6	19
B	1948:00–1951:05	3.5	+7	?	+6	≈0	<0.1	345	347.0	0.2	1.4	≈9.0	2.4	14

*Values averaged over entire in-cloud (ice and/or water) time.

**Greatest 5-s (≈350-m) average during cloud penetration.

†Pass A was approximately at cloud base.

22.3.2 The Mature Stage

The observations from the sailplane as it ascended from 5 to 8 km between 1919 and 1930 are shown in Fig. 22.7, and its horizontal position relative to the reflectivity structure is shown in Fig. 22.8. In Fig. 22.8 the periodic variations of vertical velocity from maxima to minima with periods of 70 to 80 s are a result of the sailplane flying from the main updraft region to the downdraft region east of the updraft (see Fig. 23.12). When compared with the measurements obtained in the first ascent (Fig. 22.3) where two peaks in updraft normally were observed during one complete sailplane spiral, it is obvious that the character of the cloud had changed significantly. The updraft and thermodynamic structure had become more organized and showed broader and more smoothly changing updrafts, although the maximum updraft speeds were comparable during the two ascents. Note that the measured values of θ_e and of the J-W liquid water content were for the most part close to their adiabatic values. Thus

some of this air had ascended unmixed from cloud base. This will be discussed more in Section 22.4.

The most striking feature seen in Fig. 22.7 is the existence of water drops and large ice particles even in the core of the updraft at temperatures as warm as −10°C. The water drops were 0.5 to 1 mm in diameter and some graupel were as large as 6 mm. Based on the growth rates and discussion in Section 22.2 it is clear that the graupel must have nucleated and grown to an appreciable fraction of their observed size elsewhere in the cloud and then been somehow "recirculated" back into the updraft region. By "recirculation" we mean that the particles were transferred from other regions of the cloud. Later we will show that organized cellular motions, sedimentation, and perhaps turbulent transfer could have occurred. Photographs from the sailplane's camera (Cannon, 1974) of some particles (Fig. 22.5) show that there were water drops, graupel (i,m,n), glossy ice particles that appeared to have been partially melted (e,f), recently frozen water drops (j), and large ice particles growing wet (g).

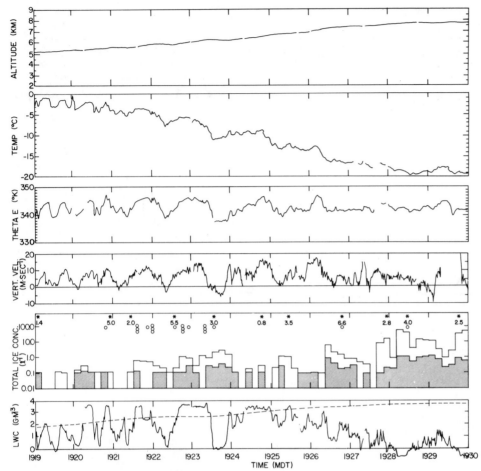

Figure 22.7 Same as Fig. 22.3, except for the sailplane's second ascent from 1919 to 1930. Each water drop photographed by the camera is indicated by an open circle in the appropriate 10-s interval near the top of the "Total Ice Concentration" segment of the figure. The time of observation (shown by an asterisk) and diameter (in millimeters) of the largest ice particle observed during each 1-min period is shown at the top of that segment. The dashed line in the LWC plot is the adiabatic value.

The observations along the horizontal path of N10UW during passes 3, 4, and 5 ($T = -3\,°C$ to $-5\,°C$) help to identify the region of recirculated particles. All three passes showed the strongest updrafts and highest, broadest regions of liquid water to be located on the south-southwest side of the storm. The measurements made from N10UW on passes 3 and 4 are shown in Figs. 22.9 and 22.10, with the corresponding positions relative to reflectivity contours shown in Figs. 22.8 and 22.11 respectively. With the 2-D probe the distinction between water drops and various types of ice particles is based solely on shape. Consequently, the particles identified as water drops by their circular images may or may not be frozen, but if so, they must have frozen only recently. Precipitation size "drops" were detected on all three passes in or on the fringes of updrafts and near the reflectivity core. The larger raindrops and graupel observed on these passes by N10UW could not have grown to the observed sizes in a single transit from cloud base. Thus, again, recirculation is suggested.

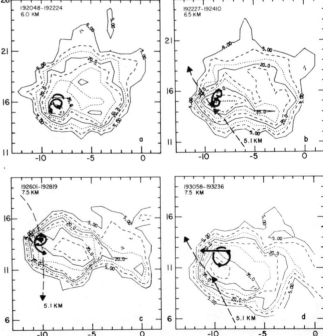

Figure 22.8 Similar to Fig. 22.4, except the sailplane tracks are for the time periods (a) 1920-1922, (b) 1922-1924, (c) 1926-1928, and (d) 1930-1933. N10UW tracks (passes 3, 4, and 5) at about the 5.1-km level ($-4\,°C$) are shown by heavy dashed lines and are from the following time periods: (b) 1922-1924, (c) 1927-1928, and (d) 1931-1933, respectively.

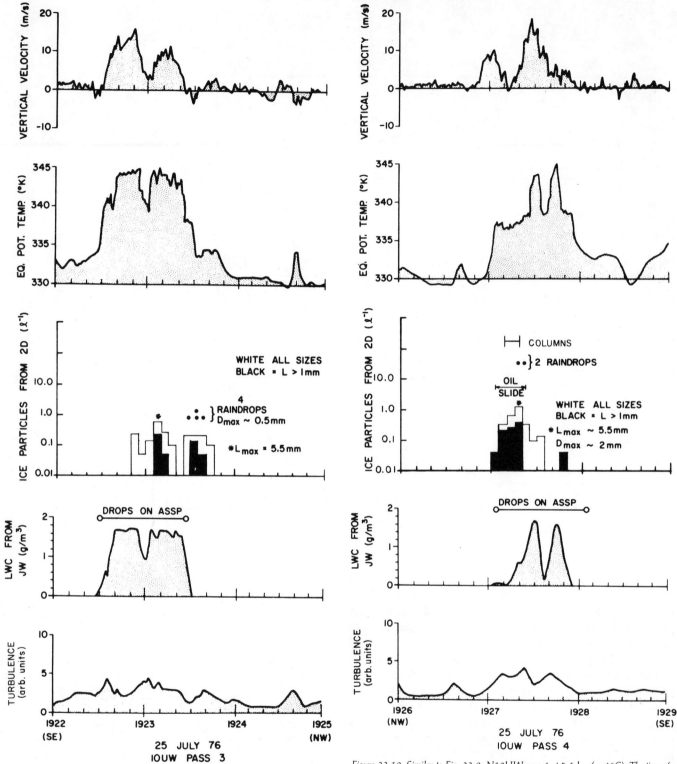

Figure 22.9 Similar to Fig. 22.6: N10UW pass 3 at 5.1 km (−4°C). Each water drop detected by the PMS 2-D probe is shown in the appropriate 5-s interval at the top of the "Total Ice Concentration" segment of the figure.

Figure 22.10 Similar to Fig. 22.9: N10UW pass 4 at 5.1 km (−4°C). The times for which columns were observed by the PMS 2-D probe are also shown.

Column crystals were seen in the 2-D data and oil slides taken by the N10UW crew during the fourth (Fig. 22.11) and fifth passes. In both cases the crystals were observed at temperatures of −4°C to −5°C near the reflectivity core. The crystals had predominantly clf shapes (Magono and Lee, 1966). A photograph of two clf crystals collected on the oil

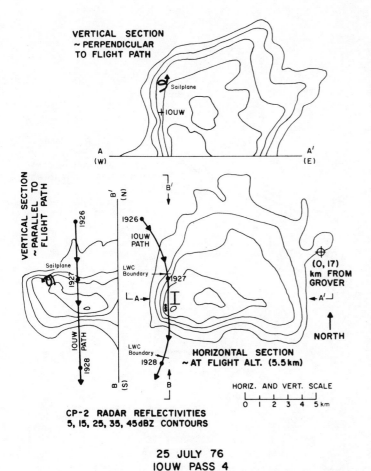

VERTICAL SECTION
~ PERPENDICULAR
TO FLIGHT PATH

CP-2 RADAR REFLECTIVITIES
5, 15, 25, 35, 45 dBZ CONTOURS

25 JULY 76
IOUW PASS 4

Figure 22.11 The flight track for N10UW during pass 4 at 5.1 km (−4°C), superimposed on horizontal and vertical reflectivity contours. The locations where column |———| and drops (●) were observed are shown along the flight track.

Figure 22.12 A photograph of columns captured on an oil-coated slide exposed from 1932:02 to 1932:27 during pass 5 of N10UW. The large, heavily rimed column was about 640 μm long. Some of the irregular pieces might have been chunks of ice shed from the walls of the sampler.

slide exposed during pass 5 is shown in Fig. 22.12. Some of the crystals were as long as 640 μm along the c-axis and had been accreting cloud droplets, while others on the same slides were only 90 μm long. Axial ratios ranged from about 1 to 8 with an average of about 5. The 2-D images also indicated columns of these sizes in concentrations of 0.2 - 0.3 ℓ^{-1}.

In the higher portions of the cloud, from 7 to 8 km, the sailplane ice particle photographs show aggregates and crystals with varying degrees of riming (photographs k, l, o in Fig. 22.5) and graupel (k, m, n in Fig. 22.5) up to 6 mm in diameter. In regions with low liquid water contents and high ice particle concentrations, aggregates were observed in concentrations of 1-10 ℓ^{-1} and were a significant fraction (10%-50%) of the precipitation particles seen in the upper levels. Starting at about 1928 (−18°C, 7.5 km), a significant number of small particles ≤ 0.2 mm in diameter began to appear in concentrations of 1 ℓ^{-1} or more and increased to nearly 10 ℓ^{-1} in the downdrafts near 1930. Before this time, particles this small were very seldom observed. The presence of small ice particles coexisting with supercooled water indicates very recent origin, and suggests that considerable nucleation had occurred in these parcels of air. By this time both the J-W

and FSSP were not operating properly so that good measurements of liquid water were not avialable during most of the period from 1928:30 until 1931.

After the penetrations at the −5°C level, N10UW made two passes (1940-1946 and 1948-1951) to investigate the microphysical and clear-air properties near cloud base and to collect some bag samples of air for aerosol analysis. The observations taken during pass A are shown in Fig. 22.13, and the position of the aircraft relative to reflectivity contours is shown in Fig. 22.14. Both cloud base passes gave similar results: updrafts of 5 to (briefly) 10 m s^{-1} on the western edge of the storm, small hail 8 or 9 mm in diameter, concentrations of particles ≥ 1 mm of about 1 ℓ^{-1}, and θ_e values of about 346 K.

In summary, during the mature stage of the storm both the sailplane and N10UW found millimetric ice particles and water drops in the updraft, including the core between 5 and

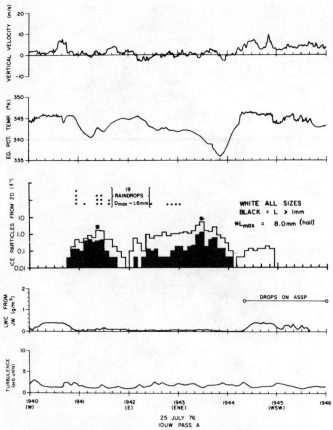

Figure 22.13 Similar to Fig 22.9: cloud base pass A at about 3.8 km (+5°C).

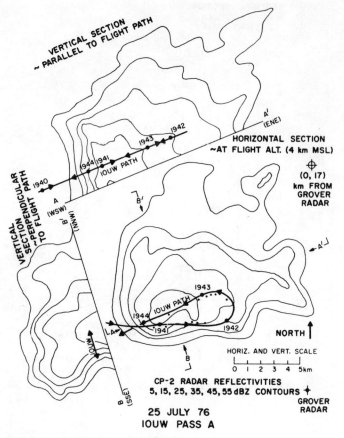

Figure 22.14 Similar to Fig. 22.11: cloud base pass A.

6 km. The particle photographs suggest that some of the particles had slightly but not totally melted and were refreezing. Column crystals in concentrations of 0.3 ℓ^{-1} and sizes up to 640 μm were observed in isolated regions of the cloud near updraft-downdraft boundaries.

22.4 DISCUSSION AND INTERPRETATION OF OBSERVATIONS

22.4.1 Thermodynamic Structure

Comparisons between measurements made during the first and second ascents of the sailplane (Figs. 22.3 and 22.7) show differences in the scale of variability of the cloud properties at the two different times. This was particularly true between 5 and 6.5 km. During the first ascent large variations in θ_e, updraft velocity, and liquid water content were occurring on scales of 800 m (20 s of flight) and less. During the second ascent these variations, while still present, were less pronounced and more regular. In fact, the minima in updraft speed, LWC, and θ_e seen between 1921:10 and 1925:10 in Fig. 22.7 are due to the fact that the sailplane's spiral extended from the updraft region into the mixed downdraft region behind the main updraft. These measurements suggest

a more organized structure on scales of 2-3 km, in contrast to those shown in Fig. 22.3, where the structure is more chaotic. Above 6.5 km (after 1925 in Fig. 22.7) the structure of the cloud during the second ascent appeared to be less organized and more turbulent.

During both ascents of the sailplane, measured values of θ_e frequently equaled the cloud base values of 346 to 347 K, and the LWC was frequently near the adiabatic value (within the error of the measurement). Furthermore, these values occurred within updrafts that were comparable with those expected from adiabatic unresisted ascent of a parcel. Heymsfield et al. (1978) have shown that the presence of these adiabatic regions is a common feature of the clouds investigated by the sailplane in northeastern Colorado. Two of the cases they present, the ascents of the sailplane on 25 July 1976, are among the better examples of adiabatic parcel ascent.

While adiabatic regions were detected during both ascents the adiabatic area was much larger at the later time, when the storm was larger, more vigorous, and more organized. During the first ascent of the sailplane and the early stage of cloud development the regions of adiabatic values (best delineated by the sailplane measurements of θ_e shown in Fig. 22.3) extended over distances of only 200 to 400 m (5 to 10 s). From 1920 to 1925 the unmixed region was more on the scale of 1 to 2 km and more organized. This is best seen from N10UW's observations during pass 3 (Fig. 22.9). It is also apparent in the sailplane data presented in Fig. 22.7.

The penetrations of N10UW were deliberately coordinated so as to be directly below the sailplane. Using the measurements from the sailplane and N10UW we constructed a composite of the thermodynamic structure of the cloud from 5.5 to 6.5 km during the period 1921 to 1927 (Fig. 22.15), superimposed on the reflectivity pattern for the 1923-1925 scan. According to the Doppler data analysis in

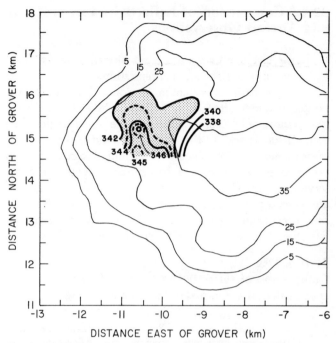

Figure 22.16 *Similar to Fig. 22.15, except using measurements from the sailplane between 6.5 and 7.5 km (1925-1928), with the reflectivity section taken at 7 km from the 1927-1931 NOAA-D scan.*

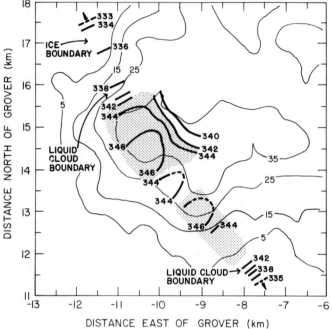

Figure 22.15 *A composite of equivalent potential temperature measurements (heavy lines) made during the sailplane's second ascent from 5.5 to 6.5 km (1921-1925) and N10UW's third pass (1922-1924) at 5.5 km, superimposed on reflectivity contours for the 6-km horizontal sections made from the 1923-1926 scan of the NOAA-D 3-cm radar. To generate this composite, aircraft positions have been corrected for storm motion.*

Chapter 23, as well as the aircraft data, the sharp gradient in θ_e to the northeast was a region of enhanced mixing behind the updraft core where air that had been diverted around the updraft ("obstacle flow") converged to form a downdraft.

A composite of sailplane measurements of θ_e obtained as it climbed from 6.5 to 7.5 km is presented in Fig. 22.16. At 7 km the unmixed region had diminished to only two small areas, even though the updraft speeds were comparable to those at 6 km.

Although measured values of LWC were not as reliable as those of temperature they show a similar pattern: adiabatic or near adiabatic values in the updraft core at 6 km with decreasing values downwind of the core and at 7 km. The N10UW measurements show a sharp boundary of LWC on both the northwest and southeast sides of the cloud, changing from zero to $>1.5\,\mathrm{g\,m^{-3}}$ in less than 500 m. The only measurements available that show east to west variations are those made when the sailplane exited the cloud at 7 km at 1933. In this location the liquid cloud boundary was also very

distinct.

Although adiabatic regions did exist in this cloud there were also large regions in which θ_e and LWC were not adiabatic. Paluch (1979) used data from the Butler storm to deduce the source of entrained air. Her study, which examined the variations of the wet equivalent potential temperature as a function of total water mixing ratio, was limited to the early stage of cloud development (the sailplane's first ascent). During this stage she concluded that much of the entrainment was from vertical mixing via penetrative downdrafts, as first proposed by Squires (1958). Her analysis of sailplane data from the time period 1904 to 1905 suggests the 7- to 8-km region as the source of much of the entrained air in the early stages of development of this reflectivity core.

Paluch showed that air entrained at the 7- to 8-km level could, by evaporative cooling, penetrate down to the 5- or 6-km level in the cloud, even against updrafts of 10 m s⁻¹. These negatively buoyant bubbles eventually descend low enough to reach an equilibrium level, and then, as more cloud air mixes with these parcels, they may become positively buoyant and begin to rise again. Thus, penetrative downdrafts can be an effective mechanism in transporting ice particles from near the top of the growing cloud into the interior of the cloud. Furthermore, if ice crystals in these parcels are in a region containing some liquid water they can continue to grow by diffusion and accretion as they fall. The sailplane observations in Fig. 22.3 from 1905 to 1911 show ice particles predominantly in downdrafts or in regions of weak vertical motion where liquid water is still available for growth.

22.4.2 Comparison of Reflectivities from Radar and Aircraft Observations

Microphysical observations from aircraft give us direct evidence of the size and concentration of some of the precipitation particles present in the cloud. However, limited sample volumes often do not allow statistically significant observations of low concentrations of the large particles. One method of checking the statistical validity of in-cloud measurements is to compare reflectivities calculated from observed particle sizes and concentrations with those observed with the CP-2 radar. The results of such comparisons are presented in Table 22.3 for the times during which significant precipitation particle concentrations were measured with the sailplane camera. The reflectivities were calculated

TABLE 22.3
Comparison of reflectivities measured by radar with those calculated from sailplane camera images

Time	Measured radar reflectivity (est. ave. in parens.)	Reflectivity from images*	Maximum particle size (mm)
190500–191000	0–18 (≈10)	−5.2	1.3
191120–191235	20–27 (≈24)	30.8	4.0
191400–191510	0–30 (≈20)	11.3	2.0
191540–191640	25–40 (≈32)	10.3	2.0
191641–191715	15–35 (≈25)	23.7	3.0
191920–192110	35–40 (≈37)	35.1	5.0
192111–192220	35–40 (≈37)	12.4	1.3
192221–192340	35–45 (≈37)	33.5	4.6
192341–192500	25–40 (≈30)	6.9	2.0
192501–192615	25–35 (≈27)	24.9	3.5
192616–192815	25–45 (≈38)	42.6	6.6
192830–193030	40–45 (≈44)	40.5	6.0
193031–193230	25–40 (≈35)	31.6	4.0
192000–192500	35–45 (≈37)	37.4	5.0
192500–192800	25–45 (≈38)	42.7	6.6

*The reflectivity from imaged particles was calculated from $Z_e = 3.003 \times 10^{-3} \sum_i n_i D_i^{7.06}$, where Z_e is the reflectivity based on the melted equivalent diameter and n_i is the concentration (m^{-3}) of ice particles of diameter D_i (mm), which were assumed to be dry.

by use of the melted equivalent diameter for each particle, based on the mass-diameter relationship reported by Heymsfield (1978). This introduces uncertainty of a few decibels in the calculated values of Z_e because of the departure of individual particles from the assumed mass-diameter relationship. Similar reflectivity comparisons were made with N10UW 2-D image data and showed similar results.

It is apparent that for many time periods the particle observations, when averaged over periods of a few minutes, can account for the reflectivity observed by radar. In particular, in the high reflectivity region during the period 1920-1930, both sailplane and N10UW data agree well with the radar reflectivity. These regions of high reflectivity are the regions in which wet and apparently refrozen, recirculated ice particles were found. When particles several millimeters in

diameter were observed by the sailplane or N10UW, the agreement with radar reflectivity was better than when the largest particles were only 1 or 2 mm in diameter. At other times (e.g., 1911:20-1912:35 for the sailplane) when a large particle was observed, the reflectivity computed from the camera data was greater than the radar-observed reflectivity. These fluctuations are caused by the statistical noise inherent with the limited sample volume of the camera on the sailplane and the 2-D probe on N10UW. When averaged over longer time periods (e.g., 1920-1925 and 1925-1930 for the sailplane) the statistical variation was less.

During the development of the first echo at the time of the sailplane's first ascent between 1905 and 1910 the observed particles were either too small or to few to explain the reflectivity measured by the radar in the vicinity of the sailplane. From 1901 to 1905 the sailplane was in a region having reflectivity between about 0 and 5 dBZ$_e$. Knowing the sample volume and detection limits of the camera and the reflectivity observed by the radar, we can place an upper limit on the particle concentration and a lower limit on the size of particles needed to give the observed reflectivity (see Dye et al., 1980). It was found that if the particle concentration was not over 10 m^{-3}, the graupel diameter would have to be at least 1 mm, in order to explain the 0- to 5-dBZ$_e$ reflectivity in the vicinity of the sailplane. Similarly, sailplane measurements from 1905 to 1910 suggest that in this region (see Fig. 22.4) with reflectivities between 10 and 15 dBZ$_e$, graupel particles at least 2 mm in diameter (or 1.1-mm water drops) in concentrations less than ≈ 10 m^{-3} were necessary to account for the observed reflectivity. These estimates are consistent with data from N10UW, which in a total sample volume of about 0.5 m^3, found two graupel particles larger than 2 mm during its first two passes. The concentration of these particles estimated from the 2-D probe was 4 m^{-3}; however, this estimate is quite uncertain statistically.

22.4.3 Early Stages of Precipitation Formation

One goal of the first echo studies was to understand the origin of the precipitation particles that made up the first echo. The sailplane started investigation of the Butler storm 8 min after detection of the −5-dBZ$_e$ return (see Fig. 21.17) and was in or near the highest reflectivity of the developing echo from 1901 to 1905. As discussed in the previous subsection, it was concluded that millimetric particles in concentrations of ≈ 10 m^{-3} would account for the observed reflectivity. It is important to note also that during the period 1901 to 1911 there were not only very few large particles but also very few small particles (50-200 μm). The average concentration of all particles from 1905 to 1911 was only 10 m^{-3}. Based on the time required for diffusional growth to 200 μm discussed in Section 22.2 it appears that very little nucleation was occurring in the updraft at temperatures warmer than −5°C. The absence of small particles in the downdraft at temperatures of $-10\,°C$ to $-12\,°C$ from 1911:30 to 1912:30 also suggest that little nucleation had occurred even in that region at somewhat colder temperatures. Thus, in the region examined by the sailplane, which as seen in Fig. 22.4 was not in the highest reflectivities but always quite close to them, the cloud contained only a small number of ice particles, and it appears that the echo was caused by a small concentration of large particles.

The origin of these particles is uncertain. Based on the relatively long particle growth times and the fact that almost all ice particles observed between 1905 and 1910 were in near-zero updraft or downdraft, we conclude that the particles causing the first echo at about $-15\,°C$ were not formed by nucleation followed by growth during one ascent in the main updraft. The particles probably grew in the fringes of the updraft or through a process whereby particles formed high in the cloud could reenter the updraft and continue growing. This contention will be supported by Doppler radar measurements of vertical air motion (Chapter 23).

The sailplane's first ascent was located within a relative minimum of reflectivity, with higher reflectivities to the west, south, and east. Doppler-derived vertical velocities showed this to be a region of updraft with weak downdrafts or nearly zero upward motion bordering it on the west and south. A composite of the sailplane and Doppler radar vertical motion measurements for about 1910 are shown superimposed on 3-cm reflectivity contours in Fig. 22.17.

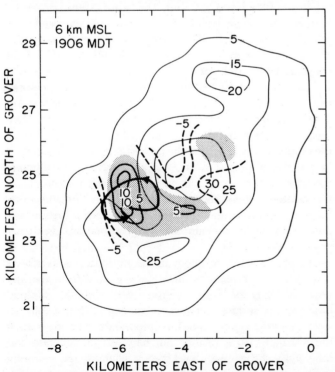

Figure 22.17 A composite of vertical motions at 6 km based on the sailplane measurements and Doppler radar results of Chapter 23. Areas of upward motion are shaded and are also shown by moderately heavy solid lines with a 2.5 m s⁻¹ contour interval. Downward motions are indicated by dashed lines. The light solid lines are reflectivity contours as labeled from the NOAA-D 3-cm radar data. The heaviest solid line shows the sailplane track from 1905 to 1907.

Note that higher reflectivities are south of the updraft. As seen in Fig. 22.4 these higher reflectivities remained on the south side of the developing echo from first echo formation until past the time of this composite. Therefore, it is reasonable to assume that the pattern of updraft-downdraft seen in Fig. 22.17 was also present at the time of first echo formation. In fact, Fig. 21.20, the high-resolution MINA II CP-2 radar data were used to show the presence of a void or minimum in the very early development of this echo. This void was in the region where, several minutes later, the sailplane found the updraft it used for the first ascent in cloud from 1903 to 1911.

It appears that the developing echo and precipitation were in a region of downdraft or near-zero vertical motion, and that the active updraft region was relatively free of precipitation and therefore had a minimum of reflectivity. Such a configuration is similar to weak echo regions found in severe storms, where intense updrafts are devoid of precipitation because there has been insufficient time for precipitation particle development (see, e.g., Marwitz, 1972). Considering the previous discussion on the time needed for precipitation particle growth, it is not at all surprising to see a miniature version of this feature in developing echoes with updrafts of 5-15 m s⁻¹, where the ice process is the dominant precipitation mechanism.

By 1910 a new updraft cell was detected by the Doppler radars (see Fig. 23.10). This new updraft penetrated a region within the southern edge of reflectivity from an earlier cell (called B_0 in Chapter 23). When first detected, reflectivities in the updraft core were already ≈ 15 dBZ$_e$. A previous Doppler scan at 1905 indicated no updraft in this area, so that hydrometeors were large enough to produce 15-dBZ$_e$ reflectivities in less than 5 min. Using Auer's (1972) graupel concentrations, this value of reflectivity implies sizes as big as 1.5 mm. Growth from nuclei to this size would have required 10 min of diffusional growth plus 3 to 7 min of accretional growth. This reflectivity core was, therefore, a consequence of growth from particles already big enough to start riming as they were engulfed in the new updraft cell of fresh cloud material.

22.4.4 Column Crystal Observations—Growth in the Fringes of the Updraft

One purpose of this case study was to elucidate how individual particles can have sufficient time to grow to precipitation size. The observations of column crystals from N10UW demonstrate one mechanism by which particles can have longer residence times than would be possible if they remained in the main updrafts. Column crystals as large as 640 μm mixed with ones as small as 90 μm provide strong evidence that growing ice particles remained in a narrow altitude region of the cloud for appreciable lengths of time. As seen in Fig. 22.12, clf shapes (hollow columns) were predominant, with *c*- to *a*-axis ratios of about 5 to 1. Hallett and Mason (1958) show that hollow column growth occurs in the −5°C to −8°C region in supersaturated conditions. Furthermore, for an axial ratio of 5:1, the growth rate studies of Ryan et al. (1976) imply that growth would have been at −5°C to −7°C and that it would have taken at least 8 min to grow a 640-μm column at a pressure of 1000 mb. At the pressure of observation (≈530 mb) the minimum time required for growth to this size would be reduced by 30%-40% (see Heymsfield et al., 1979) to about 5 min.

Most of the growth of these columns must have occurred in the −4°C to −8°C temperature range or else the crystals would have exhibited altered growth habits. This 4°C span corresponds to an altitude interval of roughly 500 m. Thus, these particles spent at least 5 min in a relatively narrow range of altitudes, implying that the average vertical velocity of the particles during this time was less than 1.5 m s⁻¹. The location of the columns observed during pass 4 is shown in Fig. 22.18 by an asterisk superimposed on the Doppler-measured horizontal motion field and vertical motion contours. Note that the columns were observed in a region of weak updraft (<2 m s⁻¹) in weak horizontal motion, and moving in a direction that could keep them in weak updrafts for several minutes more. This trajectory would have eventually carried these columns into a region of stronger downdraft to the north. The important point is that there are localized regions within a cloud in which particle growth can proceed with little vertical excursion of the particles. The fact

that some of the columns were fairly heavily rimed demonstrates that riming as well as diffusional growth was occurring.

22.4.5 Precipitation Particles in the Updraft Core

The microphysical observations from 5 and 6 km presented in Section 22.3 showed the presence of a mixture of millimetric precipitation particles in the updraft core. As early as 1914, pass 2 of N10UW showed that 2-mm particles were present in the updraft. From 1920 to 1925 sailplane observations in the updraft showed 4- to 5-mm graupel, 1-mm raindrops, recently frozen drops, and particles that appeared to have partially melted and were refreezing. Considering particle growth rates presented in Section 22.2 it is clear that these particles had insufficient time to grow during a single ascent in the updraft in which they were observed and therefore must have had more complicated trajectories. A close comparison of the location of these particles with respect to other parameters such as θ_e, liquid water content, and updraft speed show that these recirculated particles were distributed throughout the updraft region and in the unmixed core. The most probable mechanism by which these particles got into the unmixed core was by sedimentation. Any other mechanism such as turbulent mixing or large-scale entrainment would have diluted the equivalent potential temperature, liquid water content, and updraft speed. The particles observed in the mixed updrafts may also have arrived there by sedimenting or mixing with air that did not originate from cloud base.

Referring back to the comparison of the radar-measured reflectivities with those calculated from particle observations, it is important to note the agreement between the two measurements between 1920 and 1925. During this period the sailplane was in or near the highest reflectivities observed in the storm by the radar, and the aircraft and Doppler measurements show that these were updraft regions. Thus, recirculated particles caused the high reflectivities, and consequently the recirculation mechanism was, in this case, an important factor in the amount of precipitation eventually reaching the ground.

The precipitation efficiency of this storm during its mature stage was rather high, as might be expected from the low shear. By using the Doppler-derived vertical motion at cloud base and the surface water vapor mixing ratio to obtain an estimate of the water vapor flux into the storm, and by using the Martner (1975) relationship between rainfall rate and reflectivity at the lowest level scan, we determined nearly instantaneous precipitation efficiencies for the storm at different times during its history. In the early stages the efficiency was low, but it reached a peak of 60% at 1920. This peak efficiency fits well with the curve of precipitation efficiency versus shear compiled by Marwitz (1972), based on data for

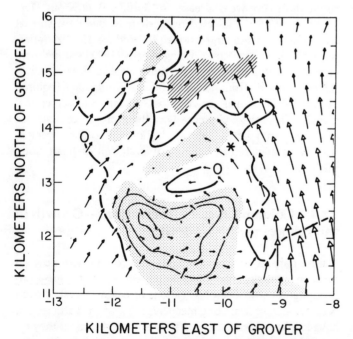

Figure 22.18 The location (shown by an asterisk) where column crystals were observed by N10UW during pass 4 relative to the horizontal and vertical motions derived from the Doppler radar measurements. Upward motions of ≥ 2.5 m s⁻¹ are shown lightly shaded, downward motions of ≤ 2.5 m s⁻¹ are shown by hatching, and the zero vertical motion boundary by a heavy solid line. The horizontal wind motions are shown relative to the storm; i.e., storm motion has been subtracted.

thunderstorms at several locations. It is important to note, however, that this was the peak efficiency and that an average for the entire storm history was much lower.

The near-vertical structure of the updraft profile derived from Doppler radar suggests that gravitational loading of the updraft might have occurred. Ice water contents derived from sailplane data for the period 1920 to 1930 are presented in Fig. 22.19. Because 2- to 4-mm diameter graupel contributed most of the mass to the ice water content, the density of the ice particles was assumed to be 0.2 g cm^{-3}, based on the graupel measurements reported by Heymsfield (1978). Prior to 1928 there is considerable fluctuation of the ice water content, much of which is due to the statistical error associated with the limited sample volume of the camera. From 1928 to 1930 the ice water content (10-s average values) ranged from 0.3 to 3 g m^{-3}, with an average value of about 1.1 g m^{-3} and a standard deviation of 0.8. Based on the 3°C temperature difference between the environmental temperature and the in-cloud adiabatic temperature at these levels (7-8 km, 400 mb), the specific buoyancy was calculated as 7.5 g m^{-3}.* Thus, even the high concentrations of ice and resulting ice water contents observed from 1928 to 1930 caused negligible loading of the updraft. Because some of the particles may have recirculated and may therefore have had higher deviation than 0.2, the ice water contents presented in Fig. 22.19 may be low. But even if the actual densities were a factor of three higher (0.6 g cm^{-3}), the average ice water content from 1928 to 1930 would be less than half the specific buoyancy.

Based on the ice particle sizes observed from 1920 to 1922, the time required to deplete the liquid water by $1 - 1/e$ (63 %) was about 800 s. The sailplane data collected between 1920 and 1930 and presented earlier in Fig. 22.7 clearly showed a

decrease in liquid water content. Some of this decrease may have been due to mixing, but clearly depletion was also acting.

22.5 CONCLUSION

During the early stages of the storm history the thermodynamic measurements in the updraft showed small adiabatic regions on scales of a few hundred meters and a lot of variability in all parameters on these same scales. Much of this variability appears to have been caused by the mixing of air descending in penetrative downdrafts from above the observation level with air ascending from cloud base. These penetrative downdrafts probably transported ice crystals nucleated high in the cloud down to lower levels.

From the coordinated radar and microphysical observations it was inferred that the first radar return in one of the major turrets of the storm was caused by ice particles larger than 2 mm in diameter at a concentration of less than 15 m^{-3}. The origin and history of these few particles responsible for the first echo are unknown. However, based on the observed updraft speeds, liquid water contents, and temperatures, it is doubtful that these particles formed during one vertical transit within the updraft. Even these first radar-detectable particles appear to have had a complicated growth history.

In the mature stages of the storm millimetric water drops, ice particles, and partially frozen ice particles were observed in the main updraft of the storm in the 0°C to -5°C region. Since some of these particles were observed in unmixed as well as mixed portions of the updraft, sedimentation as well as recirculation must have been responsible for the transfer of the particles into this region. These recirculating particles

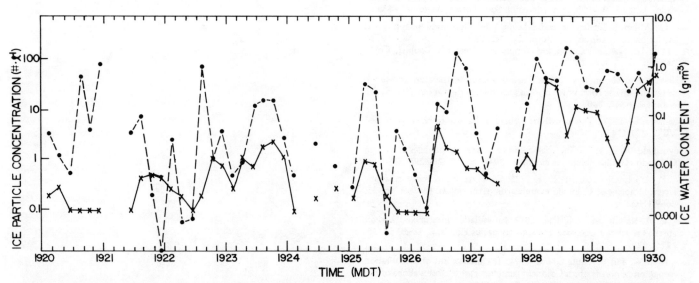

Figure 22.19. Measured total ice particle concentration (solid) and derived ice water content (dashed) are plotted against time for the sailpane's second ascent in the Butler storm.

*This assumes that differences in pressure between the cloud and the environment were negligible and that differences in temperature can be related to differences in density by means of the ideal gas law.

were observed in the regions of maximum reflectivity of the storm. Thus, in this storm the recirculation mechanism was an important factor in the development of precipitation. In other regions of the storm column crystals provided evidence of particles residing for at least 5 min in a narrow vertical extent of the cloud. From the microphysical observations alone, it is obvious that the observed precipitation particles had a rather complicated growth histoy. This history will be discussed in more detail in Chapter 23, in which the kinematic and microphysical observations are combined and a conceptual model of this storm is presented.

References

Auer, A. H. Jr., 1972: Distribution of graupel and hail with size. *Mon. Weather Rev.* 100, 325-328.

Auer, A. H., J. D. Marwitz, G. Vali, and D. L. Veal, 1971: Final report to the National Science Foundation. Dept. Atmos. Resources, Univ. of Wyoming, Laramie, Wyo., 94 pp.

Cannon, T. W., 1974: A camera for photography of atmospheric particles from aircraft. *Rev. Sci. Instrum.* 45, 1448-1455.

-----, J. E. Dye and V. Toutenhoofd, 1974: The mechanism of precipitation formation in northeastern Colorado cumulus: II. Sailplane measurements. *J. Atmos. Sci.* 31, 2148-2151.

Dye, J. E., and V. Toutenhoofd, 1973: Measurements of the vertical velocity of air inside cumulus congestus clouds. Prep. 8th Conf. on Severe Local Storms, Denver, Colo., Oct. 15-17, 1973, Am. Meteorol. Soc., Boston, Mass., 33-34.

-----, J. E., C. A. Knight, V. Toutenhoofd, and T. W. Cannon, 1974: The mechanism of precipitation formation in northeastern Colorado cumulus: III. Coordinated microphysical and radar observations and summary. *J. Atmos. Sci.* 31, 2152-2159.

-----, C. A. Knight, P. N. Johnson, T. W. Cannon, and V. Toutenhoofd, 1976: Observations of the development of precipitation-sized ice particles in NE Colorado thunderstorms. Prep. Intl. Conf. on Cloud Physics, Boulder, Colo., July 26-30, 1976, Am. Meteorol. Soc., Boston, Mass., 478-483.

-----, L. J. Miller, B. E. Martner, and Z. Levin, 1980: Dynamical-microphysical evolution of a convective storm. NCAR Technical Note NCAR/TN-151+STR, National Center for Atmospheric Research, Boulder, Colo., 248 pp.

Hallett, J., and Mason, B. J. (1958): The influence of temperature and supersaturation on the habit of ice crystals grown from the vapour. *Proc. R. Soc. of London*, A247, 440.

Harimaya, T. (1981): The growth of graupel. *J. Fac. Sci. Hokkaido Univ.* Ser. VII, 7, 121-134.

Heymsfield, A. J., 1978: The characteristics of graupel particles in eastern Colorado cumulus congestus clouds. *J. Atmos. Sci.* 35, 284-295.

-----, P. N. Johnson, and J. E. Dye, 1978: Observations of moist adiabatic ascent in northeast Colorado cumulus congestus clouds. *J. Atmos. Sci.* 35, 1689-1703.

-----, C. A. Knight, and J. E. Dye, 1979: Ice initiation in unmixed updraft cores in northeast Colorado cumulus congestus clouds. *J. Atmos. Sci.* 36, 2216-2229.

-----, -----, -----, and V. Toutenhoofd, 1974: The mechanism of precipitation formation in northeastern Colorado cumulus: Part I. Observations of the precipitation itself. *J. Atmos. Sci.* 31, 2142-2147.

Kyle, T. G., and W. Sand, 1973: Water content in convective storms. *Science* 180, 1274-1276.

Magono, C., and C. W. Lee, 1966: Meteorological classification of natural snow crystals. *J. Fac. Sci. Hokkaido Univ.* Ser. VII, 2, 321-335.

Martner, B. E., 1975: Z-R and Z-W$_0$ relations from drop size measurements in High Plains thunderstorms. Prep. 9th Conf. on Severe Local Storms, Norman, Okla., Oct. 21-23, 1975, Am. Meteorol. Soc., Boston, Mass., 307-310.

Marwitz, J. D., 1972: Precipitation efficiency of thunderstorms on the High Plains. *J. Rech. Atmos.* 6, 367-370.

Ono, A., 1969: The shape and riming properties of ice crystals in natural clouds. *J. Atmos. Sci.* 26, 138-147.

Paluch, I. R., 1979: The entrainment mechanism in Colorado cumuli. *J. Atmos. Sci.* 36, 2467-2478.

Ryan, B. F., E. R. Wishart, and D. E. Shaw, 1976: The growth rates and densities of ice crystals between −3 and −21°C. *J. Atmos. Sci.* 33, 842-850.

Squires, P., 1958: Penetrative downdraughts in cumuli. *Tellus* 10, 381-389.

CHAPTER 23

The 25 July 1976 Case Study: Airflow from Doppler Radar Observations and Conceptual Model of Circulation

L. J. Miller, J. E. Dye, B. E. Martner

23.1 INTRODUCTION

Some time ago Ludlam and Scorer (1953) and Malkus (1952, 1954) presented descriptions of the thermal theory of cumulus convection. They suggested that large cumulus clouds are assemblages of individual thermals, which intermittently appear as rising cloud towers. In this way a cloud is gradually built up by successive thermals, with later ones rising through and mixing with residues of previous thermals. Thus, cloud droplets and ice crystals can move to new thermals, extending their growth times over that possible in the use of a single thermal to its equilibrium level. This exchange of particles between individual convective elements obviously complicates the growth trajectories of individual particles.

Woodward (1959) found that the vortex ring motion within isolated thermals, with the center rising at about twice the speed of the cap and the perimeter descending, trapped particles within the thermal until their fallspeeds exceeded some fraction of the maximum vertical motion at the center of the vortex. This fraction decreased from 0.7 cm on the updraft axis to 0.2 cm at the thermal's edge, where sinking motions were about 0.5 times the maximum on the axis.* The small size, 500 to 2000 m, of vortex ring circulations within thermals usually prevents it from being resolved in Doppler radar measurements. This flow, however, probably brings fairly large particles into new thermals rising through precipitation debris in the wakes of earlier thermals.

We will discuss a similar flow pattern that developed in proximity to the main updraft in the 25 July 1976 storm, and suggest that this pattern enabled large particles to be carried back into the updraft and continue their growth as they rose a second time. The number of times individual particles can recirculate obviously depends on the lifetime of a favorable flow pattern and the time it takes to grow to sizes large enough to fall out. Furthermore, the largest updraft speed attained in a storm must be an important control on the biggest size of particle grown. This idea has been previously discussed, for example, by Dennis and Musil (1973) and Browning and Foote (1976).

Byers and Braham (1949) defined thunderstorm cells as "regions of localization of convective activity within the thunderstorm." Later Browning (1977) described a cell as "a dynamical entity characterized by a compact region of relatively strong vertical air motion which can be identified by

radar from its associated volume of relatively intense precipitation." Since most studies of storms have used only conventional radar measurements of reflectivity and limited airflow measurements, these definitions of the cell have seemed adequate. However, as more complete data sets become available the simple definition of thunderstorm cells becomes less satisfactory, as the observations to be presented will illustrate.

As discussed in Chapter 22, cells and their associated reflectivity cores were observed from the first appearance of radar echo through their intensification and dissipation phases. The storm was weakly organized, moderately intense, and produced hailstones with diameters to about 8 mm. Each cell formed ahead of the storm on the southwest flank (upshear side), developed precipitation, and became the dominant entity. This evolutionary process is similar to the one observed in feeder clouds (Goyer et al., 1966; Dennis et al., 1970) or daughter clouds (Browning et al., 1976). As suggested by Musil (1970) and discussed by Browning (1977), hailstone embryos are probably grown in such clouds, particularly for organized, multicell hailstorms.

23.2 METHODS OF DOPPLER RADAR OBSERVATION AND DATA REDUCTION

23.2.1 Doppler Radar Characteristics and Observations

Four pulsed Doppler radars, two from NOAA's Wave Propagation Laboratory (WPL) and two from NCAR's Field Observing Facility (FOF), were operated during the 1976 NHRE summer field program. The 25 July case study included observations from only three of these radars, NOAA-C, NOAA-D, and CP-3. Figure 21.1 has shown the locations of the radars, Grover field headquarters (CP-2), and the storm of interest. The storm moved south-southwestward and was located about 25 km west of the centroid of the three Doppler radars. Because of this storm location, all radar measurements were taken at relatively low elevation angles; the consequences of this are discussed in the next section.

The radars made several coordinated scans of the storm to study the reflectivity and internal airflow patterns. All radar scans started together at the lowest elevation angles and moved upward to echo top, taking about 2 to 3 min. At any grid point all samples used to determine horizontal airflow

---

*Woodward reported these values as fractions of the rise rate of the cap, which she stated was about one-half the maximum upward speed on the thermal's axis.

were taken within 10 to 60 s. Radar characteristics and sampling increments are summarized in Table 23.1.

TABLE 23.1
Characteristics of pulsed Doppler radars

	Radars		
	NOAA-C	NOAA-D	NCAR CP-3
Wavelength (cm)	3.22	3.22	5.45
Beamwidth (deg)	0.8	0.8	1.2
Pulse length (m)	105	105	150
Number of time samples	64	64	128
Gate spacing (m)	450(1–3)*	450(1)	260
	375(4–7)	300(3–7)	
Average azimuth angle increment (deg)	0.5	1(1–3)	0.4
		0.7(4–7)	
Average elevation angle increment (deg)	0.5	1.5(1)	1.0
		1.0(4–7)	
Range to storm** center (km)	49–56	22–36	36–27
Azimuth sampling increment† (m)	428–489	334–440	251–188
Elevation sampling increment† (m)	428–489	576–628	628–471

*Numbers in parentheses refer to volume scans.

**First range applies to volume scan 1 and increases or decreases, respectively, to range at scan 7.

†First increment applies to scan 1 and second applies to scan 7.

Backscattered radar signals from pulse volumes, defined by pulse length and beamwidth, were analyzed for reflectivity-weighted mean radial velocity and equivalent radar reflectivity factor. These pulse volumes were located along the beam direction every 260 to 450 m (gate spacing) and were separated by 250 to 630 m (linear sampling increments)in the azimuth and elevation directions. At these sampling increments, scale sizes larger than 0.5 to 1.25 km were resolved, but with only 50% to 75% of their true amplitude because of the pulse-volume averaging. However, since divergence of horizontal airflow was approximated by finite difference techniques, divergence estimates approached their true values only for scales larger than 1.0 to 2.5 km. The FOF radar was equipped with hardware processing, whereas both WPL radars recorded the digitized time series of return signal onto magnetic tape for later software processing. All mean radial velocities were determined with the covariance or "pulse-pair" algorithm (e.g., Rummler, 1968; Miller and Rochwarger, 1972; Berger and Groginsky, 1973). Following Berger and Groginsky, random errors in the radial velocity estimates ranged from 0.2 to 1.0 m s[1] for the signal-to-noise power ratios (SNR > −7dB), numbers of time series samples (64 or 128), and estimated widths ($\approx$ 1 to 2 m s[-1]) of velocity spectra involved. Because these errors depend on true spectral widths and per-pulse SNR, which are only approximately known, point-by-point error analysis is not possible.

23.2.2 Data Reduction

Radial velocity and reflectivity fields in each of the radar sampling spaces (range, azimuth angle, and elevation angle) were transformed to a common Cartesian analysis grid under the assumption that each grid point estimate was well represented by a weighted spatial average of its neighboring samples. No attempt was made to account for temporal evolution during the time necessary for a radar to sample the entire storm volume. All original samples within 500 m of an analysis grid point were linearly weighted to obtain interpolated values (more details are given by Miller in an NCAR technical report, Dye et al., 1980). For such a weighting or filtering function applied over a 500-m radius sphere, scales larger than about 1 to 1.5 km were passed with more than 50% amplitude and appear in the final fields. The number of original samples within a sphere of influence varied over the analysis grid so that filtering reduced the random errors in grid point estimates of radial velocity by 0.3 to 0.6 times their original values.

Bohne and Srivastava (1976) presented a detailed description of the triple Doppler radar data analysis scheme and the attendant random errors. Dye et al. (1980) explain the scheme used here and a source of bias errors inherent in mean radial velocities measured by radar. Briefly, interpolated radial velocities that are normally non-orthogonal were combined linearly to yield three orthogonal components of precipitation motion. Large errors in the derived vertical components caused by the low scanning angles precluded using them to obtain vertical air motion. The horizontal components were used instead in integrating the anelastic form of the mass continuity equation to obtain vertical air motion. Because the lowest-level divergence estimates were in a layer centered 400 m above the ground, vertical air motion at this level was approximated by the product of these density-weighted divergences and this height increment. This approximation gave results at cloud base that were consistent with aircraft observations of vertical motions of 1 to 3 m s[-1].

The Doppler-derived horizontal air motion generally had random errors of about 0.1 to 0.5 m s[-1]. (Bias errors may have caused additional errors of about 0.5 m s[-1]. See Appendix A.) Therefore, horizontal divergences estimated by centered finite differences on a 250-m grid had errors of 0.4 to 2.0 × 10[-3] s[-1] (4 × 10[-3] s[-1] if horizontal components had errors as large as 1 m s[-1]). The larger errors in divergence generally were in regions of less than 13-dB signal-to-noise power ratio. Forcing the horizontal flow to approach the near-environmental flow as the signal power went to zero significantly reduced these errors. The details of this procedure are outlined in Dye et al. (1980). Errors in the divergence used in the integration were generally less than 1 × 10[-3] s[-1]. Vertical air motion estimates aloft still did not agree well with aircraft measurements. However, if the mass continuity equation is integrated upward, vertical speed tends to increase, usually because of propagation of errors in the density-weighted divergence and the lower boundary conditions.

To minimize propagation of errors from integration a variational scheme to adjust vertical motion estimates was used that was similar to the one formulated by O'Brien (1970). The adjusted vertical motions then agreed reasonably with aircraft-measured values of vertical air motion at all heights. Figure 23.1 presents a comparison of a penetrating aircraft's measured vertical velocities and those derived from the Doppler analysis. Line profiles of vertical velocity measured by the penetrating aircraft (N10UW) and inferred from the Doppler analysis are shown. For passes 1 and 2, the aircraft tracks were adjusted to account for movement of the storm during the time between aircraft traverse and Doppler radar scan. There generally is reasonable agreement, considering that radar measurements of velocity were volume averages and vertical air motion was derived from an integration divergence estimates obtained over a period of 2 to 3 min, while aircraft measurements of vertical velocity were line averages. Small sizes of updraft cores were comparable to spatial error in aircraft position relative to the "instantaneous," radar-derived motions.

23.3 SUMMARY OF ENVIRONMENTAL AND RADAR ECHO OBSERVATIONS

A very brief summary of environmental observations and echo history of the Butler storm, presented in detail in Chapter 21, is given here as a more wieldy context for the discussion of Doppler radar-measured airflow to follow.

23.3.1 Environmental Characteristics

The first radar echo from the Butler storm was detected in southeastern Wyoming above a low-level echo associated with a convergence line in the surface winds. Since horizontal flow was nearly parallel to the radial direction from one Doppler radar (NOAA-D), radial variation of its radial velocity could be used to obtain convergence values, giving $2 \times 10^{-3} \, s^{-1}$. A combination of orographic and mountain lee trough, low-level mechanical lifting forced moist air with a mixing ratio of 9 to 9.5 $g \, kg^{-1}$ out of the subcloud neutral layer. After reaching cloud base at 4 km, if this air had undergone undiluted parcel ascent, it would have experienced a maximum of 3°C to 4°C of thermal buoyancy in the layer between 600 and 250 mb. There was little or no temperature deficit at cloud base, in contrast to 1°C to 3°C deficits of more intense convection in the northeastern Colorado area (Marwitz, 1973; Foote and Fankhauser, 1973; Browning and Foote, 1976). Surface winds were northerly with speeds of 4 to 5 $m \, s^{-1}$ and backed to westerly at 9 km. Wind shear through the cloud layer (4 to 10 km) was weak

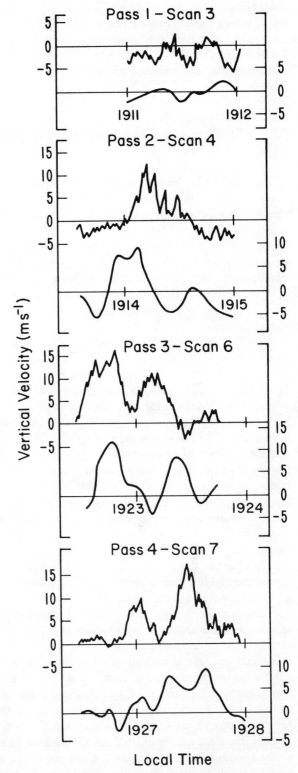

Figure 23.1 Line profiles of vertical velocity measured by N10UW (top) and inferred from Doppler analysis (bottom) for the N10UW passes and radar scans shown.

($2 \times 10^{-3} \, s^{-1}$) with upper-level wind speeds of only 10 to 12 $m \, s^{-1}$. Cloud bases were relatively warm (7°C) and cloud tops during the most intense phase reached temperatures colder than $-50\,°C$ at ≈ 13 km. Two minor temperature inversions ($\approx 1°$ to $2°$) existed, one at 5.2 km and the other at 7.6 km. Air with the lowest equivalent potential temperature (≈ 330 K) was found at about 5 km. These environmental conditions play an important role in the evolution of the air motion structure.

23.3.2 Structure and Evolution of the Radar Echo

The radar echo moved south-southwestward (200°) at 9 m s⁻¹. This movement consisted of simple advection of identifiable radar cells (Browning, 1977), fine-scale reflectivity structure (Barge and Bergwall, 1976), and discrete propagation as new cells formed on the storm's forward and right flanks. Because the radar echo is an integrated effect of the dynamics, it can be misleading in specifying the kinematical structure unless a storm is steady (e.g., Browning and Ludlam, 1962; Newton, 1966; Fankhauser, 1971; Browning and Foote, 1976).

Reflectivity features that formed as discrete elements, separated from other echo by a closed contour line at 5 dBZ_e, are defined as "reflectivity cells." "Reflectivity cores" were identifiable only as local maxima in reflectivity and often originated within the 5-dBZ_e boundary of a cell. In this storm both reflectivity features were initially associated with updraft regions. The sampling interval for the Doppler radars was about 5 min, whereas the interval for the CP-2 radar was less than 2 min. Updraft cells could not always be tracked from scan to scan in the Doppler data, but their associated reflectivity could be tracked using the 2-min reflectivity scans. The only apparent difference between cells and cores was their distance from existing echo maxima when they formed; cores were within 2 to 3 km, whereas cells were usually farther than 5 km. Barge and Bergwall (1976) speculated that fine-scale reflectivity structure (which appears to be equivalent to what we call cores) was related to hail-generation regions. They further noted that such fine structure is very often associated with both multicell and supercell storms.

Reflectivity data at a constant elevation angle from the CP-2 radar were used to reconstruct a history of the echo near 6 km, shown in Fig. 23.2. Cell A already existed when detailed scans of the storm sector began at 1852. The times of formation of cells B_0, C, and D_0 are shown alongside their tracks. The areas labeled B_1, B_2, and D_1 were all cores that formed within the 5-dBZ_e reflectivity contour of the associated cell. Most of these cores lasted more than 10 min. Except for cell D_0 and core D_1, which formed northwest of cell B_0 as a gust front propagated through the area (see Chapter 21), all the formation positions of cells and cores were along a nearly straight line, taken as the direction of storm propagation. Individual cells and cores moved to the left of this direction, parallel to the mean winds in the lower

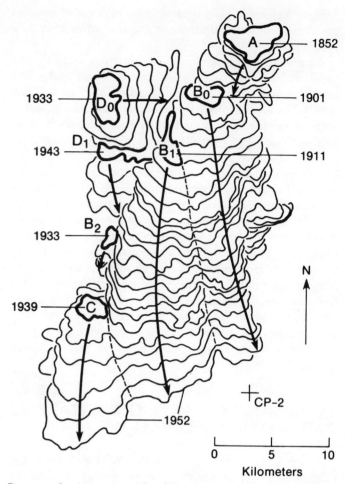

Figure 23.2 Successive positions of the 5-dBZ_e reflectivity contour near 6 km. The times of first detection of radar echo from cells and cores A through D are shown alongside the storm track. A (+) marks the location of the CP-2 radar.

layer (below ≈ 8 km) of the storm. The vector sum of propagation and advection together gave an echo motion that was to the left of the echo motion of propagation alone.

23.4 DOPPLER RADAR RESULTS

23.4.1 Overview and Time Evolution

Density-weighted horizontal convergence greater than $1.5 \times 10^{-1} s^{-1}$ at cloud base (4 km) is depicted in Fig. 23.3 along with major cell tracks. A lack of significant convergence east of the broad convergence zone on the western side of the storm implies that the reflectivity cells and cores did not have roots to cloud base for their entire lifetimes. Moist air from the sub-cloud layer ascended into the cloud layer on the western side of the storm. As this air rose, latent heat was released and precipitation was produced, which fell out as cloud elements drifted downstream, following the

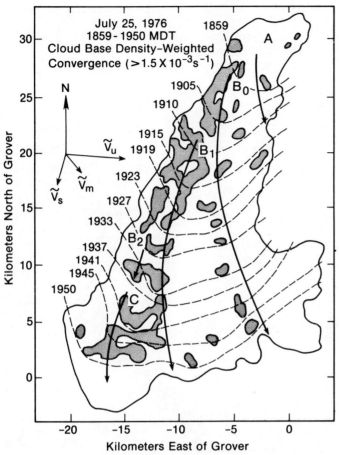

Figure 23.3 *The movement of regions of convergence at cloud base. Areas of density-weighted convergence > 1.5 × 10⁻³ s⁻¹ are shaded. The outer heavy line is the envelope of 5-dBZₑ reflectivity for the period 1859-1955. Dashed lines are isochrones of echo leading edge. Letters indicate cell or core positions when first detected. Insert shows winds at the surface (Vₛ), middle (Vₘ), and upper (Vᵤ), levels, with vectors scaled 2.5 m s⁻¹ per kilometer.*

cloud layer winds. This storm had no significant downdrafts and strong outflow so that the dynamics of sub-cloud and cloud layers were not strongly coupled. More intense storms have characteristic flow patterns that couple the cloud layer to the sub-cloud layer, for example, by intrusion of middle-level air and its subsequent downward flow into the planetary boundary layer (Browning and Ludlam 1962; Kropfli and Miller, 1976; Browning and Foote, 1976).

The movement of the Butler storm's major updraft region, deduced from Doppler radar scans described in Table 23.2, is depicted in Fig. 23.4. The horizontal sections were taken near the level of maximum updraft and slightly (≈ 500 m) above the flight level of N10UW. Updrafts were weak and sailplane measurements (Chapter 22) showed them to be disorganized in the earliest stages. At 1910 (scan 3) multiple peaks remained within the updraft region and were becoming more intense. Shortly thereafter the updraft region shifted relative to the core reflectivity, from the west-southwest side more toward the southwest side. This shift was accompanied by continued invigoration and a double-peaked updraft structure, particularly at 1923 (scan 6). Two updraft cores remained until 1940 (scan 9) when the updraft began to dissipate rapidly. Detailed data will be presented below only for scans 1 to 7.

Sailplane ascent began near the updraft region deduced from the data of scan 2 (see Fig. 23.5). Ascent continued until 1910 when the sailplane started losing lift, exited cloud, and re-entered farther south. However, this region was still north of the main updraft region and consisted mostly of downward motion, as was observed in the Doppler data (scan 4). Since it could not ascend, the sailplane exited once more and continued southward. When it entered the cloud the second time about 1919, a strong updraft region was again found, and the sailplane spiraled upward from 1920 to 1932, when it again lost lift and left the cloud. This second ascent began in one updraft core near the center of the scan 5 updraft region, which became the westernmost and most intense core detected in scan 6. At the time of scan 7 the updraft was rapidly weakening.

TABLE 23.2
Doppler radar scans (segments of sailplane and N10UW measurements that overlapped Doppler observations are also listed)

| | | Time | | Scan time | Sailplane | | N10UW | |
		Start	End	Min:Sec	Time	Altitude (km)	Time	Altitude (km)
Scan	Radars							
1	All	185901	190051	1:50				
2	NOAA-C, CP-3	190510	190743	2:33	19058	5.5		
3	All	191004	191245	2:41			191112	4.6
4	All	191502	191743	2:41	191417	6.0	191415	4.9
5	All	191900	192133	2:33	191921	5.5		
6	NOAA-C, NOAA-D	192305	192600	2:55	192527	7.0	192224	5.1
7	NOAA-C, NOAA-D	192712	193011	2:59	192729	7.5	192728	5.1
8	NOAA-C, CP-3	193303	193554	2:51				
9	All	193709	194017	3:08				

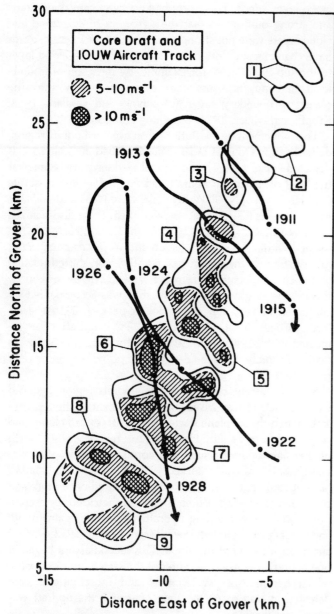

Figure 23.4 *Core updraft (≥ 2 m s⁻¹) determined from Doppler analysis for indicated scan numbers (boxed, see Table 23.3 for times) and near the flight level of the penetrating Queen Air, University of Wyoming N10UW. Horizontal levels were 5 km for scans 1 through 3 and 6 km for the rest. Aircraft penetrated at 5.5 km after scan 3. Dots on the aircraft track show its position at each min.*

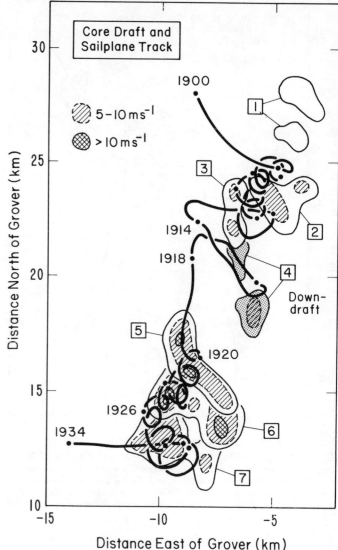

Figure 23.5 *Core updrafts near the flight level of the sailplane. During scan 4 the sailplane penetrated a downdraft region at an altitude of 6 km. All other scans depict regions of updraft ≥ 2 m s⁻¹. The sailplane's first ascent (1900-1910) was from 5- to 6.5-km altitude, and the second ascent (1920-1932) was from 5- to 8-km altitude. Dots on the track show the sailplane's position every 2 min.*

At each vertical level of each scan the updraft speeds and density-weighted convergences were averaged over the area of upward motion greater than 0.5 m s⁻¹. Time-height sections of these averages are displayed in Fig. 23.6. The updraft was associated with moderate subcloud convergence and the largest average convergence values ($\approx$ 1 to 2 × 10⁻³ s⁻¹) were consistently found at the lowest levels scanned by the radars. These convergence values are consistent with the updrafts (about 1 to 2 m s⁻¹) found by aircraft flying at cloud base (see Chapter 21) since average convergence of about 0.5 × 10⁻³ s⁻¹ over a depth of 2.5 km would give 1.25 m s⁻¹

updrafts. The maximum value of average divergence was 1 to 2 × 10⁻³ s⁻¹ and was always near echo top, indicating rapid deceleration of rising air as it approached cloud top. The level of no convergence rose from just below cloud base at 1900 to nearly 6 km during the storm's most intense phase.* Updraft average speeds increased at a rate of about 4 m s⁻¹ per 24 min and reached a maximum value of 5.5 m s⁻¹ at about 1925, 35 min after first echo detection.

*Some of this apparent rise probably resulted from the gradual filling of the subcloud convergence region with precipitation; hence, more and more of the inflow was detected. Often no detectable air-motion tracers are found in much of the low-level inflow, severely limiting the Doppler radar's ability to map the complete flow pattern in the near environment as well as in the precipitating regions. Convergence values were not adjusted with the O'Brien (1970) variational scheme. The level of no convergence must be coincident with the level of maximum updraft speed. Figure 23.6 shows that an adjusted zero-convergence level would have been consistently above cloud base but still rising from 5 km at 1900 to 7 km at 1925.

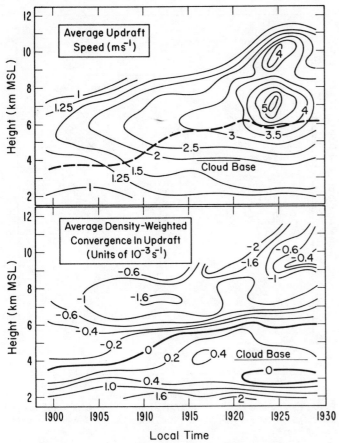

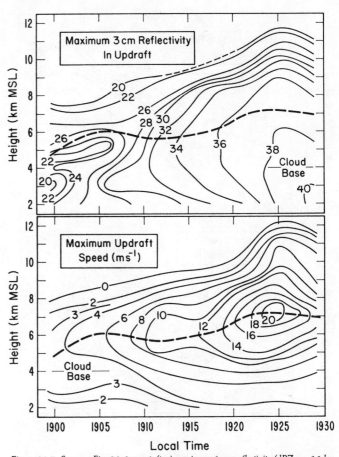

Figure 23.6 Time-height sections through the region of updraft stronger than 0.5 m s⁻¹: (top) average updraft speed (m s⁻¹) and (bottom) average density-weighted convergence (10⁻³ s⁻¹). Heavy dashed line in (top) is the level of no convergence.

Figure 23.7 Same as Fig. 23.6 except: (top) maximum 3-cm reflectivity (dBZ_e = 10 log Z_e/1 mm⁶ m⁻³) and (bottom) maximum updraft speed (m s⁻¹). Heavy dashed line is level of maximum updraft speed.

Time-height sections of maximum 3-cm reflectivity* in the updraft region and the maximum updraft speed are given in Fig. 23.7. The maximum updraft speed increased from 3 m s⁻¹ at 1900 to 20 m s⁻¹ at 1924. Except for the period before 1905, the maximum reflectivity increased at all levels until 1924, and the largest reflectivities tended to spread vertically away from the level of maximum updraft.

Figure 23.8 is a time-height plot of maximum reflectivity in the downdraft, defined as the region at each height with vertical motion less than zero, and maximum downdraft speed. There are two downdraft maxima shown by the dashed lines. One started near 4 km at 1900, rose to 5.5 km, and sank toward the ground about 1915 when a second one developed near 7 km. Before 1924, no obvious connection was apparent between the height of maximum downdraft and height of maximum reflectivity as might be expected if gravitational loading by the precipitation mass was dominant in downdraft generation. In a two-dimensional numerical model of deep moist convection, Schlesinger (1973) found that precipitation

loading acted more to curb the updraft than to initiate the downdraft.

23.4.2 Horizontal Sections

Figures 23.9 to 23.15 are horizontal sections of 3-cm reflectivity, vertical velocity, and horizontal airflow relative to the storm, which was moving toward 200° at 9 m s⁻¹. These sections show the detailed evolution of precipitation distribution and airflow pattern from the time of the first radar detection of cell B₀ (FE on 6-km frame of reflectivity for scan 1, Fig. 23.9) to the most intense phase (scan 6, Fig. 23.14) followed by the beginnings of storm dissipation (scan 7, Fig. 23.15). The compact region of precipitation northeast of FE was the remnants of an earlier cell (A, see Fig. 23.2). Horizontal positions of N10UW penetrations and sailplane ascents are displayed where appropriate on the 6-km frames of vertical velocity. Airflow for scans 2, 6, and 7 (Figs. 23.10, 23.14, and 23.15) were synthesized from data from two radars. Comparisons between two- and three-radar data analyses for the other scans indicated that two-radar analysis was sufficiently accurate to be included in this study (as might be expected from the low elevation angles of the scans).

**Reflectivities measured at 3-cm wavelength were much less than 10-cm values (compare Fig. 23.7 with Fig. 21.11) when the reflectivity exceeded about 25 dBZ. No obvious attenuation effects were found; the discrepancy was apparently caused by near-saturation in the 3-cm linear receivers. The 3-cm reflectivity values were not corrected in this chapter.

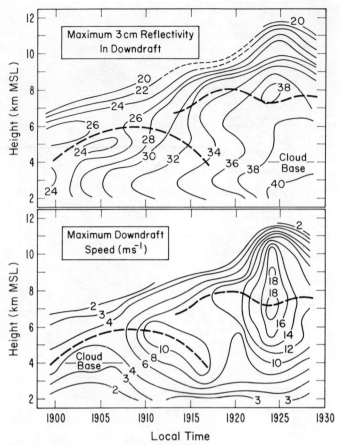

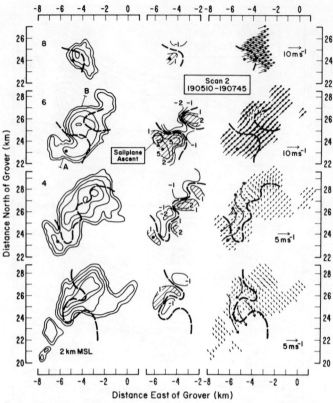

Figure 23.8 Time-height sections through the downdraft region: (top) maximum 3-cm reflectivity (dBZ$_e$) and (bottom) maximum downdraft speed (m s^{-1}).

Figure 23.10 Same as Fig. 23.9: scan 2. Reflectivity contours start at 5 dBZ$_e$ in Figs. 23.10-23.15. Location of sailplane ascent at this time is shown on the vertical air motion panel at 6 km.

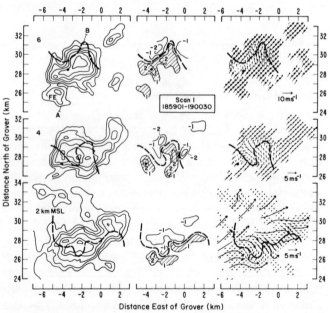

Figure 23.9 Horizontal sections of 3-cm reflectivity (left), vertical air motion (center) and horizontal airflow relative to moving storm (right) at indicated heights for scan 1. Heavy dashed lines separate regions of updraft (> 1 m s^{-1} is cross-hatched) from regions of downdraft. Reflectivity contours start at 5 dBZ$_e$ with a 5-dB increment. Values of vertical air motion (m s^{-1}) are labeled on the contours. Vector scale for 2 and 4 km is 5 m s^{-1} per km; otherwise scale is 10 m s^{-1} per km. Streamlines supplement vectors when speed is slow. Region of first echo (FE) is shown. Line AB in this and following figures is the location for the vertical sections in Fig. 23.20. Heavy dot marks position of maximum updraft.

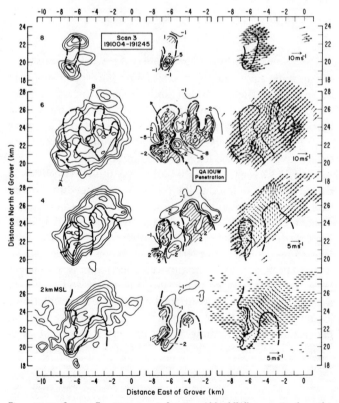

Figure 23.11 Same as Fig. 23.9: scan 3. Location of N10UW's penetration (pass 1) is shown. L marks a region of relatively low reflectivity.

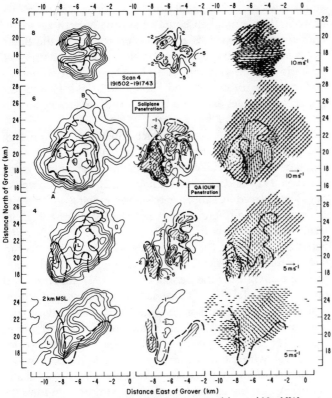

Figure 23.12 Same as Fig. 23.9: scan 4. Locations of sailplane and N10UW penetrations are shown. Position of largest (maximum dimension = 2.2 mm) ice particles observed by N10UW is marked by I on the reflectivity map at 6 km.

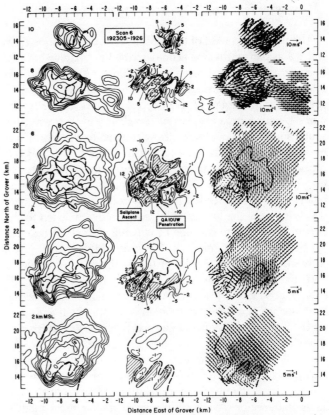

Figure 23.14 Same as Fig. 23.9: scan 6. Largest (maximum dimension = 5.5 mm) ice particles were observed by N10UW at I and rain was observed at R. Locations of N10UW penetration and sailplane ascent are shown.

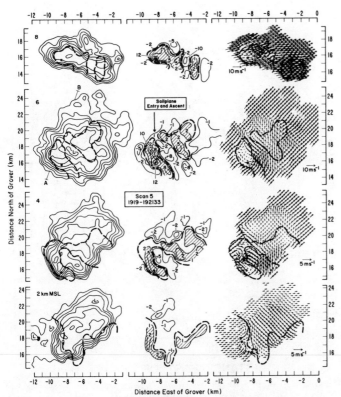

Figure 23.13 Same as Fig. 23.9: scan 5. Location of sailplane entry and ascent is shown.

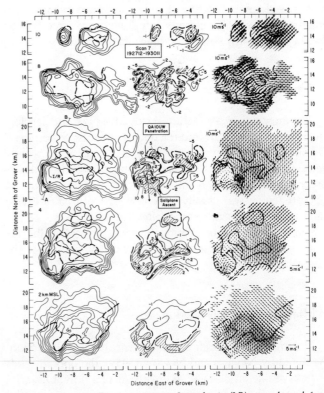

Figure 23.15 Same as Fig. 23.9: scan 7. Ice and rain (I/R) were observed from N10UW. Locations of N10UW penetration and sailplane ascent are shown.

Low-level features (2 km, ≈ 500 m above ground)

The weak echo west of the most intense precipitation from cell B was part of the line echo described in Chapter 21. This line was consistently located in the region of low-level inflow into the updraft, particularly from 1859 to 1926. Updraft speeds near the largest reflectivity values were generally 1 to 2 m s^{-1}, with a broad region of less than 1 m s^{-1} toward the south and west. Low-level inflow relative to the storm was never intense (speeds were less than 1 to 2 m s^{-1}), and this probably contributed to the unsteady, cellular character of the storm (Browning, 1977). According to this idea, strong relative winds are needed for a vigorous updraft to remain steady because the low-level inflow must be matched to the upward flow, the magnitude of which is a result of buoyancy. Moncrieff and Green (1972) expressed this by defining a convective Richardson number (Ri) as the ratio of available potential energy produced by buoyancy to available kinetic energy produced by shear. A low value of Ri favors the maintenance of a steady circulation. We estimate that $Ri \approx 2$ for the Fleming storm (Browning and Foote, 1976) which persisted for over 6 h in a quasi-steady mode. An unsteady, multicell hailstorm, the Raymer storm (Browning et al., 1976), had $Ri \approx 10$, whereas the Butler storm had $Ri \approx 50$.

Downdraft speeds were usually less than 1 m s^{-1} and downdraft regions were co-located with the most intense precipitation. Storm outflow speeds gradually increased from 1 to 2 m s^{-1} during scan 1 (Fig. 23.9) to about 3 m s^{-1} during scan 7 (Fig. 23.15). The subcloud layer was not organized with a well-defined gust front, as is usually observed in more intense storms. Both updrafts and downdrafts were formed beneath the storm within the precipitation region.

Cloud base features (4 km)

The most significant feature at cloud base was the development of a relative flow into the downwind (east or northeast) side of the main updraft, especially after scan 3. As will be seen, this flow allowed air and precipitation particles to re-enter the updraft region and be recycled. The most vigorous updraft region was usually very near or within a relatively high reflectivity region, especially after scan 5 (Fig. 23.13). An increasing mass of precipitation was present in the updraft and may have contributed to its decay, and hence the dissipation of cell B.

Middle-level features (6 km)

The maximum updraft speed was encountered near 6 km. As stated by Newton and Newton (1959), the updraft tends to behave like an obstacle to environmental flow. Obstacle-like flow in the middle troposphere has been observed in the wind field of supercell storms (e.g., Fankhauser, 1971; Brown

and Crawford, 1972; Fujita and Grandoso, 1968). While obstacle flow around an updraft has usually been associated with severe storms, it is an important feature of the Butler storm as well, only at a different scale. Flow around supercell updrafts may be as strong as 30 to 40 m s^{-1} past an updraft 40 km across (Browning and Foote, 1976); the flow in the Butler storm never exceeded 10 m s^{-1} and was around an area no more than 5 to 8 km across.

It was possible to reconstruct much of the field of equivalent potential temperature, θ_e, at this level from sailplane and N10UW measurements that were taken between 1919 and 1930 (see Chapter 22). Measured values of θ_e were placed in a coordinate system moving with the updraft region and, therefore, do not strictly represent an instantaneous picture of their spatial distribution. However, the persistent nature of the flow in the vicinity of the updraft justifies this time-to-space conversion. Figure 23.16a shows θ_e from the aircraft for the period 1921 to 1927 superimposed on the horizontal flow field, relative to the moving storm, derived at 6 km from scan 6 (Fig. 23.14). Reflectivity contours and the updraft and downdraft regions are shown in Fig. 23.16b, and it can be seen that the high θ_e regions correspond well with the two updraft maxima. As has been noted in Chapter 22, the two regions represented virtually adiabatic ascent. Air on the lateral boundaries and downwind (north-

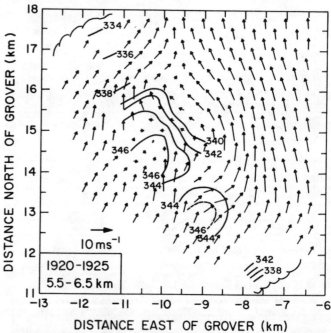

Figure 23.16a θ_e from aircraft measurements superimposed on the Doppler-derived relative flow field at 6 km during the intense phase of the storm. The visual cloud boundaries determined by N10UW penetrations are scalloped.

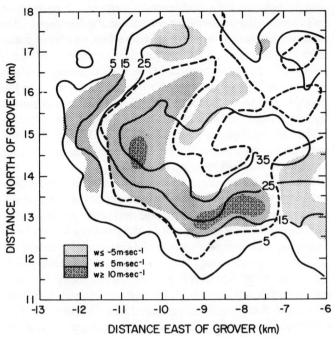

Figure 23.16b Vertical motions superimposed on radar reflectivity for the region shown in Fig. 23.16a. Contours of reflectivity are in 10-dB increments starting at 5 dBZ_e.

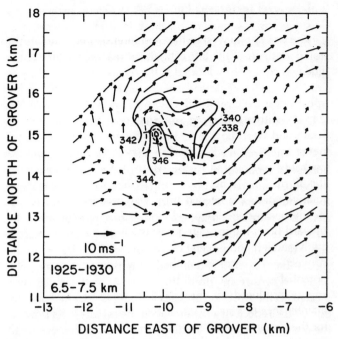

Figure 23.17a Similar to Fig. 23.16a, but at 7 km.

east) of the updraft was apparently a mixture of surface air (θ_e = 346 K) and middle-level air (θ_e = 330 K). At 7 km (Figs. 23.17a,b) it appears that the adiabatic core of the western-most updraft maximum had shrunk to a few hundred square meters even though the strong updraft area (w greater than 10 m s^{-1}) had actually increased.

As the Butler storm moved south-southwestward, it overtook middle-level (6 km) air that had a low value of θ_e. This air flowed past the updraft, eroding it and carrying fresh cloud material and precipitation particles downstream (east-northeast). The dry, potentially cold air was chilled by evaporation of cloud droplets and smaller precipitation particles and sank, forming the downdrafts on the flanks of the updraft (e.g., see Fig. 23.16b). These flanking downdrafts were not located in regions of high reflectivity, which shows that precipitation loading was not a significant factor in their generation. Measurements of θ_e (about 336 K) during the 1914-17 sailplane penetration into downdraft air (Fig. 23.12) indicated that environmental air from middle levels could have mixed with low-level air to form this downdraft in the wake of the updraft. Below middle levels, however, the sinking air and precipitation were carried back toward the updraft to be mixed with positively buoyant air. Mixing significantly reduced the negative buoyancy of downdraft air so that only weak downdrafts reached the ground.

In more intense convective storms, middle-level air often overtakes the storm from the rear and undercuts the tilted updraft (e.g., Browning and Ludlam, 1962; Kropfli and Miller, 1976). The tilt of the updraft allows it to unload precipitation efficiently into potentially cold air. In such cases, evaporation

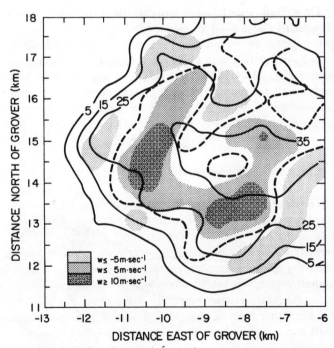

Figure 23.17b Similar to Fig. 23.16b, but at 7 km.

and precipitation loading operate in concert to form an intense downdraft, which may extend to the ground and generate a strong cold outflow that can further invigorate the inflow by mechanical lifting. Flow patterns in the plane of echo motion for an intense storm and for a weaker one, the

Butler storm, are sketched in Fig. 23.18. The weaker storm's flow pattern is similar to the flow in tropical storms (e.g., Betts, 1976; Zipser, 1977), where environmental winds are also weakly sheared as they were in the case of the Butler storm.

Upper-level features (8 and 10 km)

Outflow from the top of the updraft quickly assumed the horizontal momentum characteristic of the environment. As individual buoyant elements (turrets in Chapter 21) reached their equilibrium level near storm top, they were rapidly carried downstream, dropping their precipitation on the downshear (east-northeast) side of the storm. Some oscillation about the equilibrium level was observed, as indicated by the succession of updraft and downdraft regions (e.g., scan 5, Fig. 23.13) downstream from the main updraft, which was located on the west side of the cell. There is a marked orientation of the reflectivity pattern along the middle-level flow. In the discussion of turrets in Chapter 21, it was noted that small reflectivity cores emerged from the main cloud mass and moved toward the southeast along the environmental winds.

23.4.3 Temporal Evolution of Individual Updraft Cells and Cores

Histories of the largest reflectivity values as a function of height within updraft cells are shown in Figs. 23.19a-d. The updraft cells were tracked by use of associated reflectivity cells or cores that were better resolved in time by the shorter sampling interval of the CP-2 radar. Vertical profiles of maximum speed, w_m, along the updraft core at the time of maximum reflectivity aloft, usually near 6 to 7 km, are also depicted in Figs. 23.19a-c. Updraft speed increased up to the time of reflectivity maximum and then decreased, as shown by the height of maximum updraft and its magnitude, designated by the circled numbers in Figs. 23.19a-c. With the exception of core B_1 (Fig. 23.19c), the height of maximum updraft speed was nearly constant.

Coupling of the updraft and reflectivity within the cells is apparent, with the maximum reflectivity aloft and the maximum updraft speed increasing and decreasing in phase.

A similar correlation of updraft and precipitation has been found in time-dependent, one-dimensional models of cumulus convection (e.g., Srivastava, 1967; Weinstein, 1970). However, both Srivastava and Weinstein show temporal oscillations in updraft and hydrometeor content with two maxima in these quantities separated by about 15 to 25 min, with the maximum in hydrometeor content lagging the maximum in updraft by nearly 10 min. In an axially-symmetric model, Ogura and Takahashi (1971) found no such oscillations. Such pulsations may have been a result of the one-dimensionality of the Srivastava and Weinstein models. Nevertheless, it is apparent that these simple models replicate some important aspects of the temporal coupling of dynamics and microphysics, particularly for smaller-scale convection such as occurred in the Butler storm.

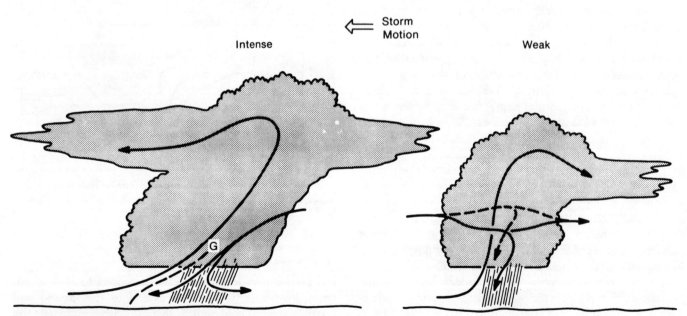

Figure 23.18 Schematic of flow in the plane of storm motion for a storm that grows in a strongly sheared wind environment (intense) and for one that grows in a weakly sheared environment (weak). The letter G denotes the gust front position.

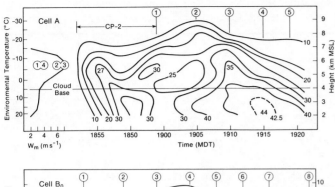

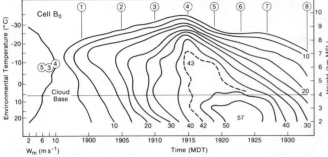

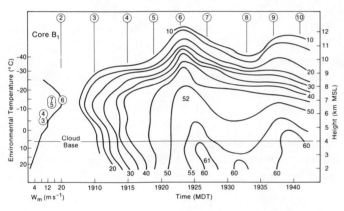

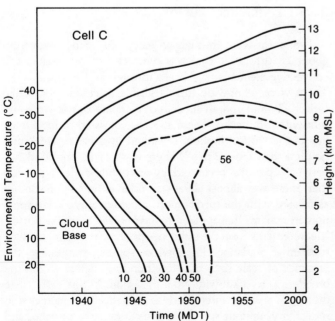

Reflectivity growth rates and greatest reflectivity achieved for each updraft cell or core observed are presented in Table 23.3. The systematic increases in growth rate and reflectivity were probably consequences of two factors: (1) each successive updraft core was more vigorous and had a higher liquid water content than the one before, and (2) each new cell, growing in proximity to earlier cells, started precipitation development upon larger particles that were ingested from these earlier cells. Chalon et al. (1976) found similar increases in the growth rates of succeeding cells in the Raymer multicell hailstorm. They speculated that this increase was due to updraft intensification and changes in the nucleus population, though the latter cause seems unlikely. The implications of cell-to-cell interaction, both microphysical and dynamical, have not been fully investigated.

TABLE 23.3
Cell or core reflectivity growth rates and maximum reflectivity and updraft achieved

Cell or core	Time (min) after first echo 5 or 10 dBZ$_e$	Maximum reflectivity (dBZ$_e$) aloft (near 6 km)	Growth rate (dBZ$_e$ min^{-1})	W$_{max}$ (m s^{-1})
A	23	35	1.1	7
B$_0$	16	43	2.0	10
B$_1$	16	52	2.3	20
C	16	56	2.9	—

23.4.4 Vertical Sections

Streamlines of relative flow in a vertical plane parallel to the echo motion are imposed over reflectivity patterns in Fig. 23.20. Time increases from right to left and top to bottom. The important feature of Fig. 23.20 is the development and persistence of an updraft-downdraft couplet that appears as a horizontal vortex behind the core reflectivity on the southwest side of the storm. This nearly-closed circulation was located near the melting level. As was discussed in Chapter 22, some precipitation particles photographed by the sailplane rising in the updraft had features that strongly suggested that they had been partly or wholly melted and refrozen. Apparently, precipitation particles recycled vertically in the wake of the most intense updraft on the forward side of the storm. Flow was generally into the plane of Fig. 23.20 to the right of the leading updraft from about 4 to 7 km; at the top of the updraft, flow was out of the plane. Precipitation particles within such a flow field could easily

Figure 23.19 Time-height sections of the maximum reflectivity in (a) cell A, (b) cell B$_0$, (c) core B$_1$, and (d) cell C. Vertical profiles of the largest upward motions at the time of strongest updraft are depicted for cells A and B$_0$ and core B$_1$. Numbers in circles denote the scan numbers of the Doppler radars. Circled numbers shown on the updraft profiles locate the heights and values of the largest updraft speed within the updraft cell at the time of the Doppler scans.

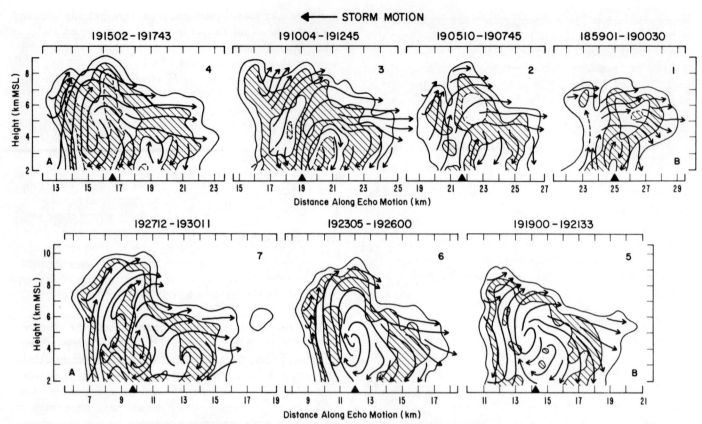

Figure 23.20 Vertical sections in the plane of echo motion of reflectivity isopleths at 10-dB intervals from 5dBZ_e and relative flow (heavy streamlines). Triangle (▲) on horizontal axis marks reference point that remains fixed as storm moves (right to left). Cloud base was at 4 km and melting level (0°C) was nearly 5 km. Scan numbers (1 to 7) are shown in upper right-hand part of panels.

have reentered the updraft to be recycled, and some of the largest particles with fallspeeds of more than about 5 m s⁻¹ probably fell back through the updraft region after having been carried above the updraft maximum. Arguments for the existence of the suggested recycling and sedimentation into the main updraft based upon observations from aircraft and calculated growth times required have been given in Chapter 22, and appear to be very strong. It would be desirable to confirm these processes by detailed growth and trajectory calculations that include the evolution of the airflow. However, the temporal and spatial sampling (about 5 min and 500 m), combined with the small sizes of important features in this storm (about 2 to 4 km) make it very unlikely that such calculations would be realistic enough to be useful.

23.5 SUMMARY AND CONCLUDING DISCUSSION

Measurements of reflectivity structure, internal airflow, precipitation particles, thermodynamics, and environmental characteristics have been examined to document the evolution of a rainshower of moderate intensity. These data were used to explore the initiation of precipitation and the production of millimeter-size particles in the cloud, especially in the

updrafts. Some hail as big as 8 mm was seen at cloud base where the temperature was 7°C, but these small sizes would melt before reaching the ground (Ludlam,, 1958).

23.5.1 Dynamical History

The storm went through an early organizing stage from about 1850 to 1920 that was characterized by small, weak convective elements or subcells throughout the cloud layer. There were 2-km-wide turrets visible at cloud tops, while reflectivity and updraft subcells of comparable size were present within the echo. These subcells consisted of local maxima in the vertical motion, separated by areas of weak motion less than about 2 m s⁻¹. There was good correspondence between updraft subcells and regions of high reflectivity. Further, there was almost always a midlevel updraft maximum associated with the turrets aloft, although there were some updraft maxima that apparently did not lead to visible turrets. Measurements from the sailplane spiraling in updraft showed substantial variation in upward speeds, suggesting the presence of scale sizes of a few hundred meters within the broader, 2- to 3-km-wide updraft region. In this early stage the storm steadily intensified, as evidenced by increases in reflectivity, updraft speed, cross-sectional area of echo, and height of echo top. The build-up of this storm by a succession of buoyant elements closely followed the thermal theory of convection (Ludlam and Scorer, 1953).

After the organizing stage, the reflectivity structure stayed fairly constant, particularly near cloud base, for the next 30 to 40 min. However, turrets persisted near cloud top and updraft subcells recurred throughout the cloud layer. The greatest upward speeds from radar measurements increased from 3 m s^{-1} to 20 m s^{-1} in the period 1900 to 1925, but the distance between subcells was still about 3 km, suggesting that no significant changes in the sizes of the major convective elements had occurred. Upward motions measured by the sailplane, however, suggest that the updrafts were smoother now than in the early stage. Doppler radar measurements did not adequately represent sizes smaller than about 3 km, so that the small-scale fluctuations found earlier in the aircraft measurements would not have been seen in Doppler winds.

The main updraft region was located toward the southwest side of the most intense precipitation shaft; relative to the direction of travel this placed the bulk of the echo downstream of the updraft area. However, the core of the largest reflectivity values was nearly centered on the main updraft region. Sub-cloud air entered the storm from the southwest, rose in the updraft, and exited and the cloud toward the east. Low-level inflow was not strong enough to match the upward flow, so that a continuous upward stream could not be maintained, which contrasts with the steadier stream of air in supercell updrafts (Marwitz, 1972; Browning, 1977).

23.5.2 Microphysical History

The steadily increasing reflectivity and size of the largest precipitation particles correlate in time with the increase in dynamical intensity, typified by the updraft speed. Maximum particle sizes in the 5- to 8-km layer in the updrafts increased from 0.2 to 6.6 mm in the period 1905 to 1927 (see Chapter 22, sailplane penetrations). Lower in the cloud, near 5 km, the largest sizes increased from 2.4 mm to 5.5 mm, with 8-mm particles present at cloud base after 1941 (see Chapter 22, N10UW penetrations).

23.5.3 Microphysical-Dynamical Interactions: Mechanisms for Recycling Particles

As demonstrated in Chapter 22, particles found in the updrafts between altitudes of 5 and 8 km needed 15 to 25 min to grow to their observed sizes. However, if all their growth had occurred during one vertical transit, at the speeds measured in the updrafts there would have been only 5 to 10 min after particles nucleated at about −5 °C before they reached the middle levels. Some means was necessary,

therefore, to recycle particles through the updrafts in order to extend their growth times. Five ways this could have happened are:

(1) Penetrative downdrafts. Squires (1958) first discussed the possibility that dry air entering clouds from above would be cooled by evaporation and penetrate several kilometers down into a growing cloud. Paluch (1979) showed that in the Butler storm such downdrafts were probably present as turrets rose through the 6- to 8-km layers. These downdrafts could transport ice crystals nucleated at cold temperatures near cloud top down into the body of the cloud updrafts where these small crystals could rise again. Downward and weak upward motions around the updraft cores at cloud top found in both the Doppler-derived and aircraft measurements were probably a result of vertical mixing of dry environmental air with cloud air.

(2) Growth in the fringes of updrafts. Since the observed column crystals required about 6 min growth time, they must have remained in nearly constant conditions, especially temperature, for this long. They were found in weak updrafts and in a region of slow horizontal motion, which would have kept them at about −6 °C for several minutes.

(3) Sedimentation. This is one mechanism that can explain the presence of millimeter-size ice particles in adiabatic regions of the updraft (Chapter 22). Transport or mixing processes that involved environmental air and undilute air rising from cloud base would reduce equivalent potential temperature and liquid water content from their adiabatic values. Cloud air containing large particles, however, could mix horizontally into the updrafts without diminishing the observed adiabatic values. Alternatively, unresolved temporal fluctuations in updraft speed such as those expected in thermals could enable large particles to fall out of one convective element into another (Ludlam, 1952). The visual appearance of cloud top as well as the structure of radar echo certainly supports the existence of small turrets aloft (Chapter 21). Evidence for bubble-like convection in other storms also comes from vertically-pointing Doppler radar observations reported by Battan (1975) and Strauch and Merrem (1976).

(4) Unorganized or turbulent motions. While these motions are surely important in storm dynamics, their effects on precipitation growth are unclear. No doubt turbulent motions, by their random nature, can increase the chances of some particles being transferred from one growth region to another. In the Butler storm, motions at scales of a few hundred meters or less were observed by penetrating aircraft.

(5) Organized, large-scale recirculation. Doppler-derived wind fields between 1900 and 1935 showed a closed loop circulation, 4 to 6 km deep, which could have carried particles down and back into the updraft. Although this was less likely than suggested by Fig. 23.20, since horizontal motion normal to the vertical plane of the figure moved particles away from the updrafts, there is little doubt that some recirculation was occurring in this storm. Ice particles found near the horizontal vortex were too large to have grown to the observed size during only one vertical transit in the updraft. Additionally, particle growth trajectories calculated from the wind field for the 1920-1924 scan, assumed to apply for 30 min, showed that some particles put into the downdraft areas northeast of the main updraft (in its wake) were carried down and back into the updraft near cloud base and rose again. Some particles even went below the melting level, consistent with photographs of ice particles showing that they glistened as if they had a wet coating.

All these processes had a role in extending the in-cloud residence time for growing particles; however, their relative importance is difficult to assess. The evidence is that they are important for the production of millimeter-size particles. Further, these mechanisms are still broadly consistent with the Ludlam-Scorer concept of thermal convection.

References

Barge, B. L., and F. Bergwall, 1976: Fine scale structure of convective storms associated with hail production 2nd WMO Scientific Conf. on Weather Mod., Geneva, 1976, World Meteorological Org., Geneva, Switzerland, 341-348.

Battan, L.J., 1975: Doppler radar observations of a hailstorm. *J. Appl. Meteorol.* 14, 98-108.

Berger, T. and H.L. Groginsky, 1973: Estimates of spectral moments of pulse trains. Presented at the Conf. on Inf. Theory, 1973, Inst. Elec. and Elec. Eng., Tel-Aviv, Israel.

Betts, A.K., 1976: The thermodynamic transformation of the tropical sub-cloud layer by precipitation and downdrafts. *J. Atmos. Sci.* 53, 1008-1020.

Bohne, A.R., and R.C. Srivastava, 1976: Random errors in wind and precipitation fallspeed measurements by a triple Doppler radar system. 17th Radar Meteorol. Conf., Seattle, Wash., 1976, Am. Meteorol. Soc., Boston, Mass., 7-14.

Brown, R.A., and K.C. Crawford, 1972: Doppler radar evidence of severe storm high-reflectivity cores acting as obstacles to airflow. 15th Radar Meteorol. Conf., Champaign-Urbana, Ill., 1972, Am. Meteorol. Soc., Boston, Mass.,16-21.

Browning, K. A., 1977: The structure and mechanisms of hailstorms. In *Hail: A Review of Hail Science and Hail Suppression, Meteorol. Monogr.* No. 38, 1-43.

-----, and G.B. Foote, 1976: Airflow and hail growth in supercell storms and some implications for hail suppression. *Q. J. R. Meteorol. Soc.* 102, 499-533.

-----, and F.H. Ludlam, 1962: Airflow in convective storms. *Q. J. R. Meteorol. Soc.* 88, 117-135.

-----, J.C. Fankhauser, J-P. Chalon, P.J. Eccles, R.G. Strauch, F.H. Merrem, D.J. Musil, E.L. May, and W.R. Sand, 1976: Structure of an evolving hailstorm: Part V. Synthesis and implications for hail growth and hail suppression. *Mon. Weather Rev.* 104, 603-610.

Byers, H.R., and R. R. Braham, Jr., 1949: *The Thunderstorm.* U.S. Govt. Printing Office, Wash., D.C., 287 pp.

Chalon, J-P., J.C. Fankhauser, and P. J. Eccles, 1976: Structure of an evolving hailstorm: Part I. General characteristics and cellular structure. *Mon. Weather Rev.* 104, 564-575.

Dennis, A.S., and D.J. Musil, 1973: Calculations of hailstone growth and trajectories in a simple cloud model. *J. Atmos. Sci.* 30, 278-288.

-----, C.A. Schock, and A. Koscielski, 1970: Characteristics of hailstorms of western South Dakota. *J. Appl. Meteorol.* 9, 127-135.

Dye, J. E., L. J. Miller, B. E. Martner, and Z. Levin, 1980: Dynamical-microphysical evolution of a convective storm. NCAR Tech. Note NCAR/TN-151+STR, National Center for Atmospheric Research, Boulder, Colo., 248 pp.

Fankhauser, J. C., 1971: Thunderstorm-environment interactions determined from aircraft and radar observations. *Mon. Weather Rev.* 99, 171-192.

Foote, G.B., and J.C. Fankhauser, 1973: Airflow and moisture budget beneath a northeast Colorado hailstorm. *J. Appl. Meteorol.* 12, 1330-1353.

Fujita, T., and H. Grandoso, 1968: Split of a thunderstorm into anti-cyclonic and cyclonic storms and their motion from numerical model experiments. *J. Atmos. Sci.* 25, 416-439.

Goyer, G.G., W.E. Howell, V.J. Schaefer, R.A. Schleusener, and P. Squires, 1966: Project Hailswath. *Bull. Am. Meteorol. Soc.* 47, 805-809.

Kropfli, R.A., and L. J. Miller, 1976: Kinematic structure and flux quantities in a convective storm from dual-Doppler radar observations. *J. Atmos. Sci.* 33, 520-529.

Ludlam, F. H., 1952: The production of showers by the growth of ice particles. *Q. J. R. Meteorol. Soc.* 78, 543-553.

-----, 1958: The hard problem. *Nubila* 1, 12-96.

-----, and R.S. Scorer, 1953: Convection in the atmosphere. *Q. J. R. Meteorol. Soc.* 79, 317-341.

Malkus, J.S., 1952: Recent advances in the study of convective clouds and their interaction with the environment. *Tellus* 4, 71-87.

-----, 1954: Some results of a trade cumulus cloud investigation. *J. Meteorol.* 11, 220-237.

Marwitz, J.D., 1972: The structure and motion of severe hailstorms: Part I. Supercell storms. *J. Appl. Meteorol.* 11, 180-188.

-----, 1973: Trajectories within the weak echo regions of hailstorms. *J. Appl. Meteorol.* 12, 1174-1182.

Miller, K. S., and M. M. Rochwarger, 1972: A covariance approach to spectral moment estimation. *IEEE Trans. on Inf. Theory* IT -18, 588-596.

Moncrieff, M. W., and J. S. A. Green, 1972: The propagation and transfer properties of steady convective overturning in shear. *Q. J. R. Meteorol. Soc.*, 98, 336-352.

Musil, D. J., 1970: Computer modeling of hailstone growth in feeder clouds. *J. Atmos. Sci.* 27, 474-482.

Newton, C. W. 1966: Circulations in large sheared cumulonimbus. *Tellus* 18, 699-713.

-----, and H. R. Newton, 1959: Dynamical interactions between large convective clouds and environment with vertical shear. *J. Meteorol.* 16, 483-498.

O'Brien, J. J., 1970: Alternative solutions to the classical vertical velocity problem. *J. Appl. Meteorol.* 9, 197-203.

Ogura, Y., and T. Takahashi, 1971: Numerical simulation of the life cycle of a thunderstorm cell. *Mon. Weather Rev.* 99, 895-911.

Paluch, I., 1979: The entrainment mechanism in Colorado cumuli. *J. Atmos. Sci.* 36, 2467-2478.

Rummler, W. D., 1968: Two-pulse spectral measurements. Tech. Memo. MM-68-4121-15, Bell Telephone Laboratories, Whippany, N.J.

Schlesinger, R. E., 1973: A numerical model of deep moist convection: Part II. A prototype experiment. *J. Atmos. Sci.* 30, 1374-1391.

Squires, P., 1958: Penetrative downdrafts in cumuli. *Tellus* 10, 381-389.

Srivastava, R. C., 1967: A study of the effect of precipitation on cumulus dynamics. *J. Atmos. Sci.* 24, 36-45.

Strauch, R. G., and F. H. Merrem, 1976: Structure of an evolving hailstorm: Part III. Internal structure from Doppler radar. *Mon. Weather Rev.* 104, 588-595.

Weinstein, A. I., 1970: A numerical model of cumulus dynamics and microphysics. *J. Atmos. Sci.* 27, 246-255.

Woodward, B., 1959: The motion in and around isolated thermals. *Q. J. R. Meteorol. Soc.* 85, 144-151.

Zipser, E. J., 1977: Mesoscale and convective-scale downdrafts as distinct components of squall-line structure. *Mon. Weather Rev.* 105, 1568-1589.